Handbook of
CHEMICAL PROCESS
Fundamentals and Applications

Handbook of
CHEMICAL PROCESS
Fundamentals and Applications

Contributors :
**Yuanyuan Wu,
Changhao Yan,** *et al.*

AURIS REFERENCE LTD.
London, UK

Handbook of Chemical Process: *Fundamentals and Applications*
Contributors : Yuanyuan Wu *and* Changhao Yan, *et al.*

Auris Reference Ltd., UK

www.aurisreference.com

United Kingdom

Copyright 2016

Printed in 2017 for Sale in the Indian Subcontinent

The information in this book has been obtained from highly regarded resources. The copyrights for individual articles remain with the authors, as indicated. All chapters are distributed under the terms of the Creative Commons Attribution License, which permit unrestricted use, distribution, and reproduction in any medium, provided the original author and source are credited.

Notice

Contributors, whose names have been given on the book cover, are not associated with the Publisher. The editors and the Publisher have attempted to trace the copyright holders of all material reproduced in this publication and apologise to copyright holders if permission has not been obtained. If any copyright holder has not been acknowledged, please write to us so we may rectify.

Reasonable efforts have been made to publish reliable data. The views articulated in the chapters are those of the individual contributors, and not necessarily those of the editors or the Publisher. Editors and/or the Publisher are not responsible for the accuracy of the information in the published chapters or consequences from their use. The Publisher accepts no responsibility for any damage or grievance to individual(s) or property arising out of the use of any material(s), instruction(s), methods or thoughts in the book.

No part of this publication maybe reproduced, stored in a retrieval system or transmitted in any form or by any means, electronic, mechanical, photocopying, recording, scanning or otherwise without prior written permission of the publisher.

Handbook of Chemical Process: *Fundamentals and Applications*

ISBN: 978-1-78154-508-9

British Library Cataloguing in Publication Data
A CIP record for this book is available from the British Library

Exclusively distributed by CBS Publishers & Distributors Pvt. Ltd.

Sales & Distribution Rights only for India, Pakistan, Bangladesh, Sri Lanka, Nepal and Bhutan. This book is not to be sold outside these territories.

Preface

A chemical process is a method of changing one or more chemicals or chemical compounds. Such a chemical process can occur by itself or be caused by an outside force, and involves a chemical reaction of some sort. In an "engineering" sense, a chemical process is a method intended to be used in manufacturing or on an industrial scale to change the composition of chemical(s) or material(s), usually using technology similar or related to that used in chemical plants or the chemical industry.

Often, one or more chemical reactions are involved, but other ways of changing chemical (or material) composition may be used, such as mixing or separation processes. The process steps may be sequential in time or sequential in space along a stream of flowing or moving material.

In addition to chemical plants for producing chemicals, chemical processes with similar technology and equipment are also used in oil refining and other refineries, natural gas processing, polymer and pharmaceutical manufacturing, food processing, and water and wastewater treatment.

This page left intentionally blank.

Contents

Preface	*v*
1. INTRODUCTION TO CHEMICAL PROCESS	**1-30**
Redox	2
Dehydrogenation	7
Hydrolysis	8
Hydration Reaction	11
Dehydration Reaction	13
Halogenation	13
Nitrification	17
Sulfation	22
Alkylation	23
Polymerization	27
2. CONTROL ARCHITECTURES	**31-92**
Feedback Control	31
Feed-forward Control	38
Dynamic Compensation	41
Cascade Control	46
Ratio Control	60
CCL Liquid	69
3. CONTROL SYSTEMS OF INDUSTRIAL APPLICATIONS	**93-136**
Temperature Sensors	98
Regulator Structure	108
Pressure Sensors	111
Level Sensors	124
4. FLOW CONTROL SENSORS	**137-179**
Common Types of Flow Meters	139

Ultrasonic Flow Meters	150
Composition Sensors	159
Electrometric Analysis	167
Chromatography	171

5. pH AND VISCOSITY CONTROL SENSORS — 180-224

Introduction to pH	180
Introduction to Viscosity	186
Miscellaneous Sensors	194
Valve Types Selection	195
Actuators	214
Valve and Actuator Selection Example	217
Valve Modeling	218

6. GASIFICATION INTRODUCTION — 225-253

Fundamentals	226
Gasification Technology	240
Hydrogen from Coal	245

7. MULTIPLE INPUT MULTIPLE OUTPUT CONTROL — 254-282

Decouple	254
Relative Gain Array	259
Model Predictive Control	268
Neural Networks	276

8. FACILE SYNTHESIS OF MONO-DISPERSED POLYSTYRENE (PS)/Ag COMPOSITE MICROSPHERES VIA MODIFIED CHEMICAL REDUCTION — 283-298

Wen Zhu, Yuanyuan Wu, Changhao Yan, Chengyin Wang, Ming Zhang and Zhonglian Wu

Introduction	284
Experimental Section	285
Results and Discussion	286
Conclusions	294
References	295

List of Contributors

Wen Zhu

College of Chemistry and Chemical Engineering, Yangzhou University, Yangzhou 225002, China; E-Mails: czzwen@163.com (W.Z.); 1747168992@qq.com (Y.W.); 250178276@qq.com (C.Y.)

and

College of Materials Science and Engineering, Jiangsu University of Technology, Changzhou 213001, China; E-Mail: 871765383@qq.com (Z.W.)

Yuanyuan Wu

College of Chemistry and Chemical Engineering, Yangzhou University, Yangzhou 225002, China; E-Mails: czzwen@163.com (W.Z.); 1747168992@qq.com (Y.W.); 250178276@qq.com (C.Y.)

Changhao Yan

College of Chemistry and Chemical Engineering, Yangzhou University, Yangzhou 225002, China; E-Mails: czzwen@163.com (W.Z.); 1747168992@qq.com (Y.W.); 250178276@qq.com (C.Y.)

Chengyin Wang

College of Chemistry and Chemical Engineering, Yangzhou University, Yangzhou 225002, China; E-Mails: czzwen@163.com (W.Z.); 1747168992@qq.com (Y.W.); 250178276@qq.com (C.Y.)

Ming Zhang

College of Chemistry and Chemical Engineering, Yangzhou University, Yangzhou 225002, China; E-Mails: czzwen@163.com (W.Z.); 1747168992@qq.com (Y.W.); 250178276@qq.com (C.Y.)

Zhonglian Wu

College of Materials Science and Engineering, Jiangsu University of Technology, Changzhou 213001, China; E-Mail: 871765383@qq.com (Z.W.)

This page left intentionally blank.

Chapter 1

Introduction to Chemical Process

In a scientific sense, a **chemical process** is a method or means of somehow changing one or more chemicals or chemical compounds. Such a chemical process can occur by itself or be caused by an outside force, and involves a chemical reaction of some sort. In an "engineering" sense, a **chemical process** is a method intended to be used in manufacturing or on an industrial scale (see Industrial process) to change the composition of chemical(s) or material(s), usually using technology similar or related to that used in chemical plants or the chemical industry.

Neither of these definitions is exact in the sense that one can always tell definitively what is a chemical process and what is not; they are practical definitions. There is also significant overlap in these two definition variations. Because of the inexactness of the definition, chemists and other scientists use the term "chemical process" only in a general sense or in the engineering sense. However, in the "process (engineering)" sense, the term "chemical process" is used extensively.

Although this type of chemical process may sometimes involve only one step, often multiple steps, referred to as unit operations, are involved. In a plant, each of the unit operations commonly occur in individual vessels or sections of the plant called **units**. Often, one or more chemical reactions are involved, but other ways of changing chemical (or material) composition may be used, such as mixing or separation processes. The process steps may be sequential in time or sequential in space along a stream of flowing or moving material. For a given amount of a feed (input) material or product (output) material, an expected amount of material can be determined at key steps in the process from empirical data and material balance calculations. These amounts can be scaled up or down to suit the desired capacity or operation of a particular chemical plant built for such a process. More than one chemical plant may use the same chemical process, each plant perhaps at differently scaled capacities. Chemical processes like distillation and crystallization go back to alchemy in Alexandria, Egypt.

Such chemical processes can be illustrated generally as **block flow diagrams** or in more detail as **process flow diagrams**. Block flow diagrams show the units as blocks and the streams flowing between them as connecting lines with arrowheads to show direction of flow.

In addition to chemical plants for producing chemicals, chemical processes with similar technology and equipment are also used in oil refining and other refineries, natural gas processing, polymer and pharmaceutical manufacturing, food processing, and water and wastewater treatment.

Unit Processing in Chemical Process

Unit processing is the basic processing in chemical engineering. Together with unit operations it forms the main principle of the varied chemical industries. Each genre of unit processing follows the same chemical law much as each genre of unit operations follows the same physical law.

Chemical engineering unit processing consists of the following important processes:

REDOX

Redox reactions include all chemical reactions in which atoms have their oxidation state changed; in general, redox reactions involve the transfer of electrons between species.

The term "redox" comes from two concepts involved with electron transfer: reduction and oxidation. It can be explained in simple terms:

- **Oxidation** is the *loss* of electrons or an *increase* in oxidation state by a molecule, atom, or ion.
- **Reduction** is the *gain* of electrons or a *decrease* in oxidation state by a molecule, atom, or ion.

Although oxidation reactions are commonly associated with the formation of oxides from oxygen molecules, these are only specific examples of a more general concept of reactions involving electron transfer.

Redox reactions, or oxidation-reduction reactions, have a number of similarities to acid–base reactions. Like acid–base reactions, redox reactions are a matched set, that is, there cannot be an oxidation reaction without a reduction reaction happening simultaneously. The oxidation alone and the reduction alone are each called a *half-reaction*, because two half-reactions always occur together to form a whole reaction. When writing half-reactions, the gained or lost electrons are typically included explicitly in order that the half-reaction be balanced with respect to electric charge.

Though sufficient for many purposes, these descriptions are not precisely correct. Oxidation and reduction properly refer to *a change in oxidation state* — the actual transfer of electrons may never occur. The oxidation state of an atom is the fictitious charge that an atom would have if all bonds between atoms of different elements were 100% ionic. Thus, oxidation is better defined as an *increase in oxida-*

tion state, and reduction as a *decrease in oxidation state*. In practice, the transfer of electrons will always cause a change in oxidation state, but there are many reactions that are classed as "redox" even though no electron transfer occurs (such as those involving covalent bonds).

There are simple redox processes, such as the oxidation of carbon to yield carbon dioxide(CO_2) or the reduction of carbon by hydrogen to yield methane(CH_4), and more complex processes such as the oxidation of glucose($C_6H_{12}O_6$) in the human body through a series of complex electron transfer processes.

Oxidizing and Reducing Agents

In redox processes, the reductant transfers electrons to the oxidant. Thus, in the reaction, the reductant or *reducing agent* loses electrons and is oxidized, and the oxidant or *oxidizing agent* gains electrons and is reduced. The pair of an oxidizing and reducing agent that are involved in a particular reaction is called a **redox pair**. A **redox couple** is a reducing species and its corresponding oxidized form, *e.g.*, Fe^{2+}/Fe^{3+}.

Oxidizers

Substances that have the ability to **oxidize** other substances (cause them to lose electrons) are said to be **oxidative** or **oxidizing** and are known as oxidizing agents, oxidants, or oxidizers. That is, the oxidant (oxidizing agent) removes electrons from another substance, and is thus itself reduced. And, because it "accepts" electrons, the oxidizing agent is also called an electron acceptor. Oxygen is the quintessential oxidizer.

Oxidants are usually chemical substances with elements in high oxidation states or else highly electronegative elements (O_2, F_2, Cl_2, Br_2) that can gain extra electrons by oxidizing another substance.

Reducers

Substances that have the ability to **reduce** other substances (cause them to gain electrons) are said to be **reductive** or **reducing** and are known as reducing agents, reductants, or reducers. The reductant(reducing agent) transfers electrons to another substance, and is thus itself oxidized. And, because it "donates" electrons, the reducing agent is also called an electron donor. Electron donors can also form charge transfer complexes with electron acceptors.

Reductants in chemistry are very diverse. Electropositive elemental metals, such as lithium, sodium, magnesium, iron, zinc, and aluminium, are good reducing agents. These metals donate or *give away* electrons readily. *Hydride transfer reagents*, such as $NaBH_4$ and $LiAlH_4$, are widely used in organic chemistry, primarily in the reduction of carbonyl compounds to alcohols. Another method of reduction involves the use of hydrogen gas (H_2) with a palladium, platinum, or nickelcatalyst. These *catalytic reductions* are used primarily in the reduction of carbon-carbon double or triple bonds.

Standard Electrode Potentials (Reduction Potentials)

Each half-reaction has a *standard electrode potential*(E^0_{cell}), which is equal to the potential difference (or voltage)(E^0_{cell}) at equilibrium under standard conditions of an electrochemical cell in which the cathode reaction is the half-reaction considered, and the anode is a standard hydrogen electrode where hydrogen is oxidized: $½ H_2 \rightarrow H^+ + e^-$.

The electrode potential of each half-reaction is also known as its *reduction potential* E^0_{red}, or potential when the half-reaction takes place at a cathode. The reduction potential is a measure of the tendency of the oxidizing agent to be reduced. Its value is zero for $H^+ + e^- \rightarrow ½ H_2$ by definition, positive for oxidizing agents stronger than H^+ (*e.g.*, +2.866 V for F_2) and negative for oxidizing agents that are weaker than H^+ (*e.g.*, −0.763 V for Zn^{2+}).

For a redox reaction that takes place in a cell, the potential difference $E^0_{cell} = E^0_{cathode} - E^0_{anode}$

However, the potential of the reaction at the anode was sometimes expressed as an *oxidation potential*, $E^0_{ox} = - E^0$. The oxidation potential is a measure of the tendency of the reducing agent to be oxidized, but does not represent the physical potential at an electrode. With this notation, the cell voltage equation is written with a plus sign $E^0_{cell} = E^0_{cathode} + E^0_{ox\ (anode)}$

Examples of Redox Reactions

A good example is the reaction between hydrogen and fluorine in which hydrogen is being oxidized and fluorine is being reduced:

$$H_2 + F_2 \rightarrow 2 HF$$

We can write this overall reaction as two half-reactions:

the oxidation reaction:

$$H_2 \rightarrow 2 H^+ + 2 e^-$$

and the reduction reaction:

$$F_2 + 2 e^- \rightarrow 2 F^-$$

Analyzing each half-reaction in isolation can often make the overall chemical process clearer. Because there is no net change in charge during a redox reaction, the number of electrons in excess in the oxidation reaction must equal the number consumed by the reduction reaction.

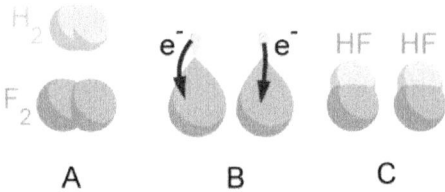

Fig. : Illustration of a redox reaction.

Elements, even in molecular form, always have an oxidation state of zero. In the first half-reaction, hydrogen is oxidized from an oxidation state of zero to an oxidation state of +1. In the second half-reaction, fluorine is reduced from an oxidation state of zero to an oxidation state of −1.

When adding the reactions together the electrons are canceled:

$$H_2 \rightarrow 2\,H^+ + 2\,e^-$$

$$F_2 + 2\,e^- \rightarrow 2\,F^-$$

$$H_2 + F_2 \rightarrow 2\,H^+ + 2\,F^-$$

And the ions combine to form hydrogen fluoride:

$$2\,H^+ + 2\,F^- \rightarrow 2\,HF$$

The overall reaction is:

$$H_2 + F_2 \rightarrow 2\,HF$$

Metal Displacement

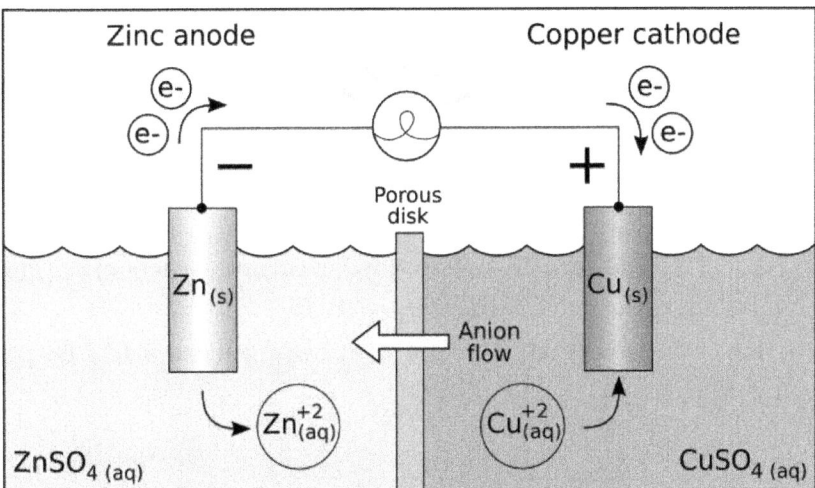

Fig. : A redox reaction is the force behind an electrochemical cell like the Galvanic cell pictured. The battery is made out of a zinc electrode in a $ZnSO_4$ solution connected with a wire and a porous disk to a copper electrode in a $CuSO_4$ solution.

In this type of reaction, a metal atom in a compound (or in a solution) is replaced by an atom of another metal. For example, copper is deposited when zinc metal is placed in a copper(II) sulfate solution:

$$Zn(s) + CuSO_4(aq) \rightarrow ZnSO_4(aq) + Cu(s)$$

In the above reaction, zinc metal displaces the copper(II) ion from copper sulfate solution and thus liberates free copper metal.

The ionic equation for this reaction is:

$$Zn + Cu^{2+} \rightarrow Zn^{2+} + Cu$$

As two half-reactions, it is seen that the zinc is oxidized:

$$Zn \rightarrow Zn^{2+} + 2e^-$$

And the copper is reduced:

$$Cu^{2+} + 2e^- \rightarrow Cu$$

Other Examples

- The reduction of nitrate to nitrogen in the presence of an acid (denitrification):

$$2\,NO_3^- + 10\,e^- + 12\,H^+ \rightarrow N_2 + 6\,H_2O$$

- The combustion of hydrocarbons, such as in an internal combustion engine, which produces water, carbon dioxide, some partially oxidized forms such as carbon monoxide, and heat energy. Complete oxidation of materials containing carbon produces carbon dioxide.
- In organic chemistry, the stepwise oxidation of a hydrocarbon by oxygen produces water and, successively, an alcohol, an aldehyde or a ketone, a carboxylic acid, and then a peroxide.

Corrosion and Rusting

- The term corrosion refers to the electrochemical oxidation of metals in reaction with an oxidant such as oxygen. Rusting, the formation of iron oxides, is a well-known example of electrochemical corrosion; it forms as a result of the oxidation of iron metal. Common rust often refers to iron(III) oxide, formed in the following chemical reaction:

$$4Fe + 3O_2 \rightarrow 2Fe_2O_3$$

- The oxidation of iron(II) to iron(III) by hydrogen peroxide in the presence of an acid:

$$Fe^{2+} \rightarrow Fe^{3+} + e^-$$
$$H_2O_2 + 2\,e^- \rightarrow 2\,OH^-$$

Overall equation:

$$2\,Fe^{2+} + H_2O_2 + 2\,H^+ \rightarrow 2\,Fe^{3+} + 2\,H_2O$$

Redox Reactions in Industry

Cathodic protection is a technique used to control the corrosion of a metal surface by making it the cathode of an electrochemical cell. A simple method of protection connects protected metal to a more easily corroded "sacrificial anode" to act as the anode. The sacrificial metal instead of the protected metal, then, corrodes. A common application of cathodic protection is in galvanized steel, in which a sacrificial coating of zinc on steel parts protects them from rust.

The primary process of reducing ore at high temperature to produce metals is known as smelting.

Oxidation is used in a wide variety of industries such as in the production of cleaning products and oxidizing ammonia to produce nitric acid, which is used in most fertilizers.

Redox reactions are the foundation of electrochemical cells.

The process of electroplating uses redox reactions to coat objects with a thin layer of a material, as in chrome-platedautomotive parts, silver platingcutlery, and gold-platedjewelry.

The production of compact discs depends on a redox reaction, which coats the disc with a thin layer of metal film.

DEHYDROGENATION

Dehydrogenation is a chemical reaction that involves the removal of hydrogen from a molecule. It is the reverse process of hydrogenation. Dehydrogenation reactions are conducted both on industrial and laboratory scales. Dehydrogenation converts saturated fats to unsaturated fats. Enzymes that catalyze dehydrogenation are called dehydrogenases. Dehydrogenation processes are used extensively to produce styrene in the fine chemicals, oleochemicals, petrochemicals, and detergents industries.

Classes of the Reaction

A variety of dehydrogenation processes have been described, especially for organic compounds:

- In typical aromatization, six-membered alicyclic rings, *e.g.* cyclohexene, can be aromatized in the presence of hydrogenation acceptors. The elements sulphur and selenium promote this process. Quinones, especially 2,3-Dichloro-5,6-dicyano-1,4-benzoquinone are effective.
- Oxidation of alcohols to ketones or aldehydes can be effected by metal catalysts such as copper chromite. In the Oppenauer oxidation, hydrogen is transferred from an alcohol to an aldehyde or ketone to bring about the oxidation.
- Dehydrogenation of amines to nitriles using a variety of reagents, such as Iodine pentafluoride(IF5).
- Dehydrogenation of paraffins and olefins — paraffins such as *n*-pentane and isopentane can be converted to pentene and isopentene using chromium(III) oxide as a catalyst at 500 °C.

Examples

One of the largest scale dehydrogenation reactions is the production of styrene by dehydrogenation of ethylbenzene. Typical dehydrogenation catalysts are based on iron(III) oxide, promoted by several percent potassium oxide or potassium carbonate.

$$C_6H_5CH_2CH_3 \rightarrow C_6H_5CH=CH_2 + H_2$$

Formaldehyde is produced industrially by the catalytic oxidation of methanol, which can also be viewed as a dehydrogenation using O_2 as the acceptor. The most common catalysts are silver metal or a mixture of an iron and molybdenum or vanadiumoxides. In the commonly used formox process, methanol and oxygen react at ca. 250–400 °C in presence of iron oxide in combination with molybdenum and/or vanadium to produce formaldehyde according to the chemical equation:

$$2\ CH_3OH + O_2 \rightarrow 2\ CH_2O + 2\ H_2O$$

HYDROLYSIS

Hydrolysis usually means the cleavage of chemical bonds by the addition of water. When a carbohydrate is broken into its component sugar molecules by hydrolysis (*e.g.* sucrose being broken down into glucose and fructose), this is termed **saccharification**. Generally, hydrolysis or saccharification is a step in the degradation of a substance.

Types

Usually hydrolysis is a chemical process in which a molecule of water is added to a substance. Sometimes this addition causes both substance and water molecule to split into two parts. In such reactions, one fragment of the target molecule (or parent molecule) gains a hydrogen ion.

Salts

A common kind of hydrolysis occurs when a salt of a weak acid or weak base(or both) is dissolved in water. Water spontaneously ionizes into hydroxide anions and hydronium cations. The salt also dissociates into its constituent anions and cations. For example, sodium acetate dissociates in water into sodium and acetate ions. Sodium ions react very little with the hydroxide ions whereas the acetate ions combine with hydronium ions to produce acetic acid. In this case the net result is a relative excess of hydroxide ions, yielding a basic solution.

Strong acids also undergo hydrolysis. For example, dissolving sulfuric acid(H_2SO_4) in water is accompanied by hydrolysis to give hydronium and bisulfate, the sulfuric acid's conjugate base. For a more technical discussion of what occurs during such a hydrolysis.

Esters and Amides

Acid–base-catalysed hydrolyses are very common; one example is the hydrolysis of amides or esters. Their hydrolysis occurs when the nucleophile(a nucleus-seeking agent, *e.g.*, water or hydroxyl ion) attacks the carbon of the carbonyl group of the ester or amide. In an aqueous base, hydroxyl ions are better nucleophiles than polar molecules such as water. In acids, the carbonyl group becomes protonated, and this leads to a much easier nucleophilic attack. The products for both hydrolyses are compounds with carboxylic acid groups.

Perhaps the oldest commercially practiced example of ester hydrolysis is saponification(formation of soap). It is the hydrolysis of a triglyceride(fat) with an aqueous base such as sodium hydroxide(NaOH). During the process, glycerol is formed, and the fatty acids react with the base, converting them to salts. These salts are called soaps, commonly used in households.

In addition, in living systems, most biochemical reactions (including ATP hydrolysis) take place during the catalysis of enzymes. The catalytic action of enzymes allows the hydrolysis of proteins, fats, oils, and carbohydrates. As an example, one may consider proteases (enzymes that aid digestion by causing hydrolysis of peptide bonds in proteins). They catalyse the hydrolysis of interior peptide bonds in peptide chains, as opposed to exopeptidases (another class of enzymes, that catalyse the hydrolysis of terminal peptide bonds, liberating one free amino acid at a time).

However, proteases do not catalyse the hydrolysis of all kinds of proteins. Their action is stereo-selective: Only proteins with a certain tertiary structure are targeted as some kind of orienting force is needed to place the amide group in the proper position for catalysis. The necessary contacts between an enzyme and its substrates (proteins) are created because the enzyme folds in such a way as to form a crevice into which the substrate fits; the crevice also contains the catalytic groups. Therefore, proteins that do not fit into the crevice will not undergo hydrolysis. This specificity preserves the integrity of other proteins such as hormones, and therefore the biological system continues to function normally.

Upon hydrolysis, an amide converts into a carboxylic acid and an amine or ammonia. The carboxylic acid has a hydroxyl group derived from a water molecule and the amine (or ammonia) gains the hydrogen ion. The hydrolysis of peptides gives amino acids.

$$R^1\text{-CO-NH-}R^2 \xrightarrow{H_2O} R^1\text{-COOH} + H_2N\text{-}R^2$$

Many polyamide polymers such as nylon 6,6 hydrolyse in the presence of strong acids. The process leads to depolymerization. For this reason nylon products fail by fracturing when exposed to small amounts of acidic water. Polyesters are also susceptible to similar polymer degradation reactions. The problem is known as stress corrosion cracking.

ATP

Hydrolysis is related to energy metabolism and storage. All living cells require a continual supply of energy for two main purposes: the biosynthesis of micro and macromolecules, and the active transport of ions and molecules across cell membranes. The energy derived from the oxidation of nutrients is not used directly but, by means of a complex and long sequence of reactions, it is channelled into a special energy-storage molecule, adenosine triphosphate(ATP). The ATP molecule contains pyrophosphate linkages (bonds formed when two phosphate units are combined together) that release energy when needed. ATP can undergo hydrolysis in two ways: the removal of terminal phosphate to form adenosine diphosphate(ADP) and inorganic phosphate, or the removal of a terminal diphosphate to yield adenosine monophosphate(AMP) and pyrophosphate. The latter usually undergoes further cleavage into its two constituent phosphates. This results in biosynthesis reactions, which usually occur in chains, that can be driven in the direction of synthesis when the phosphate bonds have undergone hydrolysis.

Polysaccharides

Fig. : Sucrose. The glycoside bond is represented by the central oxygen atom, which holds the two monosaccharide units together.

Monosaccharides can be linked together by glycosidic bonds, which can be cleaved by hydrolysis. Two, three, several or many monosaccharides thus linked form disaccharides, trisaccharides, oligosaccharides or polysaccharides, respectively. Enzymes that hydrolyseglycosidic bonds are called "glycoside hydrolases" or "glycosidases".

The best-known disaccharide is sucrose(table sugar). Hydrolysis of sucrose yields glucose and fructose. Invertase is a sucrase used industrially for the hydrolysis of sucrose to so-called invert sugar. Lactase is essential for digestive hydrolysis of lactose in milk; many adult humans do not produce lactase and cannot digest the lactose in milk (not a disorder).

The hydrolysis of polysaccharides to soluble sugars is called "saccharification". Malt made from barley is used as a source of β-amylase to break down starch into the disaccharide maltose, which can be used by yeast to produce beer. Other amylase enzymes may convert starch to glucose or to oligosaccharides. Cellulose is first hydrolyzed to cellobiose by cellulase and then cellobiose is further hydrolyzed to glucose by beta-glucosidase. Animals such as cows (ruminants) are

Metal Aqua Ions

Metal ions are Lewis acids, and in aqueous solution they form metal aqua ions of the general formula $M(H_2O)_n^{m+}$. The aqua ions undergo hydrolysis, to a greater or lesser extent. The first hydrolysis step is given generically as

$$M(H_2O)_n^{m+} + H_2O \rightleftharpoons M(H_2O)_{n-1}(OH)^{(m-1)+} + H_3O^+$$

Thus the aqua cations behave as acids in terms of Brønsted-Lowry acid-base theory. This effect is easily explained by considering the inductive effect of the positively charged metal ion, which weakens the O-H bond of an attached water molecule, making the liberation of a proton relatively easy.

The dissociation constant, pK_a, for this reaction is more or less linearly related to the charge-to-size ratio of the metal ion. Ions with low charges, such as Na^+ are very weak acids with almost imperceptible hydrolysis. Large divalent ions such as Ca^{2+}, Zn^{2+}, Sn^{2+} and Pb^{2+} have a pK_a of 6 or more and would not normally be classed as acids, but small divalent ions such as Be^{2+} undergo extensive hydrolysis. Trivalent ions like Al^{3+} and Fe^{3+} are weak acids whose pK_a is comparable to that of acetic acid. Solutions of salts such as $BeCl_2$ or $Al(NO_3)_3$ in water are noticeably acidic; the hydrolysis can be suppressed by adding an acid such as nitric acid, making the solution more acidic.

Hydrolysis may proceed beyond the first step, often with the formation of polynuclear species via the process of olation. Some "exotic" species such as $Sn_3(OH)_4^{2+}$ are well characterized. Hydrolysis tends to proceed as pH rises leading, in many cases, to the precipitation of a hydroxide such as $Al(OH)_3$ or $AlO(OH)$. These substances, major constituents of bauxite, are known as laterites and are formed by leaching from rocks of most of the ions other than aluminium and iron and subsequent hydrolysis of the remaining aluminium and iron.

HYDRATION REACTION

In chemistry, a hydration reaction is a chemical reaction in which a substance combines with water. In organic chemistry, water is added to an unsaturated substrate, which is usually an alkene or an alkyne. This type of reaction is employed industrially to produce ethanol, isopropanol, and 2-butanol. Organic chemistry

Epoxides to Glycol

Several billion kilograms of ethylene glycol is produced annually by the hydration of ethylene oxide:

$$C_2H_4O + H_2O \rightarrow HO-CH_2CH_2-OH$$

Acid catalysts are typically used.

Alkenes

For the hydration of alkenes, the general chemical equation of the reaction is the following:

$$RRC=CH_2 \text{ in } H_2O/acid \rightarrow RRC(-OH)-CH_3$$

A hydroxyl group (OH⁻) attaches to one carbon of the double bond, and a proton(H⁺)adds to the other carbon of the double bond. The reaction is highly exothermic. In the first step, the alkene acts as a nucleophile and attacks the proton, following Markovnikov's rule. In the second step an H_2O molecule bonds to the other, more highly substituted carbon. The oxygen atom at this point has three bonds and carries a positive charge (*i.e.*, the molecule is an oxonium). Another water molecule comes along and takes up the extra proton. This reaction tends to yield many undesirable side products, (for example diethyl ether in the process of creating Ethanol) and in its simple form described here is not considered very useful for the production of alcohol.

Two approaches are taken. Traditionally the alkene is treated with sulfuric acid to give alkyl sulfate esters. In the case of ethanol production, this step can be written:

$$H_2SO_4 + C_2H_4 \rightarrow C_2H_5\text{-}O\text{-}SO_3H$$

Subsequently this sulfate ester is hydrolyzed to regenerate sulfuric acid and release ethanol:

$$C_2H_5\text{-}O\text{-}SO_3H + H_2O \rightarrow H_2SO_4 + C_2H_5OH$$

This two step route is called the "indirect process".

In the "direct process," the acid protonates the alkene, and water reacts with this incipient carbocation to give the alcohol. The direct process is more popular because it is simpler. The acid catalysts include phosphoric acid and several solid acids. Here an example reaction mechanism of the hydration of 1-methylcyclohexene to 1-methylcyclohexanol.

Hydration of Other Substrates

Any unsaturated organic compound is susceptible to hydration. Acetylene hydrates to give acetaldehyde: The process typically relies on mercury catalysts and has been discontinued in the West but is still practiced in China. The Hg^{2+} center binds to C≡C triple bond, which is then attacked by water. The reaction is:

$$H_2O + C_2H_2 \rightarrow CH_3CHO$$

Nitriles undergo hydration to give amides:

$$H_2O + RCN \rightarrow RC(O)NH_2$$

This reaction is employed in the production of acrylamide.

Aldehydes and to some extent even ketones, hydrate to geminaldiols. The reaction is especially dominant for formaldehyde, which, in the presence of water, exists significantly as dihydroxymethane.

Conceptually similar reactions include hydroamination and hydroalkoxylation, which involve adding amines and alcohols to alkenes.

Inorganic and Materials Chemistry

Hydration is an important process in many other applications; one example is the production of Portland cement by the crosslinking of calcium oxides and silicates that is induced by water. Hydration is the process by which desiccants function.

DEHYDRATION REACTION

In chemistry and the biological sciences, a **dehydration reaction** is usually defined as a chemical reaction that involves the loss of a water molecule from the reacting molecule. Dehydration reactions are a subset of condensation reactions. Because the hydroxyl group (–OH) is a poor leaving group, having a Brønsted acid catalyst often helps by protonating the hydroxyl group to give the better leaving group, $-OH_2^+$. The reverse of a dehydration reaction is a hydration reaction. Common dehydrating agents used in organic synthesis include concentrated sulfuric acid, concentrated phosphoric acid, hot aluminium oxide and hot ceramic.

Dehydration reactions and dehydration synthesis have the same meaning, and are often used interchangeably. Two monosaccharides, such as glucose and fructose, can be joined together (to form sucrose) using dehydration synthesis. The new molecule, consisting of two monosaccharides, is called a disaccharide.

The process of hydrolysis is the reverse reaction, meaning that the water is recombined with the two hydroxyl groups and the disaccharide reverts to being monosaccharides.

HALOGENATION

Halogenation is a chemical reaction that involves the reaction of a compound with a halogen and results in the halogen being added to the compound. Organic compounds undergo halogenation much more readily than inorganic compounds. **Dehalogenation** is the reverse of halogenation and results in the removal of a halogen from a molecule. The pathway and stoichiometry of halogenation depends on the structural features and functional groups of the organic substrate, as well as on the specific halogen. Inorganic compounds such as metals also undergo halogenation.

Halogenation of Organic Compounds: Kinds of Reactions

There are several processes for the halogenation of organic compounds, including free radical halogenation, ketone halogenation, electrophilic halogenation, and halogen addition reaction. The determining factors are the functional groups.

Free Radical Halogenation

Saturated hydrocarbons typically do not add halogens but undergo free radical halogenation, involving substitution of hydrogen atoms by halogen. The regiochemistry of the halogenation of alkanes is usually determined by the relative weakness of the available C-H bonds. The preference for reaction at tertiary and secondary positions results from greater stability of the corresponding free radicals and the transition state leading to them. Free radical halogenation is used for the industrial production of chlorinated methanes:

$$CH_4 + Cl_2 \rightarrow CH_3Cl + HCl$$

Addition of Halogens to Alkenes and Alkynes

Unsaturated compounds, especially alkenes and alkynes, *add* halogens:

$$RCH=CHR' + X_2 \rightarrow RCHX\text{-}CHXR'$$

The addition of halogens to alkenes proceeds via intermediate halonium ions. In special cases, such intermediates have been isolated.

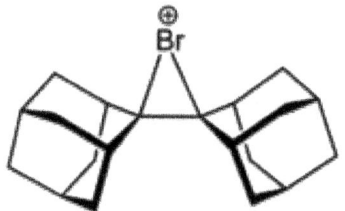

Fig. : Structure of a bromonium ion.

Halogenation of Aromatic Compounds

Aromatic compounds are subject to electrophilic halogenation:

$$RC_6H_5 + X_2 \rightarrow HX + RC_6H_4X$$

The facility of halogenation is influenced by the halogen. Fluorine and chlorine are more electrophilic and are more aggressive halogenating agents. Bromine is a weaker halogenating agent than both fluorine and chlorine, while iodine is least reactive of them all. The facility of hydrogenolysis follows the reverse trend: iodine is most easily removed from organic compounds and organofluorine compounds are highly stable.

Other Halogenation Methods

In the Hunsdiecker reaction, from carboxylic acids are converted to the chain-shortened halide. The carboxylic acid is first converted to its silver salt, which is then oxidized with halogen:

$$RCO_2Ag + Br_2 \rightarrow RBr + CO_2 + AgBr$$

The Sandmeyer reaction is used to give aryl halides from diazonium salts, which are obtained from anilines.

In the Hell–Volhard–Zelinsky halogenation, carboxylic acids are alpha-halogenated.

Halogenation of Organic Compounds: Classification by Halogen

Fluorination

Organic compounds, saturated and unsaturated alike, react readily, usually explosively, with fluorine. This process requires highly specialized conditions. In practice, organic compounds are fluorinated electrochemically. Reactions occur at an anode using hydrogen fluoride as the source of fluorine. The method is called electrochemical fluorination. Aside from F_2 and its electrochemically generated equivalent, a variety of fluorinating reagents are known such as xenon difluoride and cobalt(III) fluoride.

Chlorination

Chlorination is generally highly exothermic. Both saturated and unsaturated compounds react directly with chlorine, the former usually requiring UV light to initiate homolysis of chlorine. Chlorination is conducted on a large scale industrially; major processes include routes to 1,2-dichloroethane(a precursor to PVC), as well as various chlorinated ethanes, as solvents. Competitive with direct chlorination (use of Cl_2) is oxychlorination, which uses hydrogen chloride in combination with oxygen.

Oxychlorination

Oxychlorination is the process of chlorinating hydrocarbons, using a mixture of hydrogen chloride(HCl) and oxygen(O_2). This process is attractive industrially, because hydrogen chloride is less expensive than chlorine. The most common substrate for this reaction is ethylene:

$$CH_2=CH_2 + 2HCl + \tfrac{1}{2} O_2 \rightarrow ClCH_2CH_2Cl + H_2O$$

The reaction is initiated by copper(II) chloride($CuCl_2$), which is the most common catalyst in the production of 1,2-dichloroethane. In some cases, $CuCl_2$ is supported on silica in presence of KCl, $LaCl_3$, or $AlCl_3$ as cocatalysts. Aside from silica, a variety of supports have also been used including various types of alumina, diatomaceous earth, or pumice. Because this reaction is highly exothermic

(238 kJ/mol), the temperature is monitored, to guard against thermal degradation of the catalyst. Since the $CuCl_2$ molecule is the donor of the chloride atom, this catalyst has an important role in the formation of the hydrocarbon double bond of chlorination. The reaction is as follows:

$$CH_2=CH_2 + 2\ CuCl_2 \rightarrow 2CuCl + ClH_2C\text{-}CH_2Cl$$

The copper(II) chloride is regenerated by sequential reactions of the cuprous chloride with oxygen and then hydrogen chloride:

$$\tfrac{1}{2}\ O_2 + 2CuCl \rightarrow CuOCuCl_2$$
$$2HCl + CuOCuCl_2 \rightarrow 2\ CuCl_2 + H_2O$$

Oxychlorination is of special importance in the making of 1,2-dichloroethane, which is then converted into vinyl chloride. As can be seen in the following reaction, 1,2-dichloroethane is cracked:

$$ClCH_2CH_2Cl \rightarrow CH_2=CHCl + HCl$$
$$2HCl + CH_2=CH_2 + \tfrac{1}{2}\ O_2 \rightarrow ClCH_2CH_2Cl + H_2O$$

The HCl from this cracking process is recycled by oxychlorination. The fact that the reaction is self-supplied is one of the reasons that industry uses oxychlorination instead of direct chlorination.

Bromination

Bromination is more selective than chlorination because the reaction is less exothermic. Most commonly bromination is conducted by the addition of Br_2 to alkenes. Bromination of saturated hydrocarbons and aromatic substrates is common in nature, giving rise to a host of organobromine compounds. The usual catalyst is the bromoperoxidase which utilizes bromide in combination with oxygen as an oxidant. An example of bromination can be found in the organic synthesis of the anesthetic halothane from trichloroethylene:

Organobromine compounds are the most common organohalides in nature. Their formation is catalyzed by the enzyme bromoperoxidase. The oceans are estimated to release 1–2 million tons of bromoform and 56,000 tons of bromomethane annually.

Iodination

Iodine is the least reactive halogen and is reluctant to react with most organic compounds. The addition of iodine to alkenes is the basis of the analytical method called the iodine number, a measure of the degree of unsaturation for fats. The iodoform reaction involves degradation of methyl ketones.

Inorganic Chemistry

All elements aside from argon, neon, and helium form fluorides by direct reaction with fluorine. Chlorine is slightly more selective, but still reacts with most metals and heavier nonmetals. Following the usual trend, bromine is less reactive and iodine least of all. Of the many reactions possible, illustrative is the formation of gold(III) chloride by the chlorination of gold. The chlorination of metals is usually not very important industrially since the chlorides are more easily made from the oxides and the hydrogen halide. Where chlorination of inorganic compounds is practiced on a relatively large scale is for the production of phosphorus trichloride and sulfur monochloride.

Chemical Dehalogenation

Chemical dehalogenation is a treatment to remove halogens from harmful chemicals or contaminated areas by making them less toxic. There are two types of dehalogenation: glycolatedehalogenation and base-catalyzed decomposition

NITRIFICATION

Nitrification is the biological oxidation of ammonia or ammonium to nitrite followed by the oxidation of the nitrite to nitrate. The transformation of ammonia to nitrite is usually the rate limiting step of nitrification. Nitrification is an important step in the nitrogen cycle in soil. Nitrification is an aerobic process performed by small groups of autotrophic bacteria and archaea. This process was discovered by the Russianmicrobiologist, Sergei Winogradsky.

Microbiology and Ecology

The oxidation of ammonia into nitrite is performed by two groups of organisms, ammonia-oxidizing bacteria(**AOB**) and ammonia-oxidizing archaea(**AOA**). AOB can be found among the β-proteobacteria and gammaproteobacteria. Currently, two AOA, *Nitrosopumilusmaritimus* and *Nitrososphaeraviennensis*, have been isolated and described. In soils the most studied AOB belong to the genera *Nitrosomonas* and *Nitrosococcus*. Although in soils ammonia oxidation occurs by both AOB and AOA, AOA dominate in both soils and marine environments, suggesting that *Thaumarchaeota* may be greater contributors to ammonia oxidation in these environments.

The second step (oxidation of nitrite into nitrate) is done (mainly) by bacteria of the genus *Nitrobacter* and *Nitrospira*. Both steps are producing energy to be coupled to ATP synthesis. Nitrifying organisms are chemoautotrophs, and use carbon dioxide as their carbon source for growth. Some AOB possess the enzyme, urease, which catalyzes the conversion of the urea molecule to two ammonia molecules and one carbon dioxide molecule. *Nitrosomonaseuropaea*, as well as populations of soil-dwelling AOB, have been shown to assimilate the carbon dioxide released by the reaction to make biomass via the Calvin Cycle, and harvest energy by oxidizing ammonia (the other product of urease) to nitrite. This feature may explain enhanced growth of AOB in the presence of urea in acidic environments.

In most environments, organisms are present that will complete both steps of the process, yielding nitrate as the final product. However, it is possible to design systems in which nitrite is formed (the *Sharon process*).

Nitrification is important in agricultural systems, where fertilizer is often applied as ammonia. Conversion of this ammonia to nitrate increases nitrogen leaching because nitrate is more water-soluble than ammonia.

Nitrification also plays an important role in the removal of nitrogen from municipal wastewater. The conventional removal is nitrification, followed by denitrification. The cost of this process resides mainly in aeration(bringing oxygen in the reactor) and the addition of an external carbon source (*e.g.*, methanol) for the denitrification.

Nitrification can also occur in drinking water. In distribution systems where chloramines are used as the secondary disinfectant, the presence of free ammonia can act as a substrate for ammonia-oxidizing microorganisms. The associated reactions can lead to the depletion of the disinfectant residual in the system. The addition of chlorite ion to chloramine-treated water has been shown to control nitrification.

Together with ammonification, nitrification forms a mineralization process that refers to the complete decomposition of organic material, with the release of available nitrogen compounds. This replenishes the nitrogen cycle.

Chemistry

Nitrification is a process of nitrogen compound oxidation(effectively, loss of electrons from the nitrogen atom to the oxygen atoms):

1. $2\,NH_4^+ + 3\,O_2 \rightarrow 2\,NO_2^- + 2\,H_2O + 4\,H^+$ (Nitrosomonas)
2. $2\,NO_2^- + O_2 \rightarrow 2\,NO_3^-$ (Nitrobacter, Nitrospira)
3. $NH_3 + O_2 \rightarrow NO_2^- + 3H^+ + 2e^-$
4. $NO_2^- + H_2O \rightarrow NO_3^- + 2H^+ + 2e^-$

Nitrification in the Marine Environment

In the marine environment, nitrogen is often the limiting nutrient, so the nitrogen cycle in the ocean is of particular interest. The nitrification step of the cycle is of particular interest in the ocean because it creates nitrate, the primary form of nitrogen responsible for "new" production. Furthermore, as the ocean becomes enriched in anthropogenic CO_2, the resulting decrease in pH could lead to decreasing rates of nitrification. Nitrification could potentially become a "bottleneck" in the nitrogen cycle.

Nitrification, as stated above, is formally a two-step process; in the first step ammonia is oxidized to nitrite, and in the second step nitrite is oxidized to nitrate. Different microbes are responsible for each step in the marine environment. Several groups of ammonia-oxidizing bacteria (AOB) are known in the marine environment, including *Nitrosomonas, Nitrospira,* and *Nitrosococcus*. All contain the functional gene ammonia monooxygenase (AMO) which, as its name implies, is responsible for the oxidation of ammonia. More recent metagenomic studies have revealed that some Thaumarchaeota(formerly Crenarchaeota) possess AMO. Thaumarchaeotes are abundant in the ocean and some species have a 200 times greater affinity for ammonia than AOB, leading researchers to challenge the previous belief that AOB are primarily responsible for nitrification in the ocean. Furthermore, though nitrification is classically thought to be vertically separated from primary production because the oxidation of nitrogen by bacteria is inhibited by light, nitrification by AOA does not appear to be light inhibited, meaning that nitrification is occurring throughout the water column, challenging the classical definitions of "new" and "recycled" production.

In the second step, nitrite is oxidized to nitrate. In the oceans, this step is not as well understood as the first, but the bacteria *Nitrospina* and *Nitrobacter* are known to carry out this step in the sea.

Soil Conditions Controlling Nitrification Rates

- Substrate availability (presence of NH_4^+)
- Aeration (availability of O_2)
- Well-drained soils with 60% soil moisture
- pH(near neutral)
- Temperature (best 20-30°C) => Nitrification is seasonal, affected by land use practices

Inhibitors of Nitrification

Nitrification inhibitors are chemical compounds that slow the nitrification of ammonia, ammonium-containing, or urea-containing fertilizers, which are applied to soil as fertilizers. These inhibitors can help reduce losses of nitrogen in soil that would otherwise be used by crops. Nitrification inhibitors are used widely, being added to approximately 50% of the fall-applied anhydrous ammonia in states in the U.S., like Illinois. They are usually effective in increasing recovery of nitrogen fertilizer in row crops, but the level of effectiveness depends on external conditions and their benefits are most likely to be seen at less than optimal nitrogen rates.

The environmental concerns of nitrification also contribute to interest in the use of nitrification inhibitors: the primary product, nitrate, leaches into groundwater, producing acute toxicity in multiple species of wildlife and contributing to the eutrophication of standing water. Some inhibitors of nitrification also inhibit the production of methane, a greenhouse gas.

The inhibition of the nitrification process is primarily facilitated by the selection and inhibition/destruction of the bacteria that oxidize ammonia compounds.

A multitude of compounds that inhibit nitrification, which can be divided into the following areas: the active site of ammonia monooxygenase(AMO), mechanistic inhibitors, and the process of N-heterocyclic compounds. The process for the latter of the three is not yet widely understood, but is prominent. The presence of AMO has been confirmed on many substrates that are nitrogen inhibitors such as dicyandiamide, ammonium thiosulfate, and nitrapyrin.

The conversion of ammonia to hydroxylamine is the first step in nitrification, where AH_2 represents a range of potential electron donors.

$$NH_3 + AH_2 + O_2 \rightarrow NH_2OH + A + H_2O$$

This reaction is catalyzed by AMO. Inhibitors of this reaction bind to the active site on AMO and prevent or delay the process. The process of oxidation of ammonia by AMO is regarded with importance due to the fact that other processes require the co-oxidation of NH_3 for a supply of reducing equivalents. This is usually supplied by the compound hydroxylamine oxidoreductase(HAO) which catalyzes the reaction:

$$NH_2OH + H_2O \rightarrow NO_2^- + 5 H^+ + 4 e^-$$

The mechanism of inhibition is complicated by this requirement. Kinetic analysis of the inhibition of NH_3 oxidation has shown that the substrates of AMO have shown kinetics ranging from competitive to noncompetitive. The binding and oxidation can occur on two different locations on AMO: in competitive substrates, binding and oxidation occurs at the NH_3 site, while in noncompetitive substrates it occurs at another site.

Mechanism based inhibitors can be defined as compounds that interrupt the normal reaction catalyzed by an enzyme. This method occurs by the inactivation of the enzyme via covalent modification of the product, which ultimately inhibits nitrification. Through the process, AMO is deactivated and one or more proteins is covalently bonded to the final product. This is found to be most prominent in a broad range of sulfur or acetylenic compounds.

Sulfur-containing compounds, including ammonium thiosulfate (a popular inhibitor) are found to operate by producing volatile compounds with strong inhibitory effects such as carbon disulfide and thiourea.

In particular, thiophosphoryltriamide has been a notable addition where it has the dual purpose of inhibiting both the production of urease and nitrification. In a study of inhibitory effects of oxidation by the bacteria Nitrosomonaseuropaea, the use of thioethers resulted in the oxidation of these compounds to sulfoxides, where the S atom is the primary site of oxidation by AMO. This is most strongly correlated to the field of competitive inhibition.

Fig. : Examples of N-heterocyclic molecules.

N-heterocyclic compounds are also highly effective nitrification inhibitors and are often classified by their ring structure. The mode of action for these compounds is not well understood: while nitrapyrin, a widely-used inhibitor and substrate of AMO, is a weak mechanism-based inhibitor of said enzyme, the effects of said mechanism are unable to correlate directly with the compound's ability to inhibit nitrification. It is suggested that nitrapyrin acts against the monooxygenase enzyme within the bacteria, preventing growth and CH_4/NH_4 oxidation. Compounds containing two or three adjacent ring N atoms (pyridazine, pyrazole, indazole) tend to have a significantly higher inhibition effect than compounds containing non-adjacent N atoms or singular ring N atoms (pyridine, pyrrole). This suggests that the presence of ring N atoms is directly correlated with the inhibition effect of this class of compounds.

Methane Inhibition

Some enzymatic nitrification inhibitors, such as urease, can also inhibit the production of methane in methanotrophic bacteria. AMO shows similar kinetic turnover rates to methane monooxygenase(MMO) found in methanotrophs, indicating that MMO is a similar catalyst to AMO for the purpose of methane oxidation. Furthermore, methanotrophic bacteria share many similarities to NH_3 oxidizers such as Nitrosomonas. The inhibitor profile of particulate forms of MMO (pMMO) shows similarity to the profile of AMO, leading to similarity in properties between MMO in methanotrophs and AMO in autotrophs.

Environmental Concerns

Nitrification inhibitors are also of interest from an environmental standpoint because of the production of nitrates and nitrous oxide from the process of nitrification. Nitrous oxide (N_2O), although its atmospheric concentration is much lower than that of CO_2, has a global warming potential of about 300 times greater than carbon dioxide and contributes 6% of planetary warming due to greenhouse gases. This compound is also notable for catalyzing the breakup of ozone in the stratosphere. Nitrates, a toxic compound for wildlife and livestock and a product of nitrification, are also of concern.

Soil, consisting of polyanionic clays and silicates, generally has a net anionic charge. Consequently, ammonium (NH_4^+) binds tightly to the soil but nitrate ions (NO_3^-) do not. Because nitrate is more mobile, it leaches into groundwater supplies through agricultural runoff. Wildlife such as amphibians, freshwater fish, and insects are sensitive to nitrate levels, and have been known to cause death and developmental anomalies in affected species. In addition, because they easily leach into groundwater, contributing to eutrophication, a process in which large algal blooms reduce oxygen levels in bodies of water and lead to death in oxygen-consuming creatures due to anoxia. Nitrification is also thought to contribute to the formation of photochemical smog, ground level ozone, acid rain, changes in species diversity, and other undesirable processes. In addition, nitrification inhibitors have also been shown to suppress the oxidation of methane (CH_4), a potent greenhouse gas, to CO_2. Both nitrapyrin and acetylene are shown to be especially strong suppressors of both processes, although the modes of action distinguishing them are unclear.

SULFATION

Sulfation, sulfurylation, or sometimes incorrectly descirbed as **sulfonation**, in biochemistry is the enzyme-catalyzed conjugation of a sulfo group(not a sulfate or sulfuryl group) to another molecule. This biotransformation involves a sulfotransferase enzyme catalyzing the transfer of a sulfo group from a donor cosubstrate, usually 3'-phosphoadenosine-5'-phosphosulfate (PAPS), to a substrate molecule's hydroxyl or amine. Sulfation is involved in a variety of biological processes, including detoxification, hormone regulation, molecular recognition, cell signaling, and viral entry into cells. It is among the reactions in phase II drug metabolism, oftentimes effective in rendering a xenobiotic less active from a pharmacological and toxicological standpoint, but sometimes playing a role in the activation of xenobiotics(*e.g.* aromatic amines, methyl-substituted polycyclic aromatic hydrocarbons). Another example of biological sulfation is in the synthesis of sulfonatedglycosaminoglycans, such as **heparin, heparan sulfate, chondroitin sulfate**, and **dermatan sulfate**. Sulfation is also a possible posttranslational modification of proteins.

Tyrosine Sulfation

Tyrosine sulfation is a posttranslational modification in which a tyrosine residue of a protein is sulfated by a tyrosylproteinsulfotransferase(TPST) typically in the Golgi apparatus. Secreted proteins and extracellular parts of membrane proteins that pass through the Golgi apparatus may be sulfated. Such sulfation was first discovered by Bettelheim in bovine fibrinopeptide B in 1954 and later found be present in animals and plants but not in prokaryotes or in yeasts. Sulfation sites are tyrosine residues exposed on the surface of the protein typically surrounded by acidic residues. A detailed description of the characteristics of the sulfation site is available from PROSITE. Two types of tyrosylproteinsulfotransfe rases(TPST-1 and TPST-2) have been identified.

Function

Sulfation plays role in strengthening protein–protein interactions. Types of human proteins known to undergo tyrosine sulfation include adhesion molecules, G-protein-coupled receptors, coagulation factors, serine protease inhibitors, extracellular matrix proteins, and hormones. Tyrosine *O*-sulfate is a stable molecule and is excreted in urine in animals. No enzymatic mechanism of tyrosine sulfate desulfation is known to exist. By knock-out of *TPST* genes in mice, it may be observed that tyrosine sulfation has effects on the growth of the mice, such as body weight, fecundity, and postnatal viability.

Regulation

There is very limited evidence that the TPST genes are subject to transcriptional regulation and tyrosine O-sulfate is very stable and cannot be easily degraded by mammalian sulfatases. Tyrosine O-sulfation is an irreversible process *in vivo*. An antibody called PSG2 shows high sensitivity and specificity for epitopes containing sulfotyrosine independent of the sequence context.

ALKYLATION

Alkylation is the transfer of an alkyl group from one molecule to another. The alkyl group may be transferred as an alkyl carbocation, a free radical, a carbanion or a carbene (or their equivalents). An alkyl group is a piece of a molecule with the general formula C_nH_{2n+1}, where n is the integer depicting the number of carbons linked together. For example, a methyl group (CH_3) is a fragment of a methane molecule (CH_4); n = 1 in this case. **Alkylating agents** Utilize selective alkylation, by adding the desired aliphatic carbon chain to the previously chosen starting molecule. This is one of many known chemical syntheses.

In oil refining contexts, **alkylation** refers to a particular alkylation of isobutane with olefins. For upgrading of petroleum, alkylation produces synthetic C_7-C_8 alkylate, which is a premium blending stock for gasoline.

In medicine, alkylation of DNA is used in chemotherapy to damage the DNA of cancer cells. Alkylation is accomplished with the class of drugs called alkylating antineoplastic agents.

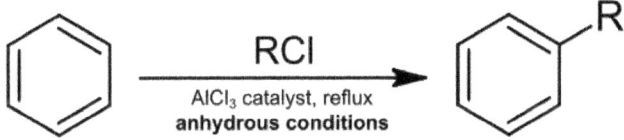

Fig. : Benzene Friedel-Crafts alkylation.

Alkylating Agents

Alkylating agents are classified according to their nucleophilic or electrophilic character.

Nucleophilic Alkylating Agents

Nucleophilic alkylating agents deliver the equivalent of an alkylanion (carbanion). Examples include the use of organometallic compounds such as Grignard (organomagnesium), organolithium, organocopper, and organosodium reagents. These compounds typically can add to an electron-deficient carbon atom such as at a carbonyl group. Nucleophilic alkylating agents can also displace halide substituents on a carbon atom. In the presence of catalysts, they also alkylate alkyl and aryl halides, as exemplified by Suzuki couplings.

Electrophilic Alkylating Agents

Electrophilic alkylating agents deliver the equivalent of an alkyl cation. Examples include the use of alkyl halides with a Lewis acid catalyst to alkylate aromatic substrates in Friedel-Crafts reactions. Alkyl halides can also react directly with amines to form C-N bonds; the same holds true for other nucleophiles such as alcohols, carboxylic acids, thiols, *etc.* Trimethyloxoniumtetrafluoroborate and triethyloxoniumtetrafluoroborate are particularly strong electrophiles due to their overt positive charge and an inert leaving group (dimethyl or diethyl ether).

Electrophilic, soluble alkylating agents are often very toxic, due to their ability to alkylate DNA. They should be handled with proper PPE. This mechanism of toxicity is also responsible for the ability of some alkylating agents to perform as anti-cancer drugs in the form of alkylating antineoplastic agents, and also as chemical weapons such as mustard gas. Alkylated DNA either does not coil or uncoil properly, or cannot be processed by information-decoding enzymes. This results in cytotoxicity with the effects of inhibition the growth of the cell, initiation of programmed cell death or apoptosis. However, mutations are also triggered, including carcinogenic mutations, explaining the higher incidence of cancer after exposure.

Alcohols and phenols can be alkylated to give alkyl ethers:

$$R\text{-}OH + R'\text{-}X \rightarrow R\text{-}O\text{-}R' + H\text{-}X$$

The produced acid HX is removed with a base, or, alternatively, the alcohol is deprotonated first to give an alkoxide or phenoxide. For example, dimethyl sulfate alkylates the sodium salt of phenol to give anisole, the methyl ether of phenol. The dimethyl sulfate is dealkylated to sodium methylsulfate.

$$Ph\text{-}O^- \, Na^+ + Me_2SO_4 \rightarrow Ph\text{-}O\text{-}Me + Na^+ \, MeSO_4^-$$

On the contrary, the alkylation of amines introduces the problem that the alkylation of an amine makes it *more* nucleophilic. Thus, when an electrophilic alkylating agent is introduced to a primary amine, it will preferentially alkylate all the way to a quaternary ammoniumcation.

$$R\text{-}NH_2 \rightarrow R\text{-}NH\text{-}R' \rightarrow R\text{-}N(R')_2 \rightarrow R\text{-}N(R')_3^+ \text{(alkylating agent omitted for clarity)}$$

If the quaternary ammonium is not the desired product, more circuitous routes such as reductive amination are necessary.

Carbene Alkylating Agents

Carbenes are extremely reactive and are known to attack even unactivated C-H bonds. Carbenes can be generated by elimination of a diazo group. Unlike electrophilic or nucleophilic alkylating agents, carbenes are neutral, and they insert into bonds rather than discard leaving groups. A metal can form a carbene equivalent called a transition metal carbene complex.

In Biology

Methylation is the most common type of alkylation, being associated with the transfer of a methyl group. Methylation is distinct from alkylation in that it is specifically the transfer of one carbon, whereas alkylation can refer to the transfer of long chain carbon groups. Methylation in nature is typically effected by vitamin B12-derived enzymes, where the methyl group is carried by cobalt. In methanogenesis, coenzyme M is methylated by tetrahydromethanopterin.

Electrophilic compounds may alkylate different nucleophiles in the body. The toxicity, carcinogenity, and paradoxically, cancer cell-killing abilities of different DNA alkylating agents are an example.

Oil Refining

Alkylation of alkenes (shown here is propene) by isobutane is a major process in refineries. It is catalysed by strong acids such as HF and sulfuric acid.

In a standard oil refinery process, isobutane is alkylated with low-molecular-weight alkenes(primarily a mixture of propene and butene) in the presence of a Bronsted acid catalyst, either sulfuric acid or hydrofluoric acid. In an oil refinery it is referred to as a sulfuric acid alkylation unit(SAAU) or a hydrofluoric alkylation unit, (HFAU). Refinery workers may simply refer to it as the alky or alky unit. The catalyst protonates the alkenes (propene, butene) to produce reactive carbocations, which alkylate isobutane. The reaction is carried out at mild temperatures (0 and 30 °C) in a two-phase reaction. Because the reaction is exothermic, cooling is needed: SAAU plants require lower temperatures so the cooling medium needs to be chilled, for HFAU normal refinery cooling water will suffice. It is important to keep a high ratio of isobutane to alkene at the point of reaction to prevent side reactions which produces a lower octane product, so the plants have a high recycle of isobutane back to feed. The phases separate spontaneously, so the acid phase is vigorously mixed with the hydrocarbon phase to create sufficient contact surface.

The product is called alkylate and is composed of a mixture of high-octane, branched-chain paraffinichydrocarbons(mostly isoheptane and isooctane). Alkylate is a premium gasoline blending stock because it has exceptional anti-knock properties and is clean burning. Alkylate is also a key component of avgas. The octane number of the alkylate depends mainly upon the kind of alkenes used and upon operating conditions. For example, isooctane results from combining butylene with isobutane and has an octane rating of 100 by definition. There are other products in the alkylate, so the octane rating will vary accordingly.

Since crude oil generally contains only 10 to 40 percent of hydrocarbon constituents in the gasoline range, refineries use a fluid catalytic cracking process to convert high molecular weight hydrocarbons into smaller and more volatile compounds, which are then converted into liquid gasoline-size hydrocarbons. Alkylation processes transform low molecular-weight alkenes and iso-paraffin molecules into larger iso-paraffins with a high octane number.

Combining cracking, polymerization, and alkylation can result in a gasoline yield representing 70 percent of the starting crude oil. More advanced processes, such as cyclicization of paraffins and dehydrogenation of naphthenes forming aromatic hydrocarbons in a catalytic reformer, have also been developed to increase the octane rating of gasoline. Modern refinery operation can be shifted to produce almost any fuel type with specified performance criteria from a single crude feedstock.

Refineries examine whether it makes sense economically to install alkylation units. Alkylation units are complex, with substantial economy of scale. In addition to a suitable quantity of feedstock, the price spread between the value of alkylate product and alternate feedstock disposition value must be large enough to justify the installation. Alternative outlets for refinery alklylationfeedstocks include sales as LPG, blending of C_4 streams directly into gasoline and feedstocks for chemical plants. Local market conditions vary widely between plants. Variation in the RVP specification for gasoline between countries and between seasons dramatically impacts the amount of butane streams that can be blended directly into gasoline. The transportation of specific types of LPG streams can be expensive so local disparities in economic conditions are often not fully mitigated by cross market movements of alkylation feedstocks.

The availability of a suitable catalyst is also an important factor in deciding whether to build an alkylation plant. If sulfuric acid is used, significant volumes are needed. Access to a suitable plant is required for the supply of fresh acid and the disposition of spent acid. If a sulfuric acid plant must be constructed specifically to support an alkylation unit, such construction will have a significant impact on both the initial requirements for capital and ongoing costs of operation. Alternatively it is possible to install a WSA Process unit to regenerate the spent acid. No drying of the gas takes place. This means that there will be no loss of acid, no acidic waste material and no heat is lost in process gas reheating. The selective condensation in the WSA condenser ensures that the regenerated fresh acid will be 98% w/w even with the humid process gas. It is possible to combine spent acid regeneration with disposal of hydrogen sulfide by using the hydrogen sulfide as internal fuel in the refinery or elsewhere.

The second main catalyst option is hydrofluoric acid. In typical alkylation plants, rates of consumption for acid are much lower than for sulfuric acid. These plants also produce alkylate with better octane rating than do sulfuric plants. However, due to its hazardous nature, HF acid is produced at very few locations and transportation must be managed rigorously.

POLYMERIZATION

In polymer chemistry, **polymerization** is a process of reacting monomer molecules together in a chemical reaction to form polymer chains or three-dimensional networks. There are many forms of polymerization and different systems exist to categorize them.

Fig. : An example of **alkene polymerization**, in which each styrene monomer's double bond reforms as a single bond plus a bond to another styrene monomer. The product is polystyrene.

In chemical compounds, polymerization occurs via a variety of reaction mechanisms that vary in complexity due to functional groups present in reacting compounds and their inherent steric effects. In more straightforward polymerization, alkenes, which are relatively stable due to σ bonding between carbon atoms, form polymers through relatively simple radical reactions; in contrast, more complex reactions such as those that involve substitution at the carbonyl group require more complex synthesis due to the way in which reacting molecules polymerize.

As alkenes can be formed in somewhat straightforward reaction mechanisms, they form useful compounds such as polyethylene and polyvinyl chloride(PVC) when undergoing radical reactions, which are produced in high tonnages each year due to their usefulness in manufacturing processes of commercial products, such as piping, insulation and packaging. In general, polymers such as PVC are referred to as "**homopolymers**," as they consist of repeated long chains or structures of the same monomer unit, whereas polymers that consist of more than one molecule are referred to as copolymers(or co-polymers).

Other monomer units, such as formaldehyde hydrates or simple aldehydes, are able to polymerize themselves at quite low temperatures (ca. −80 °C) to form trimers; molecules consisting of 3 monomer units, which can cyclize to form ring cyclic structures, or undergo further reactions to form tetramers, or 4 monomer-unit compounds. Further compounds either being referred to as oligomers in smaller molecules. Generally, because formaldehyde is an exceptionally reactive electrophile it allows nucleophillic addition of hemiacetal intermediates, which are in general short-lived and relatively unstable "mid-stage" compounds that react with other molecules present to form more stable polymeric compounds.

Polymerization that is not sufficiently moderated and proceeds at a fast rate can be very hazardous. This phenomenon is known as hazardous polymerization and can cause fires and explosions.

Step-growth

Step-growth polymers are defined as polymers formed by the stepwise reaction between functional groups of monomers, usually containing heteroatoms such as nitrogen or oxygen. Most step-growth polymers are also classified as condensation polymers, but not all step-growth polymers (like polyurethanes formed from isocyanate and alcohol bifunctional monomers) release condensates; in this case, we talk about addition polymers. Step-growth polymers increase in molecular weight at a very slow rate at lower conversions and reach moderately high molecular weights only at very high conversion (*i.e.*, >95%).

To alleviate inconsistencies in these naming methods, adjusted definitions for condensation and addition polymers have been developed. A condensation polymer is defined as a polymer that involves loss of small molecules during its synthesis, or contains heteroatoms as part of its backbone chain, or its repeat unit does not contain all the atoms present in the hypothetical monomer to which it can be degraded.

Chain-growth

Chain-growth polymerization (or addition polymerization) involves the linking together of molecules incorporating double or triple carbon-carbon bonds. These unsaturated *monomers*(the identical molecules that make up the polymers) have extra internal bonds that are able to break and link up with other monomers to form a repeating chain, whose backbone typically contains only carbon atoms. Chain-growth polymerization is involved in the manufacture of polymers such as polyethylene, polypropylene, and polyvinyl chloride(PVC). A special case of chain-growth polymerization leads to living polymerization.

In the radical polymerization of ethylene, its π bond is broken, and the two electrons rearrange to create a new propagating center like the one that attacked it. The form this propagating center takes depends on the specific type of addition mechanism. There are several mechanisms through which this can be initiated. The free radical mechanism is one of the first methods to be used. Free radicals are very reactive atoms or molecules that have unpaired electrons. Taking the polymerization of ethylene as an example, the free radical mechanism can be divided into three stages: chain initiation, chain propagation, and chain termination.

Fig. : Polymerization of ethylene.

Free radical addition polymerization of ethylene must take place at high temperatures and pressures, approximately 300 °C and 2000 atm. While most other free radical polymerizations do not require such extreme temperatures and pressures, they do tend to lack control. One effect of this lack of control is a high degree of branching. Also, as termination occurs randomly, when two chains collide, it is impossible to control the length of individual chains. A newer method of polymerization similar to free radical, but allowing more control involves the Ziegler-Natta catalyst, especially with respect to polymer branching.

Other forms of chain growth polymerization include cationic addition polymerization and anionic addition polymerization. While not used to a large extent in industry yet due to stringent reaction conditions such as lack of water and oxygen, these methods provide ways to polymerize some monomers that cannot be polymerized by free radical methods such as polypropylene. Cationic and anionic mechanisms are also more ideally suited for living polymerizations, although free radical living polymerizations have also been developed.

Esters of acrylic acid contain a carbon-carbon double bond which is conjugated to an ester group. This allows the possibility of both types of polymerization mechanism. An acrylic ester by itself can undergo chain-growth polymerization to form a homopolymer with a carbon-carbon backbone, such as poly(methyl methacrylate). Also, however, certain acrylic esters can react with diamine monomers by nucleophilic conjugate addition of amine groups to acrylic C=C bonds. In this case the polymerization proceeds by step-growth and the products are poly(beta-amino ester)copolymers, with backbones containing nitrogen (as amine) and oxygen (as ester) as well as carbon.

Physical Polymer Reaction Engineering

To produce a high-molecular-weight, uniform product, various methods are employed to better control the initiation, propagation, and termination rates during chain polymerization and also to remove excess concentrated heat during these exothermic reactions compared to polymerization of the pure monomer (also referred to as bulk polymerization). These include emulsion polymerization, solution polymerization, suspension polymerization, and precipitation polymerization. Although the polymer polydispersity and molecular weight may be improved, these methods may introduce additional processing requirements to isolate the product from a solvent.

Photopolymerization

Most **photopolymerization** reactions are chain-growth polymerizations which are initiated by the absorption of visible or ultraviolet light. The light may be absorbed either directly by the reactant monomer (*direct* photopolymerization), or else by a *photosensitizer* which absorbs the light and then transfers energy to the monomer. In general only the initiation step differs from that of the ordinary thermal polymerization of the same monomer; subsequent propagation, termination and chain transfer steps are unchanged. In step-growth photopolymerization, absorption of light triggers an addition (or condensation) reaction between two comonomers that do not react without light. A propagation cycle is not initiated because each growth step requires the assistance of light.

Photopolymerization can be used as a photographic or printing process, because polymerization only occurs in regions which have been exposed to light. Unreacted monomer can be removed from unexposed regions, leaving a relief polymeric image. Several forms of 3D printing—including layer-by-layer stereolithography and two-photon absorption 3D photopolymerization—use photopolymerization.

Chapter 2

CONTROL ARCHITECTURES

There are many different control mechanisms that can be used, both in everyday life and in chemical engineering applications. Two broad control schemes, both of which encompass each other are *feedback control* and *feed-forward control*. *Feedback control* is a control mechanism that uses information from **measurements** to manipulate a variable to achieve the desired result. *Feed-forward control*, also called anticipative control, is a control mechanism that predicts the effects of measured **disturbances** and takes corrective action to achieve the desired result. The focus of this chapter is to explain application, advantages, and disadvantages of feedback control.

Feedback control is employed in a wide variety of situations in everyday life, from simple home thermostats that maintain a specified temperature, to complex devices that maintain the position of communication satellites. Feedback control also occurs in natural situations, such as the regulation of blood-sugar levels in the body. Feedback control was even used more than 2,000 years ago by the Greeks, who manufactured such systems as the float valve which regulated water level. Today, this same idea is used to control water levels in boilers and reservoirs.

FEEDBACK CONTROL

In feedback control, the variable being controlled is measured and compared with a target value. This difference between the actual and desired value is called the error. Feedback control manipulates an input to the system to minimize this error. **Figure** shows an overview of a basic feedback control loop. The error in the system would be the *Output - Desired Output*. Feedback control reacts to the system and works to minimize this error. The desired output is generally entered into the system through a user interface. The output of the system is measured (by a flow meter, thermometer or similar instrument) and the difference is calculated. This difference is used to control the system inputs to reduce the error in the system.

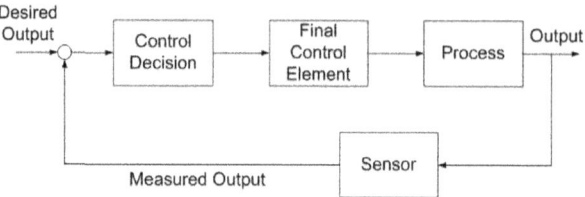

Fig. : Feedback control loop (Adapted from Lee, Newell, and Cameron 1998: 6).

To understand the principle of feedback control, consider figure. In order to bake cookies, one has to preheat an electric oven to 350°F. After setting the desired temperature, a sensor takes a reading inside the oven. If the oven is below the set temperature, a signal is sent to the heater to power on until the oven heats to the desired temperature. In this example, the variable to be controlled (oven temperature) is measured and determines how the input variable (heat into oven) should be manipulated to reach the desired value.

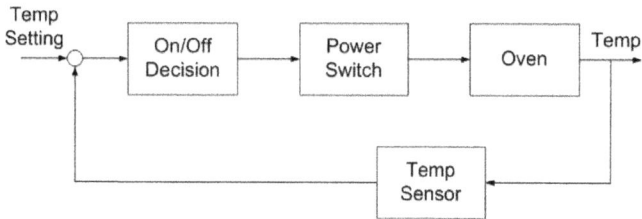

Fig. : Feedback control in an electric oven.

Feedback control can also be demonstrated with human behaviour. For example, if a person goes outside in Michigan winter, he or she will experience a temperature drop in the skin. The brain (controller) receives this signal and generates a motor action to put on a jacket. This minimizes the discrepancy between the skin temperature and the physiological set point in the person. The example is illustrated below:

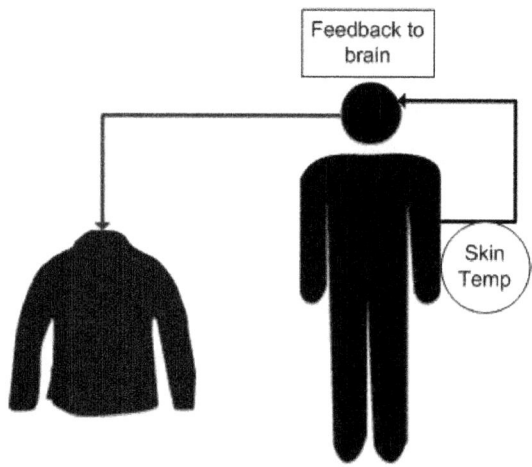

There are two types of feedback control: negative and positive. Negative feedback is the most useful control type since it typically helps a system converge toward an equilibrium state. On the other hand, positive feedback can lead a system away from an equilibrium state thus rendering it unstable, even potentially producing unexpected results. Unless stated explicitly, the term feedback control most often refers to negative feedback.

Negative Feedback

By definition, negative feedback is when a change (increase/decrease) in some variable results in an opposite change (decrease/increase) in a second variable. This is demonstrated in **Figure** where a loop represents a variation toward a plus that triggers a correction toward the minus, and vice versa. Negative feedback leads to a tight control situation whereby the corrective action taken by the controller forces the controlled variable toward the set point, thus leading the system to oscillate around equilibrium.

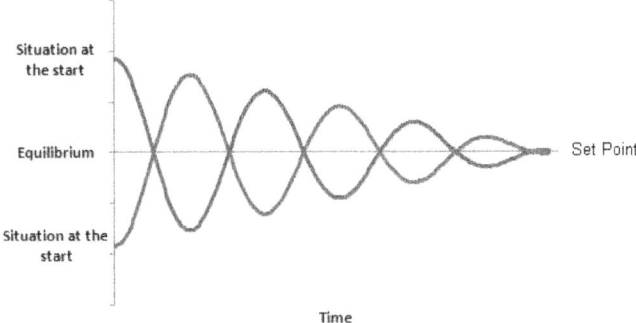

Fig. : Negative Feedback: Maintenance of equilibrium and convergence.

Figure shows the negative feedback mechanism between Duck Population and Duck Death Rate. For a given flock, we say that if the death rate increases, the duck flock will decrease. On the contrary, if the duck flock increases, the death rate of the flock will decrease.

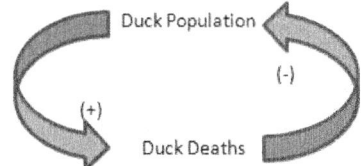

Fig. : Duck population negative feedback loop.

Positive Feedback

As opposed to negative feedback, positive feedback is when a change (increase/decrease) in some variable results in a subsequently similar change (increase/decrease) in a second variable. In some cases, positive feedback leads to an undesirable behaviour whereby the system diverges away from equilibrium. This can cause the system to either run away toward infinity, risking an expansion or even an explosion, or run away toward zero, which leads to a total blocking of activities.

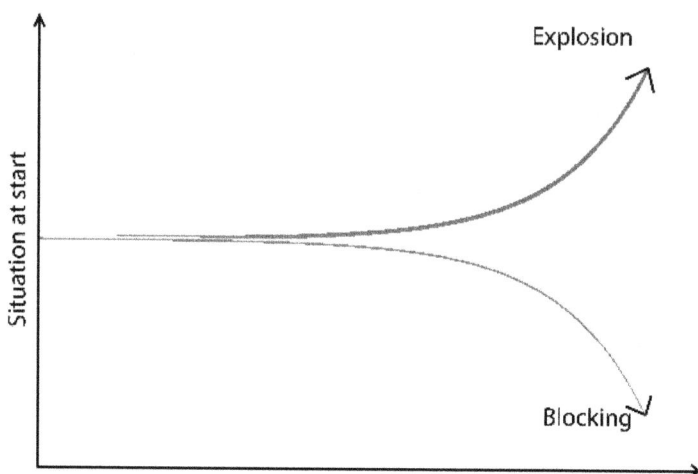

Fig. : Positive Feedback: Exponential Growth and divergent behaviour, no intermediate situation.

Figure shows the feedback mechanism responsible for the growth of a duck flock *via* births. In this example, we consider two system variables: Duck Births and Ducks Population. For a given flock, we state that if the birth rate increases, the duck flock will increase. Similarly, if the duck flock increases, the birth rate of the flock will increase.

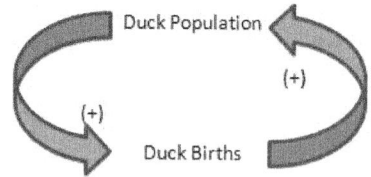

Fig. : Duck population positive feedback loop.

Applications

Control mechanisms are used to achieve a desired output or operating condition. More specifically, feedback control is a useful mechanism in numerous applications across many engineering fields. In chemical engineering, feedback control is commonly used to manipulate and stabilize the conditions of a CSTR. **Figure** shows how feedback control can be effectively used to stabilize the concentrations of reactants in a CSTR by adjusting the flow rates.

CSTR with Feedback Control

Several types of feedback control can be used to manipulate the conditions in a CSTR: positive feedback, negative feedback, or a combination of both. **Figure** illustrates each of these possible situations. As depicted below, each CSTR is equipped with two electrodes that measure the voltage of the solution contained inside the reactor. A computer adjusts the flow rates of the pump(s) in response to any changes in the voltage.

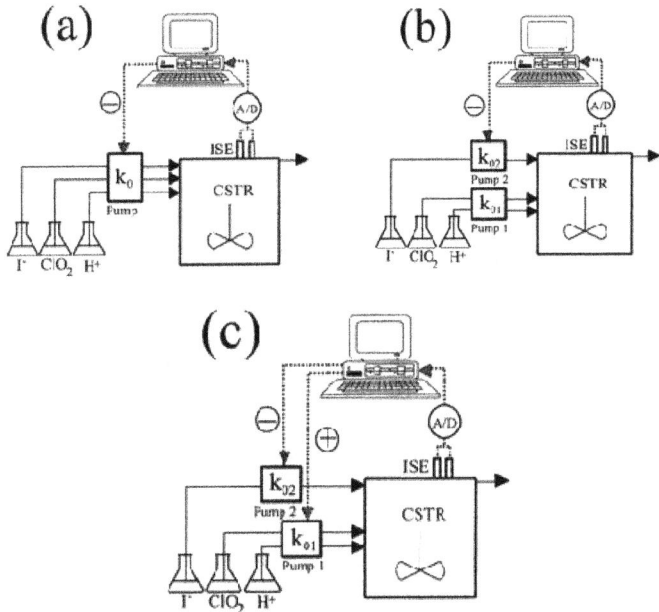

Fig. : CSTR with feedback control: equipment and control configuration.

(a) All of the reagents are pumped into the reactor by the same pump. The flow rate through the pump is adjusted constantly by a negative feedback mechanism; when level of the iodide solution is low, the computer detects the insufficiency and increases the flow rate of all the reactants.

(b) The iodide solution is pumped into the reactor by one pump, while the other two reactants are pumped in at a constant flow rate. The flow rate of the iodine solution is controlled by a negative feedback mechanism; when the computer detects an insufficient amount of iodine in the CSTR, it signals the pump. The flow rate of iodine into the CSTR is therefore increased.

(c) Two pumps are used to feed the reactor: one pump for the iodine solution and one for the ClO_2 and H^+ solutions. The flow of the iodine solution is controlled by a negative feedback mechanism; if the computer detects an iodine deficiency, it signals Pump 2 to increase the flow rate of the iodide solution. The flow rate of the ClO_2 and H^+ solutions is controlled by a positive feedback mechanism; if the computer detects an iodide deficiency, it will signal Pump 1 to decrease the flow rates of ClO_2 and H^+, thereby increasing the concentraion of iodide.

It is easy to see that by combining feedback controls, such as in **Figure**, output concentrations and operating conditions can be maintained at the desired state much more effectively than by only using one type of feedback control.

Advantages and Disadvantages

The unique architecture of the feedback control provides for many advantages and disadvantages. It is important to look at the exact application the control will be used for before determining which type of control will be the best choice.

Advantages

The advantages of feedback control lie in the fact that the feedback control obtains data at the process output. Because of this, the control takes into account unforeseen disturbances such as frictional and pressure losses. Feedback control architecture ensures the desired performance by altering the inputs immediately once deviations are observed regardless of what caused the disturbance.

An additional advantage of feedback control is that by analyzing the output of a system, unstable processes may be stabilized. Feedback controls do not require detailed knowledge of the system and, in particular, do not require a mathematical model of the process. Feedback controls can be easily duplicated from one system to another. A feedback control system consists of five basic components:

(1) input,
(2) process being controlled,
(3) output,
(4) sensing elements, and
(5) controller and actuating devices.

A final advantage of feedback control stems from the ability to track the process output and, thus, track the system's overall performance.

Disadvantages

Time lag in a system causes the main disadvantage of feedback control. With feedback control, a process deviation occurring near the beginning of the process will not be recognized until the process output. The feedback control will then have to adjust the process inputs in order to correct this deviation. This results in the possibility of substantial deviation throughout the entire process. The system could possibly miss process output disturbance and the error could continue without adjustment.

Generally, feedback controllers only take input from one sensor. This may be inefficient if there is a more direct way to control a system using multiple sensors. Operator intervention is generally required when a feedback controller proves unable to maintain stable closed-loop control. Because the control responds to the perturbation after its occurrence, perfect control of the system is theoretically impossible. Finally, feedback control does not take predictive control action towards the effects of known disturbances.

Closed Loop Control *versus* Open Loop Control

Although there are various types of controllers, most of them can be grouped into one of the two broad categories: closed loop and open loop controllers. The subsections below summarize the differentiation.

Closed Loop System

In a closed loop control system, the input variable is adjusted by the controller in order to minimize the error between the measured output variable and its set point. This control design is synonymous to feedback control, in which the deviations between the measured variable and a set point are fed back to the controller to generate appropriate control actions. The controller C takes the difference e between the reference r and the output to change the inputs u to the system. This is shown in figure below. The output of the system y is fed back to the sensor, and the measured outputs go to the reference value

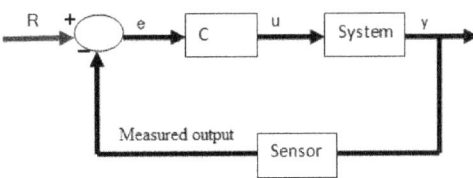

Open Loop System

On the other hand, any control system that does not use feedback information to adjust the process is classified as open loop control. In open loop control,

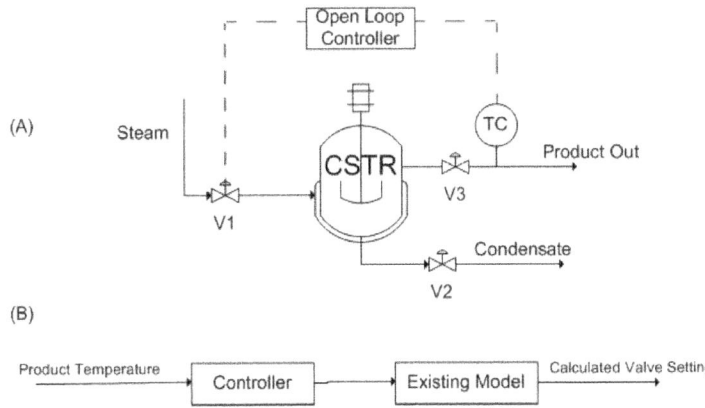

Fig. : Example of CSTR open loop controller.(A) System diagram. (B) Open control loop.

the controller takes in one or several measured variables to generate control actions based on existing equations or models. Consider a CSTR reactor that needs to maintain a set reaction temperature by means of steam flow: A temperature sensor measures the product temperature, and this information is sent to a computer for processing. But instead of outputting a valve setting by using the error in temperature, the computer (controller) simply plugs the information into a predetermined equation to reach output valve setting. In other words, the valve setting is simply a function of product temperature.

Note that the open loop controller only uses the **current state** of the measured variable (product temperature) and a model to generate its control output (valve setting), as opposed to monitoring errors that have already taken place. As the result, the quality of the control system depends entirely upon the accuracy of the implemented model, which is challenging to develop. For this reason, feedback, or closed loop, controllers are generally recognized as the more reliable control system.

Short Summary on Closed and Open Loop Controllers

Feedback Controller = Closed Loop Controller

Non-Feedback Controller = Open Loop Controller

FEED-FORWARD CONTROL

Feed-forward control is a useful tool in the field of chemical engineering when there is a known set of deviations occurring upstream of the system. This would allow engineers to account for that particular deviation within the controller and reduce the effects of the deviation on the system. An example would be a car's cruise-control system. If it has feedback control, when there is a slope and therefore a speed reduction, the feedback controller would compensate by applying additional throttle to the engine. If it uses a feed-forward controller instead, the controller would calculate the slope beforehand and throttle the engine before any speed reduction occurs. In this sense, the controller predicts the incoming deviation and compensates for it.

The following block diagram shows a feed-forward controller implimented on a idealized process with a setpoint R and load U:

Where:

$\hat{G}_P$ represents the process operator, $\hat{G}_M$ represents the measurement operator, $\hat{G}_C$ represents the controller operator, and FF is the feed-forward controller.

The perfect feed-forward controller is the inverse of the process operator, $\hat{G}_P^{-1}$

For example:

$$\hat{G}_P^{-1} = K_P \left(\tau \frac{\delta}{\delta t} + 1 \right) = \hat{G}_{FF}$$

Control Architectures

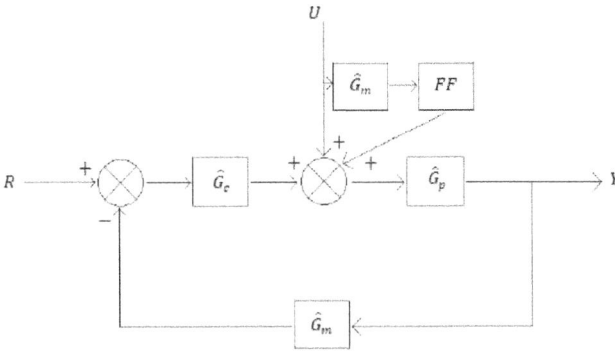

In General:

$$\hat{G}_P^{-1} Y = (U + \hat{G}_{FF} U) + \hat{G}_C (R - Y)$$

(in the language of operator or state space).

The objective of feed-forward control is to measure disturbances and compensate for them before the controlled variable deviates from the setpoint. Feed-forward control basically involves a control equation which has certain corrective terms which account for predicted disturbances entering the system. The equation is only effective for gains in a steady state process. Dynamic compensation should be used in the control equation if there are any dynamic deviations with the process response to the control action. This dynamic compensation ability will be discussed further in the next section.

One form of a feed-forward control would be a Derivative (D) control which calculates the change in error and compensates proportionately. But a D-control can't perform by itself and usually requires working in conjunction with a Proportional (P) or Proportional Integral (PI) control.

Simulated below is a typical shell and tube heat exchanger which heats up liquid water using steam.

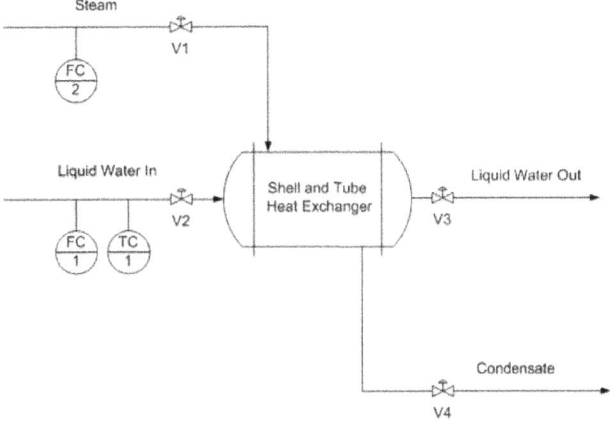

Adding a feed-forward control into the system manipulates the amount of steam required to compensate for the varying amounts of liquid feed coming in.

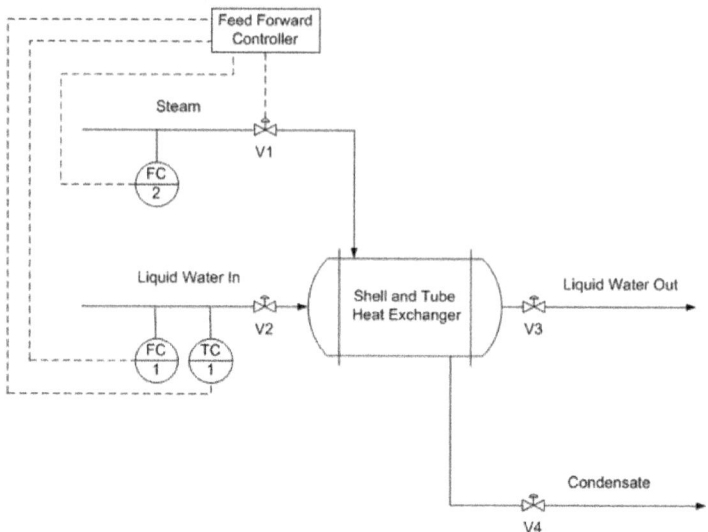

Overall heat balance:

Heat into the system is equal to the heat leaving the system.

Heat gained by liquid:

$$q_{out} = m_l\, C_p(T_2 - T_1)$$

Where:

m_l -- mass flow rate of the liquid

C_p -- the heat capacity of the liquid

T_1 -- the input temperature of the liquid

T_2 -- the desired temperature setpoint for the liquid

Heat lost by steam:

$$q_{in} = m_s \lambda$$

Where:

λ -- heat of vapourization

m_s -- mass flow rate of the steam

Therefore:

$$m_s = \frac{C_p}{\lambda} m_l (T_2 - T_1)$$

To compensate for a deviation from the desired inlet temperature of the liquid water, the amount of steam must be adjusted. Equation acts as the control equation for the feed-forward controller. This controller actuates the inlet valve of the steam and acquires temperature and flow rate data from the inlet streams. Based on the desired temperature setpoint, T_2, and the actual input temperature, T_1, we can use this equation to calculate the mass flow rate of the steam required to compensate for the temperature deviation.

Accounting for System Non-Idealities

Often, one of the most difficult tasks associated with created a functional feed forward controller is determining the necessary equations that govern the system. Even more complexities arise when the system is not, and cannot be treated as, ideal. This is the case with many real and practical systems. There is simply too much heat lost or too many unforeseen effects to safely assume ideal conditions. The pure math of the example above does not account for these effects.

The equations will output a value to a control valve (often in voltage). That voltage will be derived from some previously determined relationship between voltage, valve %open, and steam flow rate. A very simple way to begin managing the issue of non-ideality is by including a "non-ideality constant". This can be an additive or a multiplicative constant that adjusts the voltage output determined by the equations.

$V = f(m_s)$ *Voltage output is some function of the calculated required steam flow*

$V = c_N * f(m_s)$ *or* $V = c_N + f(m_s)$ *Voltage output is adjusted by some constant*

This non-ideality constant c_n often must be determined by trial. One way to accomplish this is to use manual control to determine the output voltage needed at various inlet conditions. Using the data from the manual trials, and the resulting voltage that your unadjusted feed forward controller would output, it is possible to determine by what factor your feed forward voltage needs to be adjusted. For example, if by manual control you determine that for inlet conditions X the required voltage is 300mV and your feed forward controller is only outputting 270mV for conditions X, you need some factor to adjust.

It may also happen that your "non-ideality constant" will not turn out to be constant when you begin to look at the data. In this situation, consider using a linear relationship between the non-ideality factor and some inlet condition or implementing CASE or IF statements covering discrete ranges of inlet conditions and giving different constants.

DYNAMIC COMPENSATION

Dynamic compensation is a method to account for factors such as lead and lag times when using feed-forward control. For instance, in the example above, when the feed forward controller monitors a temperature decrease in the liquid feed, it will increase the steam flow rate - the manipulated variable - to compensate and thus maintain the temperature of the exiting liquid flow - the controlled variable.

However, the steam may enter the heat exchanger faster than the liquid feed and this will cause a transient increase in the controlled variable from the setpoint.

In an ideal case the steam and liquid feed would enter the heat exchanger at the same time and no deviation from the set exiting temperature would be observed. Therefore, dynamic compensation involves predicting non-ideal behaviour and accounting for it. While perfect feed-forward control is nearly impossible, dynamic compensation is one step closer.

Open Loop System

Feed-forward control is a open loop system. In an open loop system, the controller uses current, or live, information of the system to generate appropriate actions by using predetermined models. The sensor providing the reference command to the closed loop actuator is not an error signal generated from a feedback sensor but a command based on measurements. This is the defining characteristic of an open loop system, in which the controller does not manipulate the system by trying to minimize errors in the controlled variable.

Because a feed-forward controller listens to the system and calculates adjustments without directly knowing how well the controlled variable is behaving (if it does, it is getting feedback), it is open loop in nature. In other words, the controller operates on "faith", that its mathematical models are able to accurately generate responses (valve setting, motor speed, *etc.*) that lead to expected results. Therefore, it is critical to implement good models in feed-forward systems. This is often the most difficult part of a feed-forward design, as mentioned previously.

However, not all open loop systems are feed-forward. Open loop simply means the system is not getting feedback information, in which the controlled variable tells the controller how well it is doing compared to its set point (*i.e.* error). Feed-forward control, on the other hand, takes an extra step by using non-feedback information to produce predictive actions.

This simple open loop system is an example of a open loop system that is not feed-forward. The controller simply takes the current state of the controlled variable to generate a valve setting. It is not getting a feedback, because the product temperature is not compared to a set pont. It is not feed-forward either, because there is no mechanism that produces actions that may yield expected results in the future.

Feed-forward Applications

Feed-forward control is used in many chemical engineering applications. These include heat exchangers, CSTRs, distillation columns and many other applications. A typical furnace, shown below, is heating up an input fluid using fuel gas.

One possible disturbance is the flow rate of the incoming fluid. For example: If the fluid input rate was increased by 10%, then the required heat duty should also be increased by approximately 10%. This feed-forward control strategy immediately changes the fuel gas flow rate and therefore the heat duty. The performance of feed-forward controls is limited by model uncertainty, and in practice

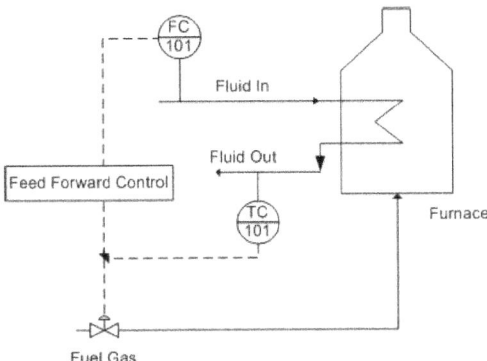

feed-forward control is combined with feedback control in order to ensure better control of the system. Feed-forward control will compensate for fluid input disturbances while feedback control will compensate for other disturbances — such as the fuel gas flow rate or the temperature of the furnace — and model uncertainty. This system can also be represented by the following block diagram.

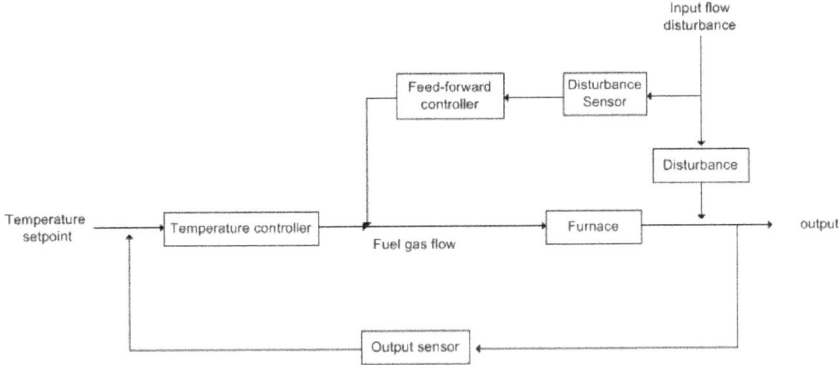

Pros and Cons of Feed-Forward Control

Different applications require different types of control strategies. Sometimes feed-forward solutions are required for proper system control; sometimes only feedback solutions are necessary. Feedback systems don't always maintain the setpoint as well because of the lag that comes with waiting for the disturbance to propagate through the system. As a result, many control systems use a combination of feed-forward and feedback strategies, such as PID controllers.

PID controllers use the Proportional-Integral control for feedback and the Derivative control for feed-forward control. This forms a system with multiple loops, otherwise known as a cascading system. A critical advantage of running both forward and backward controls is that the system is still somewhat able to adjust a variable if one mechanism fails since the two loops use different sensors. As a result, PID controllers are great for controlling processes, however, they require a number of equations to determine feed-forward and feedback correction.

Feed-forward systems work by checking the conditions of an incoming stream and adjusting it before the system is adversely affected. If the controller is told the traits of an acceptable incoming stream, then it can compare that standard to whatever is coming down the pipe. The feed-forward controller can look at this error and send a corrective signal to the automatic valve responsible for that pipe (or any other control device).

In order to have this kind of predictive ability, the controller must have explicitly defined equations that account for the effects of a disturbance on the system. In addition, these equations must also then prescribe action to counter-act the disturbance. This can become even more difficult when there are several incoming stream traits that are being observed. When there are multiple inputs, the feed-forward system will require non-linear equations, leading to the development of neural networks. Neural networks are based on "neurons", which are representations of non-linear equations. This concept is based on the brain's use of neurons to process and transmit information. The neuron is actually comprised of a set of sigmoidal equations relating inputs to outputs. Sigmoid functions are non-linear equations that take inputs and apply constants, or weights, to transform the value to make an output. Below is a picture that illustrates the function that these neurons serve.

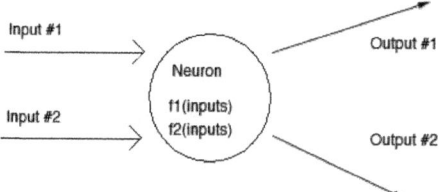

The picture above shows a multiple input-output system. Most feed-forward systems have to deal with more than one input.

Neurons can be "connected" in ways that allow the inputs to be transformed any number of times. Neurons that are connected indicate that one sigmoidal function's output becomes the input of another one. Although the concept of neurons is easily understandable, the difficulty lies in the potential complexity of real systems. For example: the number and type of inputs go to which neurons, the initial weights be, the number of neurons needed, *etc.* As you can see, there are a lot of design questions that make developing the neural network difficult.

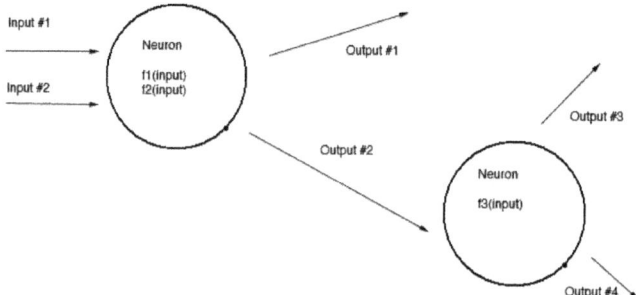

Coupled with an expected set of upstream conditions, the feed-forward system can continually adjust the method it uses to control an output variable. The system is capable of doing this by measuring sensor inputs coming into the controller, using neurons to transform the data, and comparing the resulting output(s) to a set of known or desired outputs. If the actual output is different from the desired output, the weights are altered in some fashion and the process repeats until convergence is achieved. This is how a controller is able to "learn". Learning is just discovering the weighting factors that allow the transformed ouputs to match the desired outputs.

The following table shows a list of feed-forward pros and cons:

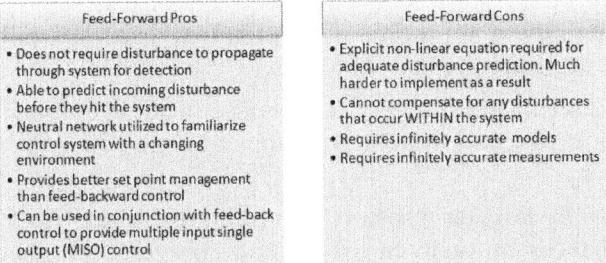

Feed-Forward Pros	Feed-Forward Cons
• Does not require disturbance to propagate through system for detection • Able to predict incoming disturbance before they hit the system • Neutral network utilized to familiarize control system with a changing environment • Provides better set point management than feed-backward control • Can be used in conjunction with feed-back control to provide multiple input single output (MISO) control	• Explicit non-linear equation required for adequate disturbance prediction. Much harder to implement as a result • Cannot compensate for any disturbances that occur WITHIN the system • Requires infinitely accurate models • Requires infinitely accurate measurements

A CSTR with a given volume with heat-exchange capability has a coolant water system to maintain a specific system temperature (368K). To maintain this temperature, the flow of coolant water oscillates. Let a temperature disturbance of 100K be introduced over a period of 10 minutes. For this simulation, the Excel spreadsheet from PID-Tuning Optimization was used to create feed-forward and feed-backward data. In the following plot, observe the difference between using feed-forward control versus feed-backward control in an example.

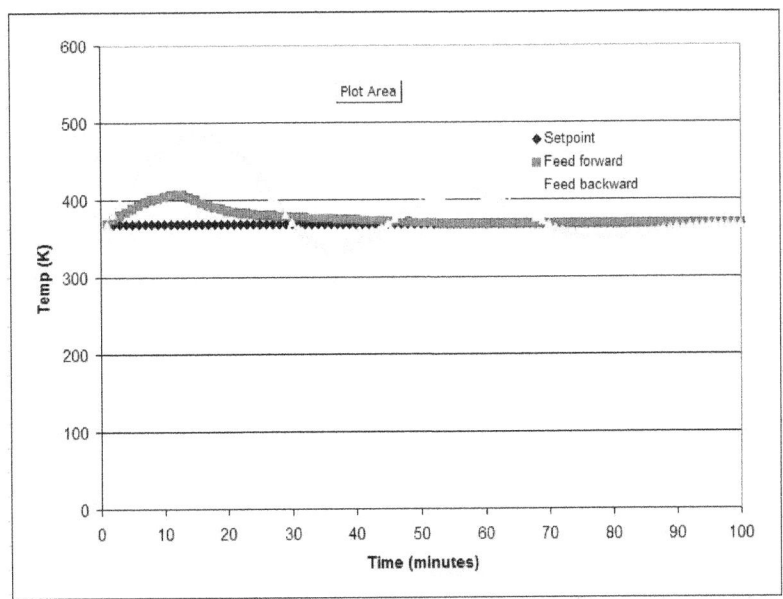

As shown, the feed-forward strategy works more effectively than the feed-backward strategy at nullifying the disturbance (over the time frame). The temperatures reached in feedback mode vary more than the ones seen in feed-forward mode. However, the controller response is solely defined by the parameters used in the Excel spreadsheet. If one assumes that the PID parameters used by the **PID-optimization group** were optimized, then we can say feed-forward would be the best option here.

Feed-Forward Design Procedure

This section provides an outline of the different steps for designing a feed-forward control strategy. The steps pay close attention to designing a feed-forward controller where there are multiple disturbances. Feed-forward design can be broken down into eight steps.

Step 1. State the control objective. This step includes defining which variable needs to be controlled and what the set point is. The setpoint should be adjustable by the operator.

Step 2. List the possible measured disturbances. This step includes identifying which disturbances are easily measured and how fast each disturbance should be expected to vary.

Step 3. State which variable is going to be manipulated by the feed-forward controller.

Step 4. The feed-forward controller consists of two parts: steady-state and dynamic compensators. Develop the steady-state compensator first. The compensator should be an equation where the manipulated variable, identified in step 3, can be calculated from the measured disturbances, identified in step 2, and the control objective (set point), identified in step 1.

Step 5. Reevaluate the list of disturbances. The effect of a disturbance on the controlled variable can be calculated from the equation. Three criteria will be used to determine which disturbance the feed-forward controller will correct: the effect the disturbance has on the controlled variable, the frequency and magnitude of variation, and the capital cost and maintenance of the sensor.

Step 6. Introduce the feedback compensation. This depends on the physical significance assigned to the feedback signal.

Step 7. Decide whether dynamic compensation, lead/lag, and/or dead time is required, and decide how to introduce it to the design.

Step 8. Draw the instrumentation diagram from the feed forward control strategy. The details of the diagram depend largely on the control system being used.

CASCADE CONTROL

In the previous chapters, only single input, single output (SISO) systems are discussed. SISO involves a single loop control that uses only one measured signal (input). This signal is then compared to a set point of the control variable (output) before being sent to an actuator (*i.e.* pump or valve) that adjusts accordingly to meet the set point. Cascade controls, in contrast, make use of multiple control loops that involve multiple signals for one manipulated variable. Utilizing cascade controls can allow a system to be more responsive to disturbances.

Before venturing further into the topic of cascade controls, the terms 'manipulated variables', 'measured variables' and 'control variables' should be clarified. The definitions of these terms commonly found in literature are often interchangeable; but, they typically refer to either the input or output signal. For the purpose, 'control variables' will refer to inputs like flow rates, pressure readings, and temperature readings. 'Manipulated variables' and 'measured variables' will refer to the output signals which are sent to the actuator.

The simplest cascade control scheme involves two control loops that use two measurement signals to control one primary variable. In such a control system, the output of the primary controller determines the set point for the secondary controller. The output of the secondary controller is used to adjust the control variable. Generally, the secondary controller changes quickly while the primary controller changes slowly. Once cascade control is implemented, disturbances from rapid changes of the secondary controller will not affect the primary controller.

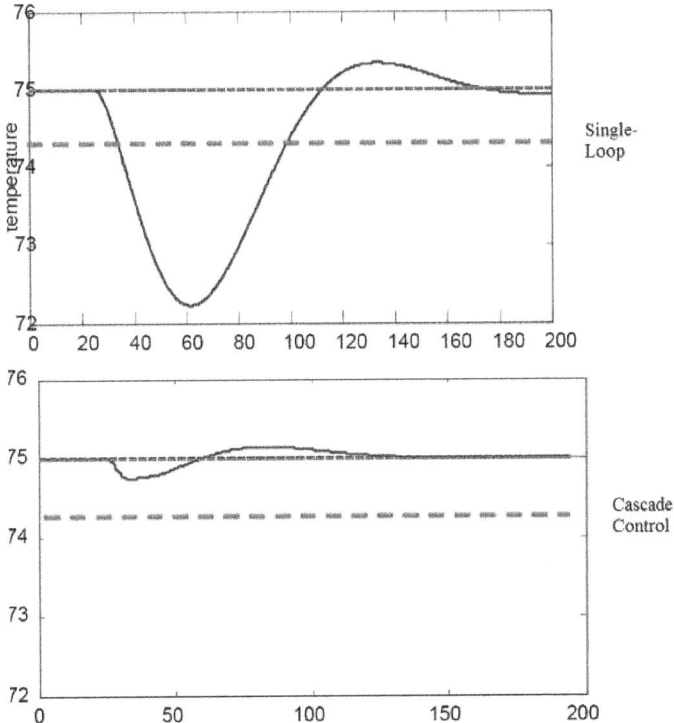

Fig. : Cascade control gives a much better performance because the disturbance in the flow is quickly corrected.

To illustrate how cascade control works and why it is used, a typical control system will be analyzed. This control system is one that is used to adjust the amount of steam used to heat up a fluid stream in a heat exchanger. Then an alternative cascade control system for the same process will be developed and compared to the typical single loop control. The figure below shows the performance of cascade control *vs.* single-loop control in CST heater.

Example of Cascade Control

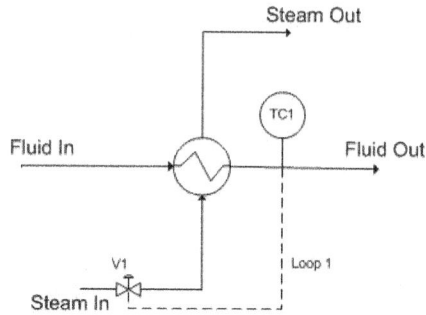

Fig. : Single loop control for a heat exchanger

In the above process, the fluid is to be heated up to a certain temperature by the steam. This process is controlled by a temperature controller (TC1) which measures the temperature of the exiting fluid and then adjusts the valve (V1) to correct the amount of steam needed by the heat exchanger to maintain the specified temperature. Figure shows the flow of information to and from the temperature controller.

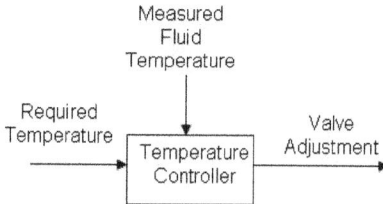

Fig. : Flow of information when single loop feedback control is used for a heat exchanger.

Initially, this process seems sufficient. However, the above control system works on the assumption that a constant flow of steam is available and that the steam to the heat exchanger is solely dependent on opening the valve to varying degrees. If the flow rate of the steam supply changes (*i.e.* pipeline leakage, clogging, drop in boiler power), the controller will not be aware of it. The controller opens the valve to the same degree expecting to get a certain flow rate of steam but will in fact be getting less than expected. The single loop control system will be unable to effectively maintain the fluid at the required temperature.

Implementing cascade control will allow us to correct for fluctuations in the flow rate of the steam going into the heat exchanger as an inner part of a grander scheme to control the temperature of the process fluid coming out of the heat exchanger. A basic cascade control uses two control loops; in the case presented below, one loop (the outer loop, or master loop, or primary loop) consists of TC1 reading the fluid out temperature, comparing it to $TC1_{set}$ (which will not change in this example) and changing $FC1_{set}$ accordingly. The other loop (the inner loop, or slave loop, or secondary loop) consists of FC1 reading the steam flow, comparing it to $FC1_{set}$ (which is controlled by the outer loop as explained above), and changing the valve opening as necessary.

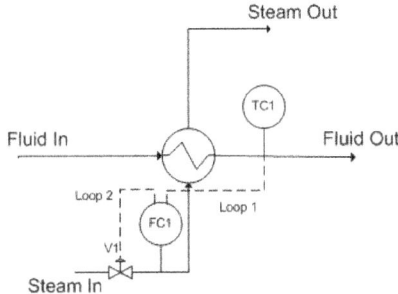

Fig. : Cascade control for a heat exchanger.

The main reason to use cascade control in this system is that the temperature has to be maintained at a specific value. The valve position does not directly affect the temperature (consider an upset in the stream input; the flow rate will be lower at the same valve setting). Thus, the **steam flow rate** is the variable that is required to maintain the process temperature.

The inner loop is chosen to be the inner loop because it is prone to higher frequency variation. The rationale behind this example is that the steam in flow can fluctuate, and if this happens, the flow measured by FC1 will change faster than the temperature measured by TC1, since it will take a finite amount of time for heat transfer to occur through the heat exchanger. Since the steam flow measured by FC1 changes at higher frequency, we chose this to be the inner loop. This way, FC1 can control the fluctuations in flow by opening and closing the valve, and TC1 can control the fluctuations in temperature by increasing or decreasing $FC1_{set}$.

Thus, the cascade control uses two inputs to control the valve and allows the system to adjust to both variable fluid flow and steam flow rates. The flow of information is shown in figure.

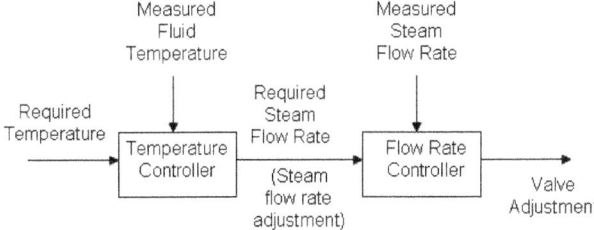

Fig. : Flow of information when cascade control is used for a heat exchanger.

In order to accomplish this, relationships between the primary and secondary loops must be defined. Generally, the primary loop is a function of the secondary loop. A possible example of such relations is:

$$TC_1 = f(FC_1); \frac{d(TC_1)}{d(FC_1)} = \ldots$$

$$FC_1 = f(V_1); \frac{d(FC_1)}{d(V_1)} = \ldots$$

$$= \frac{d(V_1)}{d(t)} = \ldots$$

Primary and Secondary Loops

In Figure, there are two separate loops. Loop 1 is known as the primary loop, outer loop, or the master, whereas loop 2 is known as the secondary loop, inner loop, or the slave. To identify the primary and secondary loops, one must identify the control variable and the manipulated variable. In this case, the control variable is the temperature and the reference variable is the steam flow rate. Hence, the primary loop (loop 1) involves the control variable and the secondary loop (loop 2) involves the reference variable. The information flow for a two loop cascade control system will typically be as shown in figure. Please note that the user sets the set point for loop 1 while the primary controller sets the set point for loop 2.

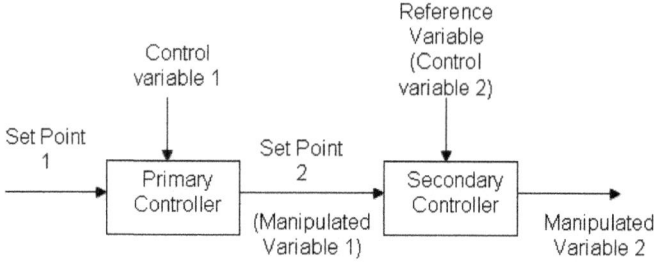

Fig. : Information flow of a two loop cascade control.

In addition to this common architecture, cascade control can have multiple secondary loops; however, there is still one primary loop and a main controlled variable. Unfortunately, with multiple inner loops, tuning the PID becomes even more challenging, making this type of cascade less common. The secondary loops can be either independent of each other, or dependent on each other, in which case each secondary loop affects the set point of the other secondary loop. When tuning such controller, the inner most loop should be tuned first.

The loop that manipulates the set point of the inner-most loop should be tuned next and so for. The figure below shows an example of using two secondary loops, independent of each other, in a fuel combustion plant. In this combustion furnace, the master controller controls the temperature in the furnace by changing the set point for the flow of fuels A and B. The secondary loops correspond to the change in the set point for the flow, by opening or closing the valves for each fuel.

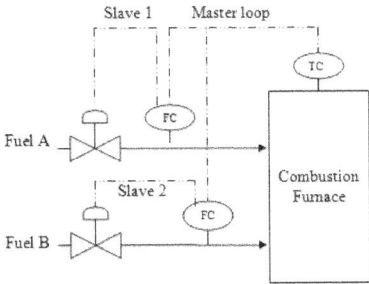

Cascade control is generally useful when

- A system error affects the primary control variable only after a long period of time as it propagates through dead time and lag time.
- A system has long dead times and long lag times.
- Multiple measurements with only one control variable are required for better response to a disturbance of a system.
- Variance occurs in multiple streams

General Cascade Control Schematic

The reactor below needs to be cooled during continuous-feed operation of an exothermic reaction. The reactor has been equipped with a cooling water jacket with the water flow rate being controlled by cold water valve. This valve is controlled by two separate temperature controllers. An "inner-loop" or "slave" (highlighted in orange) temperature transmitter communicates to the slave controller the measurement of the temperature of the jacket. The "outer-loop" or "master" (green) temperature controller uses a master temperature transmitter

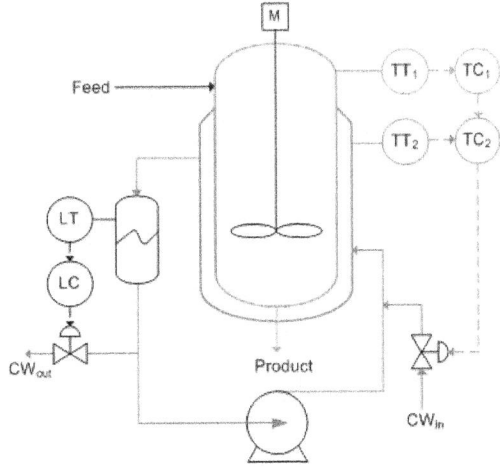

to measure the temperature of the product within the reactor. The output from the slave controller is fed into the master controller and used to adjust the cold water valve accordingly.

The cascade control loop used to control the reactor temperature can be generalized with the schematic below. We will use this main diagram to go through the formal derivation of the equations describing the behaviour of the system when there are changes in the loads U_1 and U_2 but with no change in the set point, $R_1(t)$. The general equations derived below can be used to model any type of process (first, second, third order differential equations, ect.) and use any type of control mechanism (proportional-only, PI, PD, or PID control).

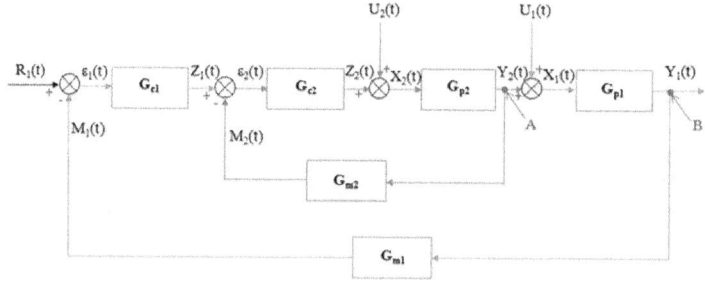

Step 1: Write down all the equations for each stage of the control loop

Master Loop

$Y_1(t) = \hat{G}_{P1} X_1(t)$

$X_1(t) = U_1(t) + Y_2(t)$

$Z_1(t) = \hat{G}_{C1} \varepsilon_1(t)$

$\varepsilon_1(t) = R_1(t) - M_1(t)$

$M_1(t) = \hat{G}_{M1} Y_1(t)$

Slave Loop

$Y_2(t) = \hat{G}_{P2} X_2(t)$

$X_2(t) = U_2(t) + Z_2(t)$

$Z_2(t) = \hat{G}_{C2} \varepsilon_2(t)$

$\varepsilon_2(t) = Z_1(t) - M_2(t)$

$M_2(t) = \hat{G}_{M2} Y_2(t)$

G_{p1} and G_{p2} are the process operators and are usually of the form:

$$\hat{G}_P^{-1} = \left(\tau_P \frac{d}{dt} + 1 \right)^n$$

Where n is a natural number.

G_{c1} and G_{c2} are the control operators and depend on the type of controller used. For PID controllers, they would be:

$$\hat{G}_C = K_C\left(1 + \frac{1}{\tau_I}\int_0^t dt' + \tau_D \frac{d}{dt}\right)$$

G_{m1} and G_{m2} are the measurement operators and usually are just equal to 1. Note that there are no equations for the "intersections" A and B shown on the diagram.

Step 2: Simplify the equations for the slave loop

$$Y_2(t) = \hat{G}_{P2} X_2(t)$$

$$Y_2(t) = \hat{G}_{P2}\{U_2(t) - Z_2(t)\}$$

$$Y_2(t) = \hat{G}_{P2}[U_2(t) - \hat{G}_{C2}\varepsilon_2(t)]$$

$$Y_2(t) = \hat{G}_{P2}[U_2(t) - \hat{G}_{C2}(Z_1(t) - M_2(t))]$$

$$Y_2(t) = \hat{G}_{P2}\left[U_2(t) - \hat{G}_{C2}\left(Z_1(t) - \hat{G}_{M2}Y_2(t)\right)\right]$$

Solve for $Y_2(t)$

$$Y_2(t) = \left(\frac{\hat{G}_{P2}}{1 + \hat{G}_{P2}\hat{G}_{C2}\hat{G}_{M2}} U_2(t) + \frac{\hat{G}_{P2}\hat{G}_{C2}}{1 + \hat{G}_{P2}\hat{G}_{C2}\hat{G}_{M2}} Z_1(t)\right)$$

Step 3: Simplify the equations for the master loop

$$Y_1(t) = \hat{G}_{P1} X_1(t)$$

$$Y_1(t) = \hat{G}_{P1}\{U_1(t) + Y_2(t)\}$$

$$Y_1(t) = \hat{G}_{P1}\left[U_1(t) + \frac{\hat{G}_{P2}}{1 + \hat{G}_{P2}\hat{G}_{C2}\hat{G}_{M2}} U_2(t) + \frac{\hat{G}_{P2}\hat{G}_{C2}}{1 + \hat{G}_{P2}\hat{G}_{C2}\hat{G}_{M2}} Z_1(t)\right]$$

$$Y_1(t) = \hat{G}_{P1}\left[U_1(t) + \frac{\hat{G}_{P2}}{1 + \hat{G}_{P2}\hat{G}_{C2}\hat{G}_{M2}} U_2(t) + \frac{\hat{G}_{P2}\hat{G}_{C2}}{1 + \hat{G}_{P2}\hat{G}_{C2}\hat{G}_{M2}} \hat{G}_{C1}\varepsilon_1(t)\right]$$

$$Y_1(t) = \hat{G}_{P1}\left[U_1(t) + \frac{\hat{G}_{P2}}{1 + \hat{G}_{P2}\hat{G}_{C2}\hat{G}_{M2}} U_2(t) + \frac{\hat{G}_{P2}\hat{G}_{C2}}{1 + \hat{G}_{P2}\hat{G}_{C2}\hat{G}_{M2}} \hat{G}_{C1}(R_1(t) - M_1(t))\right]$$

$$Y_1(t) = \hat{G}_{P1}\left[U_1(t) + \frac{\hat{G}_{P2}}{1 + \hat{G}_{P2}\hat{G}_{C2}\hat{G}_{M2}} U_2(t) + \frac{\hat{G}_{P2}\hat{G}_{C2}}{1 + \hat{G}_{P2}\hat{G}_{C2}\hat{G}_{M2}} \hat{G}_{C1}(R_1(t) - \hat{G}_{M1}Y_1(t))\right]$$

Solve for $Y_1(t)$

$$Y_1 = \left(\frac{(\hat{G}_{P1} + \hat{G}_{P1}\hat{G}_{P2}\hat{G}_{C2}\hat{G}_{M2})U_1(t) + \hat{G}_{P1}\hat{G}_{P2}U_2(t) + \hat{G}_{P1}\hat{G}_{P2}\hat{G}_{C1}\hat{G}_{C2}R_1(t)}{1 + \hat{G}_{P2}\hat{G}_{C2}\hat{G}_{M2} + \hat{G}_{P1}\hat{G}_{P2}\hat{G}_{M1}\hat{G}_{C1}\hat{G}_{C2}} \right)$$

(*Note:* here the G's are written as operators rather than Laplace transforms, and as such they shouldn't be divided. Thus, expressions in the denominator should be interpreted as inverse operators in the numerator.)

Conditions for Cascade Control

In order to have a smooth flow of information throughout the control system, a hierarchy of information must be maintained. In a double loop cascade system, the action of the secondary loop on the process should be faster than that of the primary loop. This ensures that the changes made by the primary output will be reflected quickly in the process and observed when the primary control variable is next measured. This hierarchy of information can be preserved by applying the following conditions when setting up the cascade controls.

1. There must be a clear relationship between the measured variables of the primary and secondary loops.
2. The secondary loop must have influence over the primary loop.
3. Response period of the primary loop has to be at least 4 times larger than the response period of the secondary loop.
4. The major disturbance to the system should act in the primary loop.
5. The primary loop should be able to have a large gain, Kc.

Cascade control is best when the inner loop is controlling something that happens at fairly high frequency. Cascade control is designed to allow the master controller to respond to slow changes in the system, while the slave controller controls disturbances that happen quickly. If set up in reverse order, there will be a large propagation of error. Hence, it is important to maintain the hierarchy of information. In summary, the master controller responds to SLOW changes in the system, while the slave controller responds to the high frequency, or FAST changes in the system. This also requires that the inner control scheme be tuned TIGHTLY so error is not allowed to build. Commonly, the inner loops controls a flow controller, which will reduce the effect of changes such as fluctuations in steam pressure.

Cascade Control Design Considerations

Open Loop and Closed Loop

Although cascade control generally incorporates information from several secondary loops, the system overall does not automatically become a closed loop or an open loop process. For a cascade control to be open loop, all control loops in the system should be open loop in nature. On the other hand, if any of the control loops in the system is feedback-based, the system overall is considered as closed loop. This is due to the fact that the system is getting some sort of feedback, no matter how little the "fedback" variable influences the system.

Cascade with Feed-Forward and Feedback Controls

Cascade controllers have a distinct advantage over other kinds of controller due to its ability to combine both feedback and feed-forward controls. While feed-forward loops have the potential to adjust the controlled variables to the ideal states, the feedback loops in a mixed system check deviations to make sure the system is on track. The figure below is an example of a mixed cascade control:

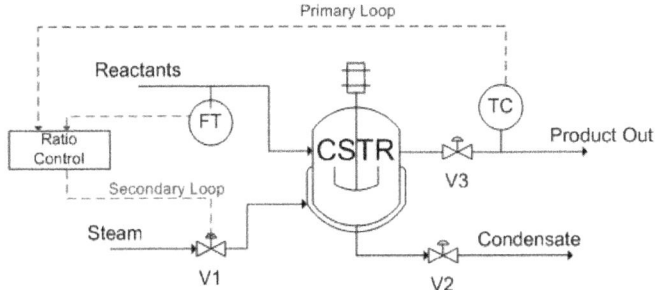

Fig. : Cascade control with both feed-forward and feedback controls.

An endothermic reaction takes place in a jacketed CSTR, in which the reactor is heated by steam. A reactant stream feeds into the CSTR, which serves as the wild stream for the ratio controller to predict a required steam flow rate (feed-forward). On the product stream side, a temperature controller manipulates the ratio setting of the ratio controller in order to minimize the product temperature errors (feedback). The temperature controller is the primary loop, whereas the ratio controller is the secondary loop.

Advantages and Disadvantages of Cascade Control

The following table shows a list of cascade control pros and cons:

Cascade Control Pros	Cacade control Cons
• Accounts for disturbances in the primary vairable more quickly and hence control the primary variable more effectively • Reduces the effects of dead time and phase lag time in the system • Can be combined with feed-forward control or other forms of control • Integrated multiple sensors readings • Improve dynamic performance and provide limits on the secondary variables	• Cascade control makes the system more complex • Cascade control requires more equipment and instrument that will drive up the cost of the process • Tuning cascade controllers is more difficult as the set point changed + more parameters

Starting up a Cascade System

A cascade system needs to be set up properly in order to function. The inner loop should be tuned before the outer loop. The following are the suggested steps for starting up a cascade system (both controllers start in automatic mode):

1. Place the primary controller in manual mode. This will break the cascade and isolate the secondary controller so that it can be tuned.
2. Tune the secondary controller as if it were the only control loop present.
3. Return the secondary controller to the remote set point and/or place the primary controller in automatic mode. This will isolate the primary controller so that it can be tuned.
4. Tune the primary control loop by manipulating the set point to the secondary controller. If the system begins to oscillate when the primary controller is placed in automatic, reduce the primary controller gain.

Cascade control systems can be tuned using conventional methods. Ziegler-Nichols tuning method can be used to tune the secondary controller. Then the parameters (depending on the system, it can be P, PI, or PID) need to be fine tuned by introducing disturbances to the system and changing the parameters accordingly. The secondary controller must be tuned tightly (no oscillations when disturbance is added), otherwise the primary controller will be responding to oscillations due to the secondary controller as well as from disturbances to the system.

This will cause the parameters for the primary controller to be inaccurate, which will cause a high controller effort. This is undesirable because it will wear down the controller. The primary controller often utilizes internal model control (IMC), which allows for improved performance by incorporating the process model into the controller setup. Another way to tune the primary controller is by using trial and error. For example, if the system has a PID primary controller, the integral and derivative gains should be set at small values. Then a small proportional gain should be introduced. The proportional gain should be tuned first then the integral and derivative gains can be added. For more information about the trial and error method refer to PID Tuning Classical section.

Startup Example

A simple example of starting up a cascade system is shown using the heat exchange system seen in figure. To begin startup the temperature controller (primary) is set to manual mode. The flow controller (secondary) is then tuned by adjusting the set point of the flow controller. This is shown in Figure below.

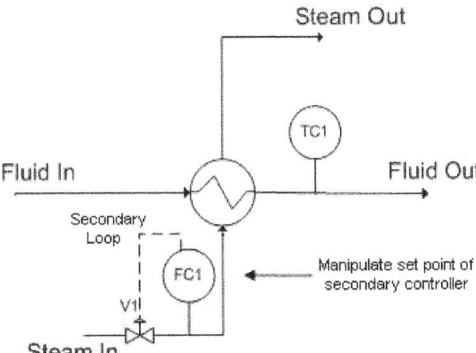

Fig. : Automatic mode for secondary controller (manual mode for primary controller).

Now the temperature controller is set to automatic mode and the flow controller is set to manual mode. In this system the temperature controller outputs a set point to the flow controller just like it would in regular cascade mode. The temperature controller is then tuned by adjusting the output to the valve. This is shown in Figure below.

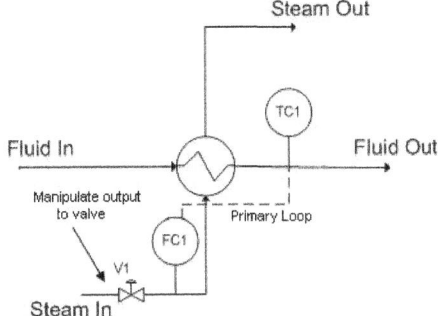

Fig. : Manual mode for secondary controller (automatic mode for primary controller)

Developing the Structure of a Cascade Algorithm

Below is a step-by-step method through which the basic structure of a Cascade Algorithm can be developed. The examples in each step refer to figure

1. **Determine the goal of your algorithm.** That is, which **ultimate output** you would like to ultimately end up changing. Also determine what **"tool"**(or aspect of the system) will most directly and physically allow you to accomplish this desired change.

This "tool" can be any physical property of the system: temperature, pressure, surface area, flow regime, flow rate, *etc.* of any part of any physical component of the system. The "tool" doesn't need to be something read by a sensor, though it can be (as it is in this example, the temperature of the heated fluid leaving the heat exchanger is read by sensor TC1).

For instance, the end goal of the process of Figure is to regulate the temperature of the fluid leaving a heat exchanger. In this case, the only way to physically accomplish this is to change the flow rate of the steam through the heat exchanger (this flow rate just so happens to be read by the sensor FC1).

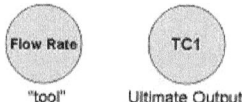

2. Determine how the "tool" physically affects the ultimate output.

For instance, an **increased** flow rate of **steam**(read by FC1) will physically **increase** the **temperature** of the fluid leaving the heat exchanger, **TC1**.

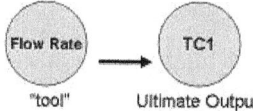

3. Determine how the ultimate output needs to affect the "tool" to achieve the desired ultimate output change.

For instance, we presume that we would like to, in general, resist any change in TC1(*i.e.* we don't want an increase in TC1 to further increase the value of TC1, that would be bad!). For this reason, we want an increase in TC1 to decrease FC1, which will then decrease TC1(negative feedback).

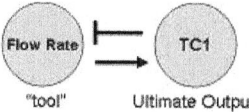

4. Determine which valve or sensor's output will *most* directly affect the "tool". Also, determine how the sensor and "tool" are related.

For instance, the setting of valve V1 will most directly affect the flow rate of steam entering the heat exchanger (it is the "closest" controllable component of the process to the "tool", the steam flow rate). An increase in V1 will increase flow rate.

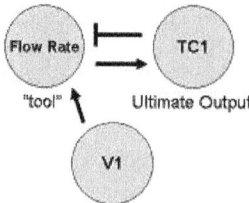

Sensor directly affecting the "tool"

5. Determine all valves/sensors in between the ultimate output and valve/sensor that will directly affect the "tool".

For instance, between the ultimate output (TC1) and the valve (V1) that will directly affect the "tool", there is only the FC1 sensor.

Control Architectures

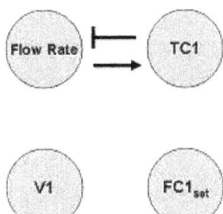

6. **Make an incidence graph**. Make sure that you use the **set points** of the valves/sensors that directly affect the "tool", because you cannot control what these sensors read (*you can only control their set points*). Confirm that the relationship between the "tool" and the ultimate output is consistent with the incidence graph!

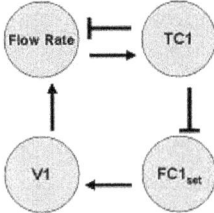

7. Use this incidence graph to **construct the algorithm**. The incidence graph is particularly helpful in determining what "sign" changes in one component should result in a the "sign" in another component. If we use incidence diagrams, and the equations for the cascade controller, we can determine a relationship between the components of the system.

For instance, if TC1 were to increase, we would want the steam flow rate to decrease, which ultimately results in a decreased temperature. An increase (a "sign" of +) in TC1 should result in a decrease in $FC1_{set}$ (a "sign of -). This can be seen from equation 2, where if $TC1_{set}$ decreases, the difference between $TC1_{set}$ and TC1 will be smaller. This causes $FC1_{set}$ to decrease relative to its previous set point. A decrease (a "sign" of -) in $FC1_{set}$ should result in a decrease in V1(a "sign of -). This is shown by equation 1, where if $FC1_{set}$ decreases while FC1 is the same, V1 will decrease relative to its previous position. A decrease (a "sign" of -) in V1 should result in a decreased flow rate (a "sign" of -). A decrease (a "sign" of -) in steam flow rate should result in a decreased TC1(a "sign of -), completing the negative-feedback mechanism initially desired.

$$V_1 = V_{1,Offset} + K_{c,FC1}(FC1_{set} - FC1) + \frac{1}{\tau_{i,FC1}} \int (FC1_{set} - FC1)dt + \tau_{d,FC1} \frac{d(FC1_{set} - FC1)}{dt}$$

$$FC1_{set} = FC1_{Offset} + K_{c,TC1}(TC1_{set} - TC1) + \frac{1}{\tau_{i,TC1}} \int (TC1_{set} - TC1)dt + \tau_{d,TC1} \frac{d(TC1_{set} - TC1)}{dt}$$

Failure

A cascade system is not to be confused with a fail-safe system. The sole purpose of implementing a cascade control system is to make the system more responsive, **not** more robust.

In a cascade system, if the master controller fails, the entire system will fail too. Just like for any other design, one should anticipate failure and build in redundancy. An example of redundancy could be having multiple controllers to be the master controller and then using selectors to choose which reading to use.

Alternatively, if the cascade system fails and has no built-in redundancy, there are a couple of ways to keep the cascade system running while the controller is being repaired. For example, the controller can be operated manually by an employee, or an average of previous readings can be used to temporarily send a constant signal to the system.

RATIO CONTROL

Ratio control architecture is used to maintain the relationship between two variables to control a third variable. Perhaps a more direct description in the context of this class is this: Ratio control architecture is used to maintain the flow rate of one (dependent *controlled feed*) stream in a process at a defined or specified proportion relative to that of another (independent *wild feed stream*) in order to control the composition of a resultant mixture.

As hinted in the definition above, ratio control architectures are most commonly used to combine two feed streams to produce a mixed flow of a desired composition or physical property downstream. Ratio controllers can also control more than two streams. Theoretically, an infinite number of streams can be controlled by the ratio controller, as long as there is one controlled feed stream. In this way, the ratio control architecture is feedfoward in nature. In this context, the ratio control architecture involves the use of an independent *wild feed stream* and a dependent stream called the *controlled feed*.

Ratio Control is the most elementary form of feed forward control. These control systems are almost exclusively applied to flow rate controls. There are many common usages of ratio controls in the context of chemical engineering. They are frequently used to control the flows on chemical reactors. In these cases, they keep the ratio of reactants entering a reaction vessel in correct proportions in order to keep the reaction conditions ideal. They are also frequently used for large scale dilutions.

Ratio Control based upon Error of a Variable Ratio

The first type of ratio control architecture uses the error of a variable ratio (R_{actual}) from a set ratio (R_{set}) to manipulate y (the *controlled stream*). R_{actual} is a ratio of the two variables *wild stream* and *controlled stream*. The controller adjusts the flow rate of stream y (*controlled stream*) in a manner appropriate for the error ($R_{actual} - R_{set}$).

The error in this system would be represented using the following equation:

$$\frac{y}{y_w} = R_{actual} \quad \text{or} \quad \frac{y_w}{y} = R_{actual}$$

$$Error = R_{actual} - R_{set}$$

CONTROL ARCHITECTURES

This error would be input to your general equation for your P, PI, or PID controller as shown below.

$$V_y = \text{Offset} + K_C(\text{Error}) + \frac{1}{\tau_I}\int(\text{Error})dt + \tau_D \frac{d(\text{Error})}{dt}$$

NOTE: Vy is used instead of y, because y is not directly adjustable. The only way to adjust y is to adjust the valve (V) that affects y.

Diagram of Ratio Dependent System

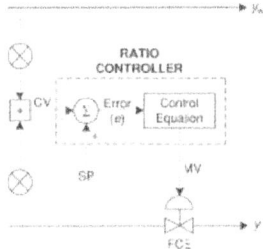

Ratio Control based upon Error of the Controlled Stream

The second type of ratio control architecture uses the error of a setpoint(y_{set}, setpoint of the control variable) from y(controlled variable) to control, once again, y. The controller adjusts the flow rate of stream y in a manner appropriate for the error (y-y_{set}).

The error in this system would be represented using the following equation:

$$y_w R_{set} = y_{set} \quad \text{or} \quad \frac{y_w}{R_{set}} = y_{set}$$

$$\text{Error} = y - y_{set}$$

This error would be input to your general equation for your P, PI, or PID controller as shown below.

$$V_y = \text{Offset} + K_C(\text{Error}) + \frac{1}{\tau_I}\int(\text{Error})dt + \tau_D \frac{d(\text{Error})}{dt}$$

Diagram of Flowrate Dependant System

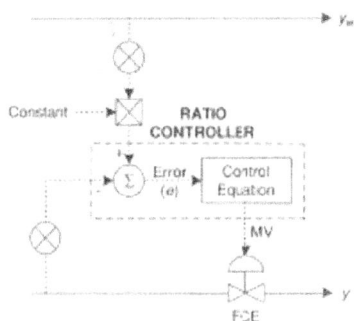

Comparing the Two Types of Ratio Control

The main difference between the two aforementioned types of ratio control architecture is that they respond to changes in the monitored variable (R_{actual} or y - the variables through which the error is determined) differently.

The first method mentioned that defines error as $R_{actual} - R_{set}$ (monitored variable of R_{actual}) responds sluggishly when y is relatively large and too quickly when y is relatively small. This is best explained by examining the equations below.

$$R = \frac{y_w}{y} \longrightarrow \frac{\partial(R)}{\partial(y_w)} = \frac{1}{y}$$

$$R = \frac{y}{y_w} \longrightarrow \frac{\partial(R)}{\partial(y_w)} = -\frac{y}{y_w^2}$$

Unlike the first method, the second method mentioned that defines error as $y - y_{set}$ (monitored variable of y) does not respond differently depending upon the relative amounts of y (or anything else, for that matter). This is best explained by examining the equations below.

$$y = y_w R_{set} \longrightarrow \frac{\partial(y)}{\partial(y_w)} = R_{set}$$

$$y = \frac{y_w}{R_{set}} \longrightarrow \frac{\partial(y)}{\partial(y_w)} = \frac{1}{R_{set}}$$

Difficulties with Ratio Controllers

Steady State Issues

A common difficulty encountered with ratio controllers occurs when the system is not at steady state. Unsteady state conditions can cause a delay in the adjustment of the manipulated flow rate and thus the desired state of the system cannot be met.

Blend Stations

To overcome the difficulty with steady state, a Blend station can be used, which takes into account both the set ratio and the wild stream. As can be seen by the diagram below, the Blend station takes in the wild stream flow rate, the set ratio, and the initial set point for the wild stream (r_1) as inputs. Using a weighting factor, called gain, the Blend station determines a new set point for the manipulated stream (r_2). The relationship that determines this new set point can be seen below.

$$r_2 = set\,ratio[\gamma * r_1 + (1-\gamma)y_w]$$

Where γ is the gain factor.

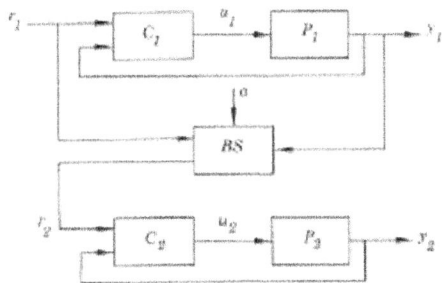

Fig. : Ratio control using the Blend station (BS).

Accuracy Issues

Another problem (which is an issue with all "feed-forward" controllers) is that the variable under control (mix ratio) is not directly measured. This requires a highly accurate characterization of the controlled stream's valves so that the desired flow rate is actually matched.

One way to address this problem is to use two levels of PID control (Cascade Control). The first level monitors the controlled streams flow rate and adjusts it to the desired set point with a valve. The outer level of control monitors the wild stream's flow rate which adjusts the set point of the controlled stream by multiplying by the desired ratio. For example, if stream A's flow rate is measured as 7gpm, and the desired ratio of A:B is 2, then the outer level of control will adjust B's set point to 3.5. Then the inner level of control will monitor B's flow rate until it achieves a flow rate of 3.5.

Ratio Control Schemes

The following subsections introduce three ratio controller schematics. Each schematic illustrates different ways that the controlled feed stream can be manipulated to account for the varying wild feed stream.

Ratio Relay Controller

The wild feed flow rate is received by the ratio relay and then multiplied by the desired mix ratio. The ratio relay outputs the calculated controlled feed flow rate which is compared to the actually flow rate of the controlled feed stream. The flow controller then adjusts the controlled feed flow rate so that it matches the set point.

The mix ratio (Fc/Fw) is not easy to access, so it requires a high level of authorization to change. This higher level of security may be an advantage so that only permitted people can change the mix ratio and decrease the chance that an accidental error occurs. A disadvantage is that if the mix ratio does need to be changed quickly, operation may be shut down while waiting for the appropriate person to change it.

Ratio Relay Controller

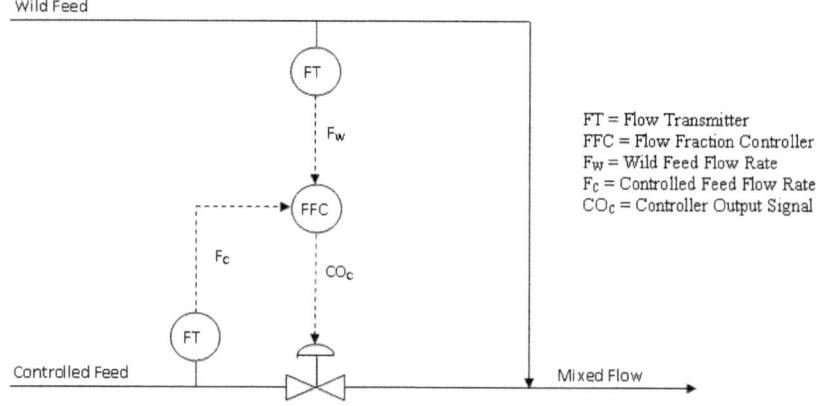

FT = Flow Transmitter
RY = Ratio Relay
FC = Flow Controller
F_W = Wild Feed Flow Rate
F_C = Controlled Feed Flow Rate
SP_C = Set Point Value
CO_C = Controller Output Signal

Another disadvantage is that linear flow signals are required. The output signals from the flow transmitters, Fw and Fc, must change linearly with a change in flow rate. Turbine flow meters provide signals that change linearly with flow rates. However, some flow meters like orifice meters require additional computations to achieve a linear relationship between the flow rate and signal.

Flow Fraction Controller

Flow fraction and ratio relay controllers are very similar except that the flow fraction controller has the advantage of being a single-input single-output controller. A flow fraction controller receives the wild feed and controlled feed flow rates directly. The desired ratio of the controlled feed to wild feed is a preconfigured option in modern computer control systems.

Flow Fraction Controller

FT = Flow Transmitter
FFC = Flow Fraction Controller
F_W = Wild Feed Flow Rate
F_C = Controlled Feed Flow Rate
CO_C = Controller Output Signal

Ratio Relay with Remote Input

The ratio relay with remote input model of a ratio controller is similar to the cascade control model in that it is part of a larger control strategy. The purpose of this model is to account for any unexpected or unmeasured changes that have occurred in the wild and controller feed streams. The diagram below illustrates how the composition of the mixed flow is used to change the relay ratio.

Ratio Relay Controller with Remote Input

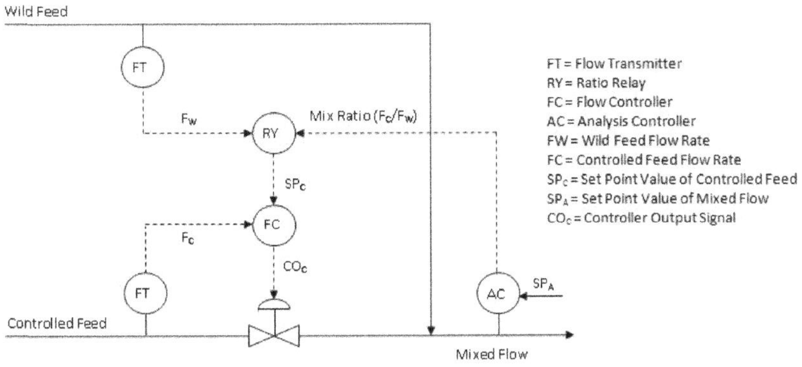

FT = Flow Transmitter
RY = Ratio Relay
FC = Flow Controller
AC = Analysis Controller
FW = Wild Feed Flow Rate
FC = Controlled Feed Flow Rate
SP_C = Set Point Value of Controlled Feed
SP_A = Set Point Value of Mixed Flow
CO_C = Controller Output Signal

The main advantage of having remote input is that the mix ratio is constantly being updated. A disadvantage of using an additional analyzer is that the analysis of the mixed stream may take a long time and decrease the control performance. The lag time depends on the type of sensor being used. For example, a pH sensor will most likely returned fast, reliable feedback while a spectrometer would require more time to analyze the sample.

Advantages and Disadvantages

There are some pros and cons to using ratio control where there is a ratio being maintained between two flow rates.

Advantages

1. Allows user to link two streams to produce and maintain a defined ratio between the streams
2. Simple to use
3. Does not require a complex model

Disadvantages

1. Often one of the flow rates is not measured directly and the controller assumes that the flows have the correct ratio through the control of only the measured flow rate

2. Requires a ratio relationship between variables that needs to be maintained
3. Not as useful for variables other than flow rates

Select Elements in Ratio Control

A select element enables further control sophistication by adding decision-making logic to the ratio control system. By doing so, a select variable can be controlled to a maximum or minimum constraint. The figure below depicts the basic action of both a low select and high select controller.

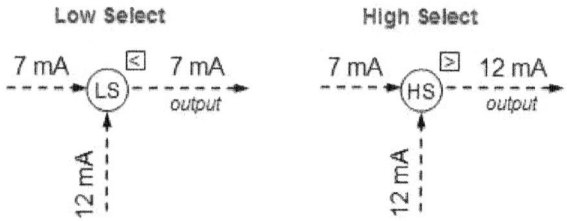

Image adapted from Houtz, Allen and Douglas Cooper

Single Select Override Control

Often times, these select elements are used as an override element in ratio control. Take, for example, ratio control system (with remote input) of a metered-air combustion process in the figure below.

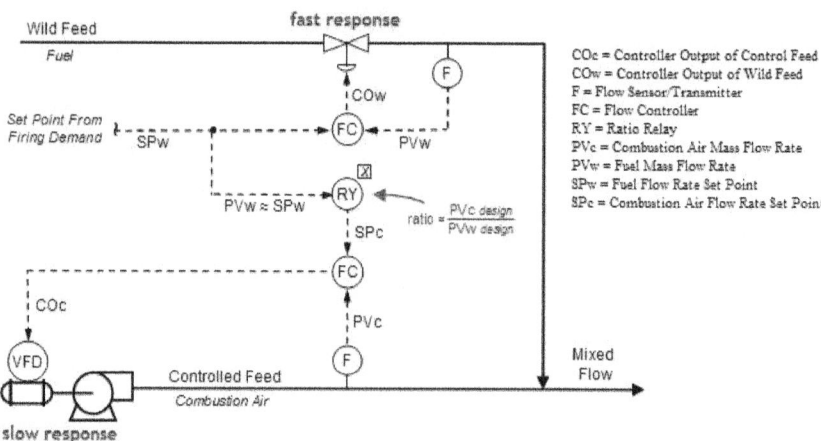

Adapted from Houtz, Allen and Douglas Cooper

In this design, the fuel set point, SPw, comes in as firing demand from a different part of the plant, so fuel flow rate cannot be adjusted freely. As the fuel set point (and therefore fuel mass flow rate, PVw) fluctuates, a ratio relay is employed to compute the combustion air set point, SPc. If the flow command outputs (COw and COc) are able to respond quickly, then the system architecture should maintain the desired air/fuel ratio despite the demand set point varying rapidly and often.

However, sometimes the final control element (such as the valve on the fuel feed stream and blower on the combustion air feed stream in the diagram) can have a slow response time. Ideally, the flow rates would fluctuate in unison to maintain a desired ratio, but the presence of a slow final control element may not allow the feed streams to be matched at that desired ratio for a significant period of time. Valves often have quick response times, however, blowers like the one controlling the combustion air feed stream can have slow response times. A solution to this problem is to add a "low select" override to the control system, as shown in the figure below.

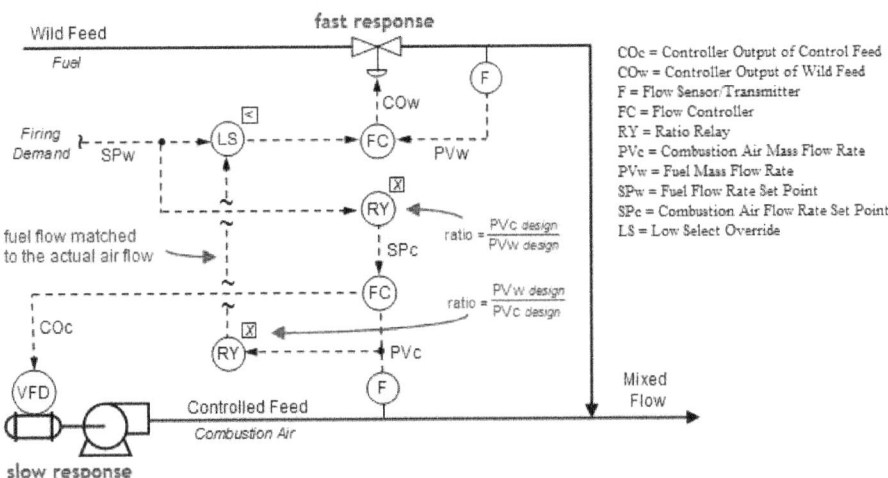

Adapted from Houtz, Allen and Douglas Cooper

The second ratio controller will receive the actual measured combustion air mass flow rate and compute a matching fuel flow rate based on the ideal design air/fuel mixture. This value is then transmitted to the low select controller, which also receives fuel flow rate set point based from the firing demand. The low select controller then has the power to pass the lower of the two input signals forward. So, if the fuel flow rate firing demand exceeds combustion air availability required to burn the fuel, then the low select controller can override the firing demand and pass along the signal of the fuel flow rate calculated based on the actual air flow (from the second ratio relay).

This low select override strategy ensures that the proper air/fuel ratio will be maintained when the firing demand rapidly increases, but does not have an effect when the firing demand is rapidly decreasing. While the low select override can help eliminate pollution when firing demand rates increase quickly, a rapid decrease in firing demand can cause incomplete combustion (as well as increased temperature) and loss of profit.

Cross-Limiting Override Control

Cross-limiting override control utilizes the benefit of multiple select controls. Most commonly, a control system using the cross-limiting override strategy will implement both a high select and low select control. As a result, the system can account for quickly increasing firing demand as well as quickly decreasing firing demand. The figure below depicts a such a scenario using the inlet combustion air and fuel for a boiler as the example.

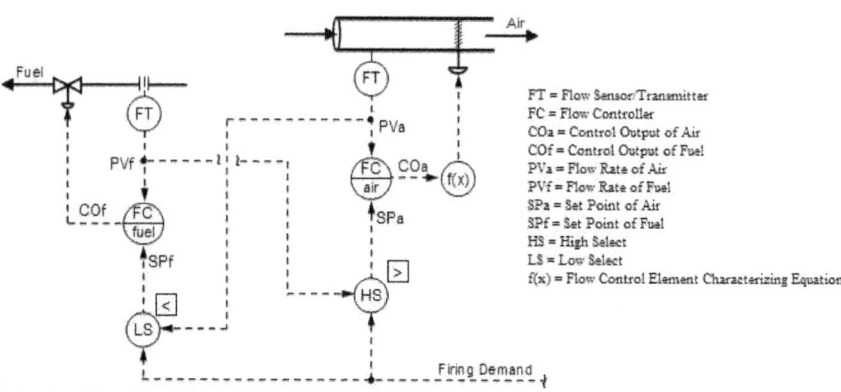

Ratio with Cross-Limiting Override Control Strategy

Adapted from Houtz, Allen and Douglas Cooper

It is assumed that same firing demand enters the high select and low select control and the flow transmitters have been calibrated so that the ideal air/fuel ratio is achieved when both signal match. As a result, the set point of air will always be the greater value of the firing demand and current fuel flow signals. Likewise, the set point of fuel will always be the lower value of the firing demand and current air flow signals. So, as the firing demand increases, the set point of air will increase.

At the same time, the low select will keep the set point of fuel to the signal set by the present flow or air. Therefore, the set point of fuel will not match the increasing firing demand, but will follow the increasing air flow rate as it is responding upwards. Similarly, if the firing demand decreases, the low select control will listen to the firing demand and the high select controller will not. As a result, the firing demand will directly cause the set point of fuel to decrease while the set point of air will follow the decreasing fuel rate as it responds downwards. The cross-limiting override strategy allows for greater balance in a ratio control system.

Summary on Control Architectures

A summary of the philosophies, advantages, and disadvantages of the different control architectures is shown below in Table.

CONTROL ARCHITECTURES   69

Table. Control architectures' philosophy, advantages, and disadvantages.

	Philosophy	Advantages	Disadvantages
Feedback	Control adjusts errors as they occur	+ simple to design + no process model needed	- corrects for error after they have already occurred - usually only takes input from one sensor
Feed Forward	Control corrects errors before they occur	+ corrects possible errors before they occur + Ideally can produce a perfect control where there is never any offset	- infinitely accurate models are needed - infinitely accurate measurements are needed
Ratio	Control connects two flows to maintain a constant ratio	+ generate defined ratio for given two streams + simple model	- the flow that you control is never truly measured and ratio model is assumed to be accurate - does not take into consideration of varying pressures
Cascade	Controls set points of sensors by controlling other sensors to integrate multiple information	+ responds quickly to high frequency changes + incorporate multiple sensor readings together (incidence graphs)	- can be very complex model - tuning model is difficult due to multiple set points and parameters
Mixed	Control integrating combinations of control architectures	+ pick the right combination of controls to accomplish main controlling objective + incorporate any number of sensors in a logical manner	- even more complex model than cascade alone - tuning model is even more complex than cascade alone, Zeigler-Nicholas tuning won't work by itself

CCL LIQUID

Liquid level and liquid pressure control are fundamental aspects of many control processes because of the reliance on the stability and convenience of liquid usage. This section will introduce the main concepts behind liquid pressure control, liquid level control, examining the loop characteristics, response, tuning and limitations. Two models are included to illustrate the application of a liquid pressure control loop and a level control loop in idealized systems.

Pressure Control Basics

Like its counterparts of temperature, level and flow, pressure is one of the most common process variables. Pressure is a key process variable as it provides a critical condition for processes such as any chemical reaction, extrusion, boiling, distillation, air conditioning and vacuuming. Poor pressure control may likely lead to problems associated with quality, safety and productivity. For example, the presence of high-pressure conditions in a sealed vessel may result in an explosion and thus it is imperative that pressure be kept under good control within its safety limits.

Liquid pressure control is one of the easier control loops in the sense that it is shares many of the same characteristics of a common flow loop. The aim of the liquid pressure loop is to control the pressure at the needed set point pressure by controlling the flow with changing process needs. Because dealing with a liquid, the flow present is that of an incompressible fluid and as a result the pressure changes very quickly leading to a fast-responding process with small dead time and capacitance. This process behaves as a fixed restriction where the change in pressure is a function of the flow through the process. Illustrated below is a diagram with liquid pressure control on the pipe leading to the process.

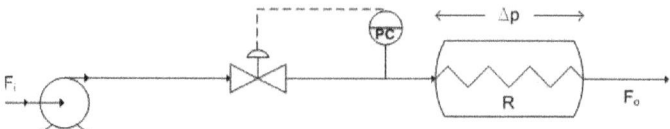

There are a few additional considerations for the aforementioned liquid pressure loop. Firstly the controller can be proportional plus integral (PI) or integral only (I-only) if $K_c<2$ otherwise a proportional only (P-only) controller is to be used. This controller is tuned similarly to the flow controller. Another consideration is that process gain is not constant, a square-root extractor or the highest loop gain should be used in tuning the controller. This highest loop gain is employed to prevent the process loop from ever becoming unstable. Finally, this liquid pressure loop is noisy like its flow loop counterpart and as a result it is recommended that derivative action in the controller not be used.

Level Control Basics

Controlling the level of liquid in a tank is one of the most basic requirements for a chemical process to run effectively. Most chemical engineering processes require some sort of liquid to be held for usage, and the most convenient way of ensuring the proper amount of liquid is available is through level control. A specific example of a process that requires level control is seen around the world. Water towers must have a designated amount of water within them to ensure sufficient water pressure to surrounding neighborhoods and hence liquid level control (LLC) is used.

Fortunately Liquid level control is one of the easier control loops available. The systems are fairly simple, a sensor reads the level of liquid within a tank, which then is conveyed to the controller. Subsequently, the controller changes a valve to increase or decrease the flowrate into or out of the tank, depending on the required action and controlled valve placement. Below is a figure with LLC on the pipe leaving the tank.

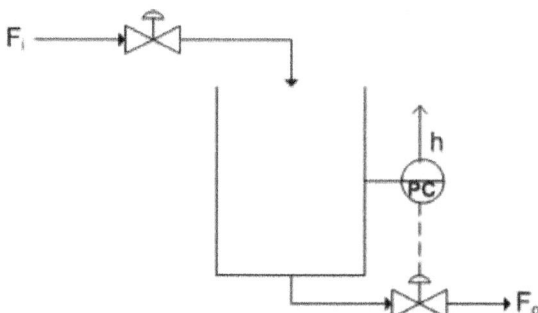

Control Architectures 71

The ease of LLC also lies in its large capacitance and nominal dead time. Usually hold-up times vary from 5-15 minutes. Problems may arise when signal noise becomes a factor, as it usually does with level controllers.

P-only Controllers

Oftentimes using a P-only controller is the best way to ensure proper level control, in most cases only a small percent error will result, and it reduces problems associated with noise. P-only controllers should only be used when gain is very small and the tank has a large capacity.

A P-only controller works off the following principle of control:

$$Output = Gain*Error + Bias$$

The output directly effects a valve to control the flowrate into or out of the tank.

Level Measurement Noise

When controlling liquid levels in drums or vessels, an important aspect to consider is that there may be noise in the level measurement due to disturbances such surface turbulence, boiling of the liquid, or agitation. Due to the existance of this noise, using a derivative action controller may not be appropiate. However, if this type of controller action is used, there are various methods that may be employed to minimize the noise.

Method 1: Use a displacer in a stilling well

Advantage: - Filters high frequency noise due to turbulence in the tank

Disadvantage: - Bobbing level due to the low-frequency movement of tank liquid into the well resulting from the formation of a U-tube between tank and well

Method 2: Use an ultrasonic level measurement with electronic filtering of the signal

Advantage: - Works well when the level response period is much lower than the noise frequency

Method 3: Use a tank weighing method

In the this method, a loading device is placed beneath the tank supports to measure the mass of the tank. Output weight values are averaged by a transmitter which are then sent to a converter which outputs the corresponding level to the controller.

Advantage: - Effectively eliminates noise in the level measurement due to turbulence

Models

The following are Excel models of Pressure and Level contol in simple systems. Proportional only control is used primarily for these models. Controller noise has been eliminated for sake of simplicity.

Liquid Pressure Control Model

The model at hand illustrates both the feed backwards system of the liquid pressure in a pipline of a process and its process gain relationships. This liquid pressure control loop is regulated by the position of the valve on the pipeline leading to the process. A P-only control is used here to change the valve position resulting in the need response in attaining the set point value. The following are any assumptions made and the equations used in modeling this process. An assumption is made that the process behaves like a fixed restriction such as an orifice plate whose Dp is a function of flow through the process. Another assumption that should be made is that the valve responds linearly to the flow-rate through it. In determining the process gain the following equations where used.

$$P = \Delta p + P_0$$

where P_0 is the downstream pressure at zero flow

$$\Delta p = \frac{F^2}{R^2}$$

where R is the process flow resistance

As a result we obtain:

$$P = \frac{F^2}{R^2} + P_0$$

The gain of the process is then determined by the following expression:

$$K_p = \frac{dP}{dF} = \frac{2F}{R^2}$$

A model for the feed backwards system for liquid pressure control introduced in this chapter is very complicated and might possibly be out of the scope of our discussion. A possible alternative model would be very similar to the liquid level model presented below with the head in the tank determining the liquid pressure in the pipeline leaving the reactor.

Liquid Level Control Model

The following model represents a feed backward system of a liquid holding tank. The control loop is regulated by the exiting valve position on the pipe leaving the tank. A maximal flowrate out is assumed and the height of water in the tank does not have any effect on the flowrate out of the tank. This assumption can be made in tanks which are sufficiently wider than the pipe exiting operating with little liquid level change, as often is the case. Another assumption made is the linear response of the valve to flowrate out.

The pipe entering the tank is set by the user at a fixed flowrate. The P-only control is used to change the exiting valve position in order to cause the desired response to obtain the setpoint value. The P control follows the equation listed above and the bias is designated by the point at which the valve is open to allow a steady state to occur. Gain in Level control systems is usually low, however this can be changed by the user as well. The setpoint for the tank is determined by the user, as are the entering liquid flowrates and the maximal exiting flowrate.

The user can input a set point change in order to view the corresponding response in the controller.

CCL Temperature

Temperature, pressure, flow, and level are the four most common process variables. Temperature is important because it provides a critical condition for processes such as combustion, chemical reactions, fermentation, drying, distillation, concentration, extrusion, crystallization, and air conditioning. Poor temperature control can cause major safety, quality, and productivity problems. Although highly desirable, it is often difficult to control the temperature because its measurement must be within a specified range of accuracy and have a specified degree of speed of response, sensitivity, and dependability. Additionally, temperature measurements must also be representative of true operating conditions in order to achieve successful automated control. The instrument selected, installation design, and location of the measuring points determine these specifications.

This chapter will serve as a guide to select the best location of measuring points to achieve the best automatic control. It will consider temperature control for three common process types: a CSTR, distillation column, and heat exchanger.

Temperature Control Loops

Before temperature control loops for specific processes are explained, we must discuss the general considerations common for all temperature control loops.

Temperature control loops can either be endothermic (requiring heat energy) or exothermic (generating heat energy). Both types are similar in that they both result in a response representing a process with a dominant capacitance plus a dead time. For both types of processes, one of the following devices is used to measure temperature:

- thermocouple
- filled thermal well system
- Resistance temperature detector (RTD)

The measurement device, or thermal well, should be selected so that is minimizes additional lag to the overall process lag. Minimizing temperature measurement lag in the temperature control loop is important in both slow and fast loops. Some general rules of thumb for reducing temperature measurement lag are:

1. Use a small-diameter bulb or thermal well to minimize the thermal resistance and thermal capacity of the measuring element.

2. Use a thermal well made from a material that minimizes thermal resistance and thermal capacity of the measuring element.
3. Use a small pipe or orfice near the measuring device to increase velocity of the passing flow. Increasing flow will increase the rate of heat transfer between the process fluid and the measuring device.
4. Place the measuring element in the liquid phase when measuring temperature in a two-phase system since thermal resistance is smaller in the liquid phase than in the vapour.
5. Use a transmitter with derivative action to cancel out some of the lag in the measuring element. Compensate for this added derivative gain in the transmitter by reducing the derivative gain in the controller.

CSTR Temperature Control

Endothermic Reactor Temperature Control Loops

Endothermic CSTR reactors are generally easier to control than exothermic CSTR reactors because when the temperature reaches a critical minimum, the reaction does not proceed until there is adequate heat. In this sense, endothermic CSTR reactors are self-regulating. A good way to think of an endothermic CSTR's controls is that of a heat exchanger being used to heat a reaction solution. This heat exchanger's response is controlled by the dead time, so typically a PI or PID controller is used.

There are two main types of temperature control methods for endothermic CSTRs: control *via* steam flow rate and control *via* steam pressure. For the steam flow rate control case, the temperature control sends a signal to the flow control, which then controls the steam inlet valve. For the steam pressure control case, the temperature control sends a signal to the pressure control, which controls the steam inlet valve. Examples of an endothermic CSTR temperature control loop can be seen below in Figures.

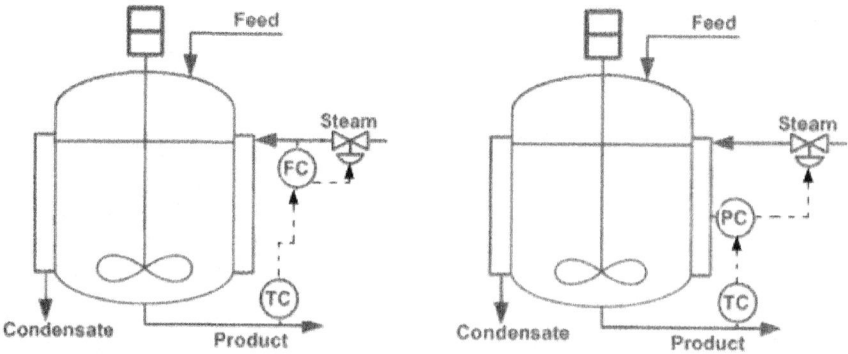

Fig.: Chemical and Bio-Process Control.

Exothermic Reactor Temperature Control Loops

In an exothermic reaction, energy is released in the form of heat. In some cases, a cooling system is require to bring the temperature back to a set point and also to ensure that the temperature does not rapidly increase, or runaway. Some steps can be taken to prevent runaway temperatures, such as reducing the feed rates or concentrations of reactants. Also, the ratio of the heat transfer area to the reactor volume can be increased to help increase the controllability of the CSTR. Typically, the temperature is controlled using a PID controller, which is described in the previous section PID Control.

Exothermic CSTRs are very difficult to control because they are very unstable and the temperature can easily runaway. The relationship between heat generation and temperature is non-linear, while the relationship between heat removal and temperature is linear. This relationship is illustrated in figure. The stability of the temperature control loops depend on the rate at which heat can be removed from the system. For example, an exothermic CSTR that generates heat at a slow steady rate is more stable than a reactor that rapidly produces heat. In this example, the rate at which the heat can be removed from the system depends on the rate at which the temperature can be changed in the cooling jacket surrounding the CSTR.

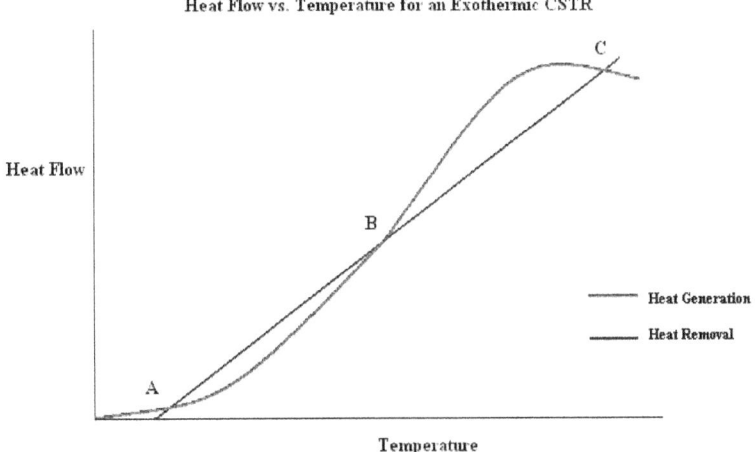

Fig. : Heat generation or removal as a function of temperature in an exothermic CSTR.

The intersections of the two curves, labeled A, B and C, represent steady state operating points. A and C are stable operation points while B is unstable. A and C are stable because as the temperature increases from this point, the rate of heat generation is less than the rate of heat removal. This means that as the temperature increases, heat is removed faster than it is generated so the temperature will be brought back down to that operating point. The same will happen if the temperature decreases from that point. As the temperature decreases, the heat generation rate is greater than the heat removal rate so the temperature will be brought back up to the operating point.

Point B is unstable because when the temperature increases from that point, the heat generation is greater than the heat removal. This means that as the temperature increases, heat is continuously added to the system and the temperature will rise until it reaches one of the stable operating points, in this case C. If the temperature decreases from point B, the heat generation rate is less than the heat removal rate so the temperature will continue to decrease until it reaches the lower, stable operating point, A.

This graph would be similar for an endothermic CSTR but there would be only one stable steady state operating point.

Very rapid exothermic reactions are the most difficult to control and they are sometimes carried out in a semi-batch reactor so the addition of reactants can be carefully controlled and runaway temperatures can be avoided.

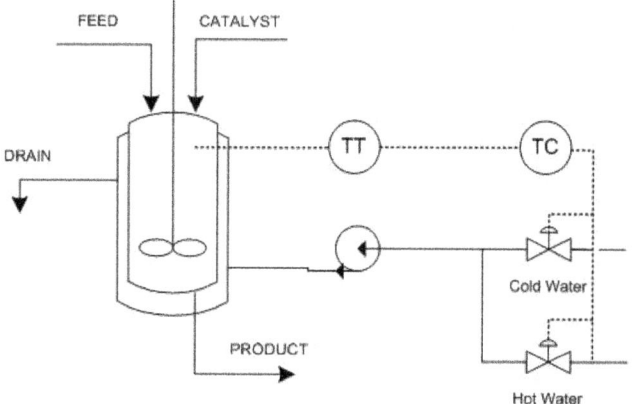

Fig. : Temperature control loops for an exothermic CSTR.

Figure illustrates a common control loop for an exothermic CSTR with a cooling jacket. A temperature transmitter (TT) sends a signal to the temperature controller (TC), which controls the hot and cold water valves on the jacket. The cold and hot water are pumped into the jacket which controls the reactor at a set temperature.

Temperature Control in Distillation

In a distillation column, temperature control is used as a means to control composition because temperature sensors are cheaper, more reliable, provide continuous measurements, and respond quicker than composition analyzers. Through equilibrium relationships, temperature measurements can be used to infer composition of the product. Inferential temperature control is only effective when the relative volatility of the components is high (greater than 2.0). Temperature controllers are used as feedback composition controllers to adjust column operation to meet production requirements. They must be able to satisfy the constraints defined by production requirements at all times, even in the face of disturbances.

Inferential Temperature Control

To control composition *via* temperature, a correlation must be made between temperature in the tray and composition of the key components. Determining a correlation is challenging because the temperature-composition relationship is affected by process nonlinearity and disturbances in feed composition, flowrates, and occurrence of entrainment or fouling. For multi-component separations, temperature does not determine a unique composition, so either an online composition analyzer or periodic lab tests must be utilized to verify composition at the temperature setpoint. Additionally, the column pressure greatly affects the measurement of tray temperature. For most systems, the following linear equation can be used to correct for variations in pressure:

$$T_{pc} = T_{mean} - K_{pr}(P - P_0)$$

Where,

T_{pc} = corrected temperature

T_{mean} = measured tray temperature

K_{pr} = pressure correction factor

P = column pressure

P_0 = reference pressure

The pressure correction factor, K_{pr}, can be estimated using a steady-state column simulator for two different operating pressures and the equation:

$$K_{pr} = \frac{T_i(P_1) - T_i(P_2)}{P_1 - P_2}$$

Where,

T_i = temperature of tray i predicted by the column simulator

Successful temperature control in the column depends on the dynamic response of measuring the tray temperature with respect to the manipulated energy source used to actuate temperature. The energy source is either the reboiler or the reflux. To have tight process control means that the equivalent dead time in the loop is small compared to the shortest time constant of a disturbance with significant amplitude. According to Svreck et al., the following observations from experimental tests are cited:

- Temperature control is made less stable by measurement lag or response times.
- The speed of response and control stability of tray temperature, when controlled by reboil heat, is the same for all tray locations.
- The speed of response and control stability of the tray temperature, when controlled by reflux, decrease in direct relation with the number of trays below the reflux tray.

- When pressure is controlled at the temperature control tray, the speed of response of the temperature control instrument can vary considerably with tray location, and is normally slower.

To achieve the best composition control, you need to determine the tray(s) whose temperature(s) show the strongest correlation with product composition. The following procedure using a steady-state column model can be used:

1. Run the column at the base conditions (x^{BC} and y^{BC}) at steady-state and record the temperature, T_i^{BC}, of each tray.
2. Increase the impurity level in the bottoms product ($x^{BC} + \Delta x$ and y^{BC}), so that Δx is about 25-50% of the impurity level for the base case. Record the temperature, $T_i^{\Delta x}$, of each tray.
3. Increase the impurity level in the overhead product (x^{BC} and + $y^{BC} + \Delta y$), so that Δy is about 25-50% of the impurity level for the base case. Record the temperature, $T_i^{\Delta y}$, of each tray.
4. The best tray for inferential temperature control of the stripping section will be the least sensitive to variations in the composition of the overhead product. This tray is the one that maximizes:

$$\Delta T_i^{net} = (T_i^{\Delta x} - T_i^{BC}) - (T_i^{\Delta y} - T_i^{BC})$$

5. The best tray for inferential temperature control of the rectifying section will be the least sensitive to variations in the composition of the bottoms product. This tray is the one that maximizes:

$$\Delta T_i^{net} = (T_i^{\Delta y} - T_i^{BC}) - (T_i^{\Delta x} - T_i^{BC})$$

6. Repeat this procedure for a range of feed compositions. Then, select the tray that maximizes the ΔT_i^{net} equations most often for the range of feed compositions.

Single Composition Control

Once you have determined the best tray to use for inferential temperature control, you must choose a manipulated variable (MV) that will have a significant gain on the tray temperature, and thus on the composition. Usually in distillation, the composition of only one product stream is controlled while the composition of the other product stream is allowed to drift. This is called single composition control. The chemical industry primarilly uses sing composition control.

When you are interested in controlling the composition of the overhead product, the reflux is used to control the purity, while the reboiler duty is held constant. The bottom composition is not directly controlled and will vary as the feed composition varies. The reflux, L, is the manipulated variable that will result in the tightest control, rather than either the distillate flowrate, D, or the reflux ratio, L/D. This is because L is the fastest-acting MV for the overhead and is the least sensitive to changes in the feed composition This scenario is depicted below in figure.

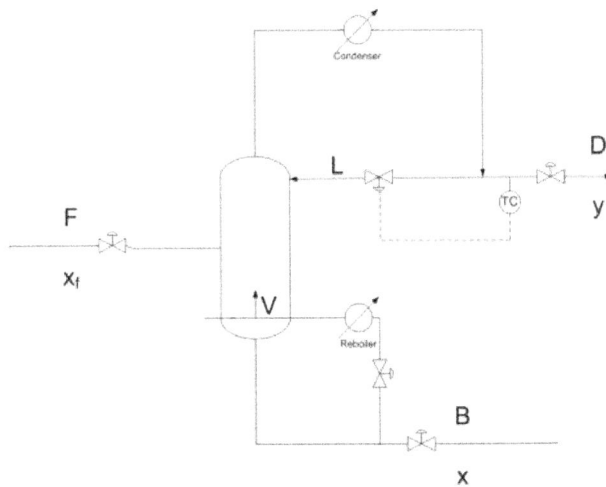

Fig. : Control diagram for inferential temperature-single composition control of overhead product stream by L.

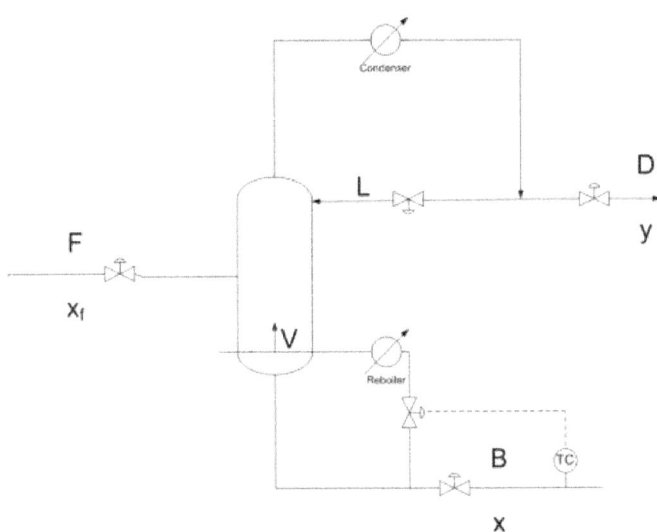

Fig. : Control diagram for inferential temperature-single composition control of bottom product stream by V.

When you are interested in controlling the composition of the bottom product, the boilup rate, V, is used to control the purity, while the reflux rate is held constant. The overhead composition is not directly controlled and will vary as the feed composition varies. V is the manipulated variable that will result in the tightest control, rather than either the bottoms product flowrate, B, or the boilup ratio, V/B. This is because V is the fastest-acting MV for the bottoms and is the least sensitive to changes in the feed composition This scenario is depicted in above figure.

Dual Composition Control

Although single composition control is most commonly used, there are some industries that need to simultaneously control the composition of both the overhead and bottom product streams. This is called dual composition control and is harder to implement, tune, and maintain than single composition control. Dual composition control increases product recovery and reduces utility costs and is used for refinery columns that generally have high associated energy usage. Industries that have refineries, such as the oil industry, use dual composition control because they are interested in using both the overhead and bottom streams as product.

In dual composition control, there are many possible control configurations. The control configuration is described by the control objective and by the MV. There are four possible control objectives: bottom product x, overhead product y, reboiler level, and accumulator level. The chosen control objectives are then paired with a MV. There are a variety of choices for the MV. These include: L, D, L/D, V, B, V/B, B/L, and D/V. Therefore, in dual composotion control, there are many possible configurations to consider, although most are not practical.

The most commonly used configuration for dual composition control is (L, V) which uses L to control y (overhead composition) and V to control x (bottom composition), because it provides good dynamic response, is the least sensitive to changes in feed composition, and is the easiest to implement. However, it is highly susceptible to coupling. In this configuration, depicted below, the setpoint for the reflux flow controller is set by the overhead composition controller, and the setpoint for the flow controller on the reboiler duty is set by the bottom composition controller. D is then used to control the accumulator level, and B is used to control the reboiler level.

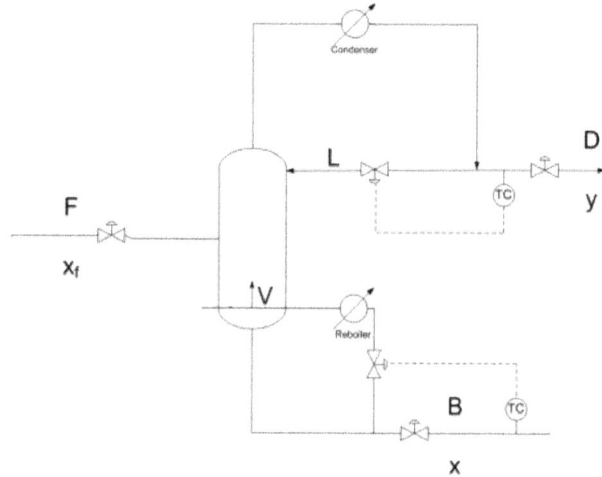

Fig. : Control diagram for inferential temperature-dual composition configuration (L,V), the control of overhead product stream by L and bottom product stream by V.

Since there are so many possibilities for configurations, there is no clear best choice for a configuration of MVs in dual composition control in distillation columns. It is impossible to know theoretically which configuration is optimum for a particular process, but there are some rules of thumb to follow to increase the possibility of choosing a good configuration. These rules are summarized in the chart below.

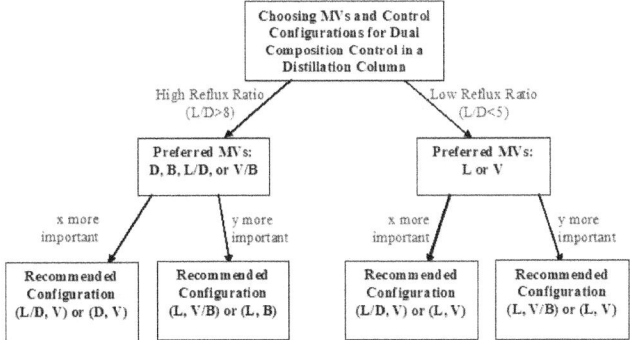

Fig. : Chart to determine a good configuration for your process when using dual composition inferential temperature control.

For the gray area where 5<L/D<8, follow the rules of thumb for the closest L/D, and use experimental data to determine the best configuration for your particular process.

Controller Tuning and Constraints

When temperature is used for inferential composition control, a PID controller is usually used because of the significant sensor lag. Common constraints on the extent of temperature control in a distillation column include:

- Capacity of reboiler power
- Capacity of overhead condenser
- Flooding or entrainment
- Maximum (undesired) column temperature

In general, temperature control can be used to control composition in a distillation column. The temperature measuring device should be chosen to minimize the lag time. A steady-state model can be used to correct for pressure changes in the column and to determine the best tray at which to measure temperature. The choice of the MV depends on if you are operating under single or dual composition control and which product composition you want to control.

Heat Exchanger Control

In heat exchanger control, the temperature of the process exit stream is the controlled variable (CV) and can be adjusted by one of four possible manipulated variables: cool side entry stream, cool side exit stream, hot side entry stream, or hot side exit stream. The selection of where control will be implemented is a combined result of three factors:

1. Whether we want to heat or cool the process stream
2. The response time of the controller
3. The capital cost

To illustrate the selection process, first it is shown that the hot stream preferred for temperature control, then two case studies are discussed, one for heating up the process stream and the other for cooling the process stream.

Controlling the Cool Side Stream

It is not advisable to place temperature control on the cool side stream. This is explained below.

The particular MV in each stream is the flow rate. It is known that the temperature of the hot stream will change once the flow rate of the cool stream is adjusted.

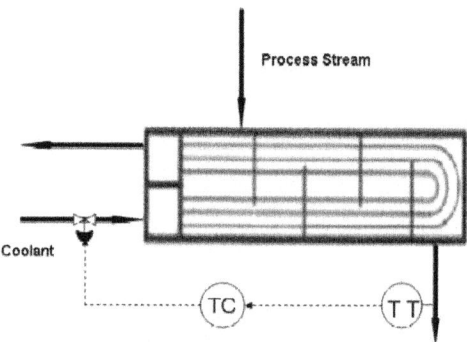

Fig. : Coolant temperature control loop for a Liquid/Liquid Heat Exchanger.

It is easier to think of the temperature change as the magnitude of the process gain ($|T_{t+1} - T_t|$). This just is the magnitude of the temperature change over a specified interval. This way the discussion can be general for both heating or cooling the process stream.

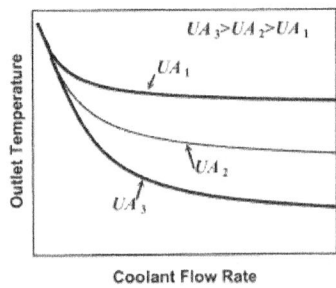

Fig. : The limit that increasing the flow rate of the coolant stream has on the exit temperature of the process stream for three heat exchanger areas.

Control Architectures

As the rate of the cool stream is increased from zero, the magnitude of the process gain continuously decreases. Above a certain rate, the gain will not be a measurable change in the process stream temperature. Because of this, *making the coolant stream the MV will make the process uncontrollable at a critical point and is not a good idea.*

Controlling the Hot Side Stream

As indicated above it is never recommended to place the temperature control on the cool side stream. This reduces controls consideration to only the hot stream. There are two cases where the control can be on the hot stream; when heating up and when cooling down the process stream.

Heating Up The Process Stream

CASE STUDY 1: Steam-Heat Heat Exchanger

In this case the process stream is the cooler of the two entering streams, the previous section states that the MV would have to on the stream used to heat up the process stream.

The temperature control loop can either be the direct steam flow or the steam pressure. Both controls the amount of steam flowing into the heat exchanger by adjusting a valve, however, the preferred of the two is the steam pressure as the MV in the temperature control loop. This is because a change in the steam supply can quickly cause the internal pressure of the heat exchanger to change hence affecting the temperature exchange. The steam pressure control responds and corrects this faster than the steam flow control. This becomes especially useful for heavy duty requirements when the change in response to change in flow rate of the process stream. The two figures below show two preferred configurations;

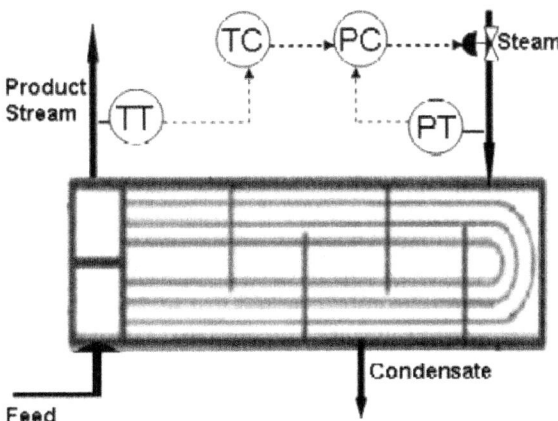

Fig. : One arrangement for temperature control for heating up a process stream.

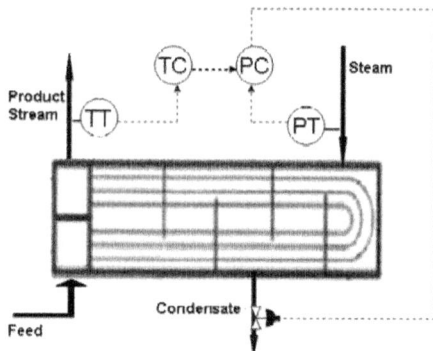

Fig. : An alternative arrangement for temperature control for heating up a process stream.

Notice that Figure shows the control valve on the condensate stream. This is an acceptable control placement, however, in most practical application the first of the configurations in Figure is preferred. In addition there are some ramifications to making the condensate flow the MV. The temperature loop in this case would not be as responsive as in the previous case. This is because the level responds slower than the pressure to the changes in the respective valves. The capitol cost involved in placing the valve in the condensate stream is much lower that that in the steam stream because the steam stream is generally larger and requires a larger, more expensive valve. In addition a steam trap is required down stream of the condensate valve, for that configuration.

Cooling Down the Process Stream

CASE STUDY 2: Liquid/Liquid Heat Exchanger

In this case the process stream is the hotter of the two entering stream so the controls will have to be on this stream. As in case one we are setting up the temperature

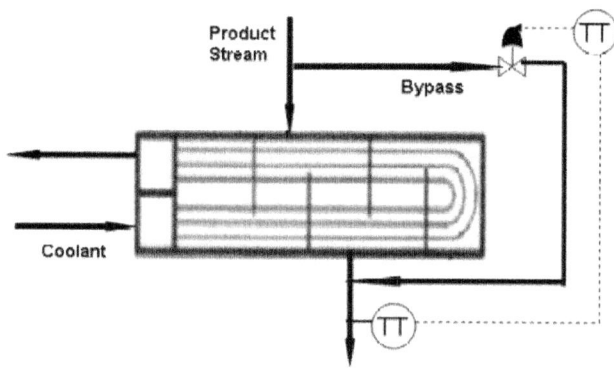

Fig. : The common arrangement for temperature control for cooling down a process stream

control loop for varying flows in the process stream because in a plant the flow rate of this stream is dependent on a process before the heat exchanger. Hence when creating control loop for cooling down the process stream the coolant flow is kept constant (as stipulated in the section "Coolant Stream Controls") and maintain the CV at the set point. The best choice for the MV in this temperature control loop is having a feed bypass stream, as illustrated in Figure below. The advantages of having this bypass stream are that the process dynamics are considerably faster with low levels of dead time and this condition is more linear, as opposed to varying the coolant flow rate. With this configuration the coolant flow rate can be maintained very high which reduced the tendency for fouling on the process fluid side of the heat-transfer surface.

CCL Reactors

Reactors are the central focus of many chemical plants. Many parameters must be controlled in a reactor for proper operation. Temperature is of great importance because it affects reaction rates and equilibrium relationships. A major challenge for temperature control is handling the nonlinear nature of temperature inside most reactors. Therefore, it is important to design an effective control architecture in order to ensure optimal operation of the reactor.

This chapter discusses the common control architectures and topologies in CSTRs. The control architectures are designed based on whether the reactor is endothermic or exothermic. However, only CSTRs will be discussed for simplicity.

Common Topologies

Here we will introduce a few of the most common control topologies which will be examined in applications to endothermic and exothermic CSTRs below.

Feedback and Feed-Forward

Feedback and feed-forward refer to the direction in which the sensor information is transferred to an actuator valve. Feedback control dictates that sensor information is "fed back" to a previous part of the process. For example, the reading from a level sensor of a filling tank can be "fed back" to the valve controlling the input to the tank. Feed-forward control means the sensor information is used to control something downstream from where the reading was taken. For example, the measured flow rate of water going into an evapourator can be used to control the heating coil inside of the evapourator.

Ratio Control

Ratio Control is used when the ratio between two measured process variables has an optimal value. In the context of two input streams with optimal flow ratio going into a reactor, one stream is designated as the control stream, and one stream is designated as the wild stream. The wild stream fluctuates, and a valve on the control stream is opened or closed to maintain the ratio between the two stream flows.

Cascade Control

Cascade control simply means that instead of one control loop found in simple control topologies where a sensor's measurement directly controls an actuator valve, multiple loops are used so that sensors' measurements can control set points for other controllers. For example, the temperature sensor measurement of process fluid exiting a reactor can be used to modify the set point of the flow controller of steam feeding the heating jacket, which then sets the steam valve. This multiple-loop system eliminates some problems caused by variable-pressure feeds, for instance.

Disturbances to CSTRs

There a few very common disturbances that CSTR may be subjected to. When designing a control architecture for a CSTR, you must investigate the possibility of all of these disturbances, determine the magnitude of each possible disturbance, and address how each will be handled.

- Changes in feed properties
    - o flow rate
    - o composition
    - o temperature
- Changes in enthalpy of heat exchange medium
- Change in heat transfer properties (ex: fouling)

Disturbances to PFRs

Plug Flow Reactors (PFRs) behave differently than CSTRs and will have different properties to consider when designing a control architecture for them. The biggest difference is that the temperature, flows, and compositions will all vary along the reactor. There is danger of exceeding a design limitation in temperature or flow in certain parts of the PFR, so control of the reactor is important along its length. To design a control architecture for a PFR, the following disturbances and changes to the system must be addressed:

- Temperature control in multiple places along reactor (hot spots can easily occur in PFRs)
- Flow control in multiple places along the reactor
- Inlet/outlet pressures
- Feed property disturbances:
    - o flow rate
    - o composition

- o temperature
- Changes in enthalpy of heat exchange medium
- Change in heat transfer properties (ex: fouling)

Endothermic Reactors

Endothermic reactors tend to be easier to control than exothermic ones because they are much less prone to runaway. There are two commonly used methods to control an endothermic CSTR. These two methods are differentiated by the variables they manipulate. In the first method, the steam pressure in the reactor jacket is in the manipulated variable, whereas in the second method, the steam flowrate is the manipulated variable. This is done to ensure that changes are being based on what is actually happening in the reactor and not what is predicted to happen. This is important in the endothermic case to ensure that the proper amount of heat is being added to the reactor to obtain a desired conversion. Feedback is more useful for control since the amount of heat needed can change quickly based on the amount or concentration of reactants being added. Generally the temperature control of the reactor is independent of the reactant feed, so the system needs a way to adjust to changes within reactant feed, and this is accomplished by using feedback control in the streams that control the temperature (*i.e.* the steam feed stream).

Controlled by Steam Pressure

Heat duty can be more effectively removed with the use of steam pressure as the manipulated variable because changes in the heat duty of the reactor will cause an immediate change in the steam pressure. By using the steam pressure, it also linearizes the temperature control system which is not the case when the steam flowrate is the manipulated variable. Using the steam pressure as the manipulated variable does not provide a direct measurement of the heat load, the amount of heat needed by the reactor. If a direct measurement is required, a steam flow indicator can be installed which would allow for the direct measurement of the applied heat load.

An added benefit, which those of you who have worked with the temperature controller in ChE460 might have noticed, is that often times the pressure of steam being provided by the utilities plant is variable. By using the steam pressure as the manipulated variable the control system will automatically adjust to this type of disturbance from the source. If a large change in steam pressure were to occur, the pressure controller would measure this change, and the control scheme would intake the change in steam pressure and translate this into an appropriate change in the valve setting.

As can be seen in the figure above, when steam pressure is used as the manipulated variable the control system is run in a feedback mode. One of two things can control the amount of steam fed into the jacket. The first is the temperature of the product, and the second is the steam pressure in the jacket. As mentioned above, a change in the heat duty required by the reactor will quickly change the steam pressure which is why it is commonly used over the steam flowrate.

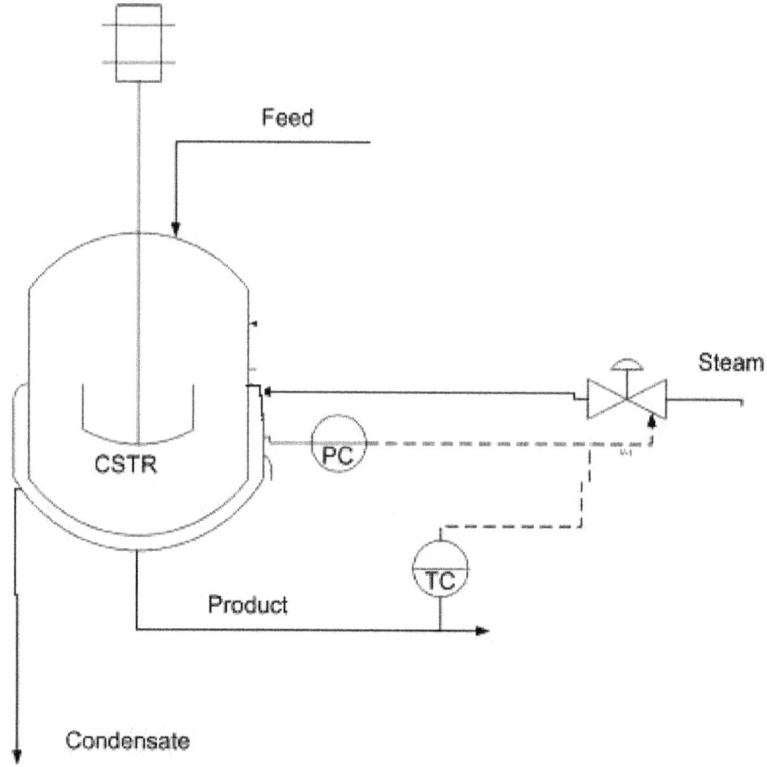

Notice that there are two controllers that are responsible for adjusting the steam valve. The pressure controller is sensitive to changes in the heat duty required by the reactor and is used to adjust the steam according to the needs of the temperature controller on the product stream. This is a prime example of cascade control. The temperature of the product stream would output a setpoint to the pressure controller for the amount of steam needed to attain the desired temperature setpoint. The pressure controller would then communicate to the valve what needs to be done in order to achieve this temperature setpoint based on the steam pressure.

Controlled by Steam Flowrate

The second method for controlling a endothermic CSTR is by manipulating the steam flow rate. Using flow rate as the manipulated variable makes the control system prone to changes in heat load and changes in the enthalpy of supplied steam. These changes require direct action of the temperature control system. One advantage of using flowrate as the manipulated variable is that heat load is directly measured, and thus, conversion is directly measured.

Control Architectures

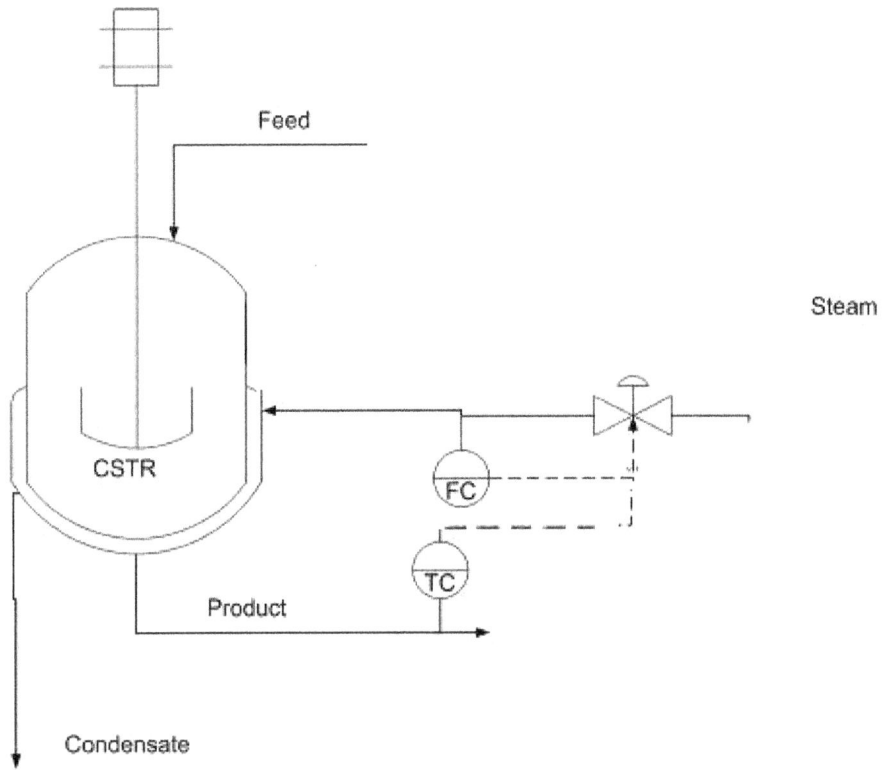

As can be seen in the figure above the system works in a feedback mode when the steam flowrate is used as the manipulated variable. The temperature of the product stream is the primary factor in adjusting the amount of steam fed into the system, and therefore this system is less responsive to changes in the amount of heat duty required for the process.

Note here that there are again two controllers used to adjust the steam valve. This setup is similar to the previous case, but now the flow controller is the "slave" controller to the temperature controller. The cascade control scheme is again at work!

Control Architecture for Endothermic Reactor		
Manipulated Variable	Advantage	Disadvantage
Steam pressure	More responsive to changes in heat duty	No direct measurement of heat load
	Linearizes temperature control system	
Steam flowrate	Heat load and conversion directly measured	Less responsive to changes in heat duty

Exothermic Reactors

Exothermic reactors are harder to control because safety is dependent on heat removal. There are two common control architectures for an exothermic CSTR based on the temperature of the coolant. The first method uses the outlet temperature of the coolant as the manipulated variable, while the second method uses the inlet temperature of the coolant. Just like the endothermic reactors, feedback control is very commonly used when controlling the temperature of exothermic reactors. The reasoning for feedback control in the exothermic case is slightly different but similar to that of endothermic reactors.

In endothermic reactors we wanted to ensure that the proper amount of heat was being supplied to our reactor, and if the amount needed changed; we needed a way to adjust our temperature accordingly. In the exothermic case we need to ensure that we are removing the proper amount of heat, not only to ensure an optimal reaction temperature, but also to prevent a runaway reaction.

Therefore, we need to know what is actually going on within the reactor instead of forecasting possible temperature changes within it. Cascade control is also used. In this case the temperature of the product stream outputs a setpoint to the temperature controller whether it be located on the outlet or inlet coolant feed. The "slave" controller can then adjust the valve position based on the temperature of the coolant water.

Controlled by Outlet Coolant Temperature

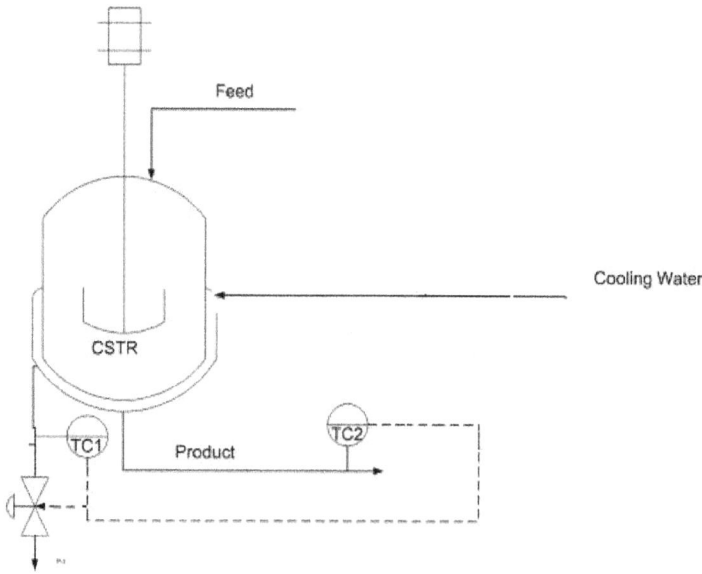

The figure above shows the control architecture for CSTR reactor temperature controlled by outlet coolant temperature in a feedback mechanism. There is a temperature control on both the product stream and outlet coolant temperature. Both of the temperature controls are used to control the valve on the inlet coolant stream. The advantage of this setup is that it responds faster to fouling on heat-transfer surfaces than the setup with the temperature controller on the inlet.

This is because in order for the inlet temperature to adjust to fouling, the fouling must first affect the temperature of the product stream. Therefore the fouling has less direct effect on the controller, but in this case fouling will immediately affect the temperature of the exiting cooling water. However, this setup has the disadvantage of responding slower to changes in the inlet coolant temperature.

As can be seen in the above diagram you will notice that again cascade control is utilized in this particular system. The temperature sensor on the product stream provides information on whether the stream needs to be cooled more or less and outputs a setpoint to the temperature controller on the recycle stream. This controller can then take into account the temperature of the recycled coolant water and make an adjustment to the amount of fresh coolant water that is added to the system.

Controlled by Inlet Coolant Temperature

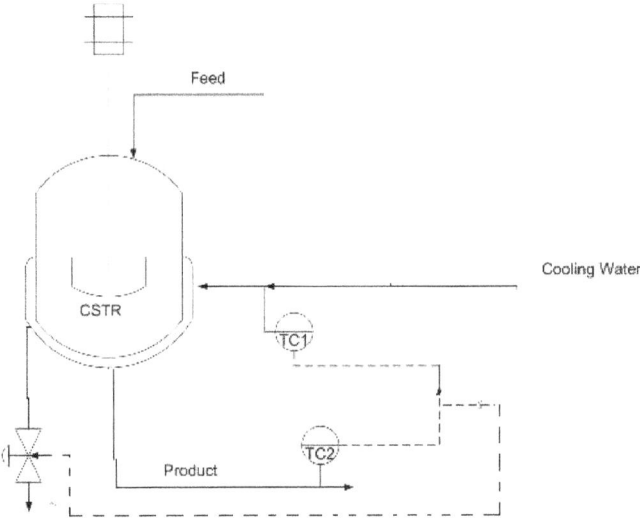

As seen in the figure above, the control configuration for CSTR reactor temperature is controlled by inlet coolant temperature in a feedforwardmechansim. There are temperature controls on both the product stream and inlet coolant temperature; both of the controls are used to control the valve on the inlet coolant stream. One advantage of this setup is that it responds faster to changes in inlet coolant temperature. A disadvantage is that it responds slower to fouling on heat-transfer surfaces.

More on Exothermic Reactors

How do you decide which exothermic reactor control architecture to use? It depends on whether a faster response to fouling on heat-transfer surfaces or a faster response to changes in inlet coolant temperature is more important.

In addition to stabilizing the temperature within an exothermic reactor, it is often a goal to maximize production rate. This can be achieved by placing a valve on the feed stream. As the temperature control adjusts the cold water valve, the cold water valve position is then communicated to the feed valve and adjusts it.

Cascade control is again utilized. In this case the secondary or "slave" controller is located in a different location, which is the only difference between the two scenarios.

Control Architecture for Exothermic Reactor		
Manipulated Variable	Advantage	Disadvantage
Outlet coolant temperature	Faster response to fouling on heat-transfer surfaces.	Slower response to changes in inlet coolant temperature
Inlet coolant temperature	Faster response to changes in inlet coolant temperature	Slower response to fouling on heat-transfer surfaces

Chapter 3

CONTROL SYSTEMS OF INDUSTRIAL APPLICATIONS

Many control systems are used today in a large number of industries consisting of applications from all kinds. The common factor of all control types is to sustain a desired outcome that may change during a chemical reaction or process. The most common control type used today in industry is a PID controller (proportional, integral, derivative), which allows the operator to apply different control techniques that can be used to achieve different settings in an experiment or process. A PID controller can be used in two main control mechanisms that include feed back and feed forward.

Temperature Control: Thermocouple

A thermocouple is a device to measure and control temperature within a system. They are used in a wide variety of industrial applications (gas turbines, chemical reactors, exhaust, chemical manufacturing *etc.*) due to their low cost and portability. The fundamental working principle for thermocouple operations is the Seebeck Effect.

Mostly, thermocouples operate in a P or a PID control mode. In order to measure temperature between two points, the thermocouple employs two metallic ends (made from different alloys). When the two conductor ends are exposed to a thermal gradient, they generate a voltage between them. This voltage drop gives rise to the temperature measurement output that a thermocouple provides.

Depending on the types of alloys in both conductor ends, and the magnitude of the thermal gradient, thermocouples can measure temperature differences between 1-23000°C. They can operate in feedback loops or feed forward loops. Thermocouples are mostly digital control units. Thermocouple prices started around $100 and cost up to $2500 for more accurate, and self-calibrated models.

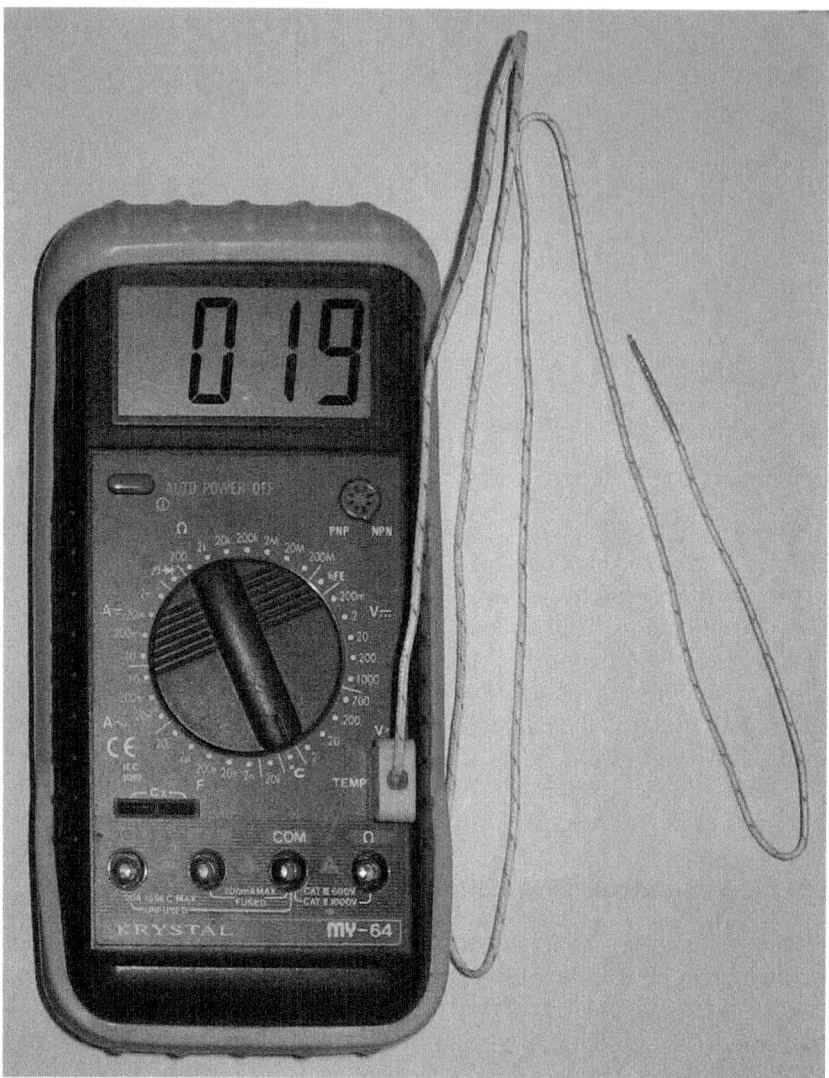

Fig. : Thermocouple is a device to measure and control temperature within a system.

Pressure Control: Pressure Switch

A pressure switch is a device that controls systems against pressure drops or pressure spikes. The most basic types of pressure switches work on an ON-OFF basis, but can also be manufactured to work in PID mode. The fundamental method of operation is to set the "Set-Pressure" to a given quantity. This deactivates the pressure switch from the circuit connecting it to the control valve upstream. If, at any point during the process, the pressure rises past the set-point, the switch is activated and completes the circuit, thus shutting off the control valve. Pressure switches can be hydraulic, or pneumatic based (air-based pressure).

Fig. : Pressure switch is a device that controls systems against pressure drops.

One common application of a pressure switch in the industry is to protect PD (positive-displacement) pumps from over-pressurization. A PD pump can generate very high pressures if not controlled by a pressure switch; thus setting a pressure switch inline with a PD pump will prevent over pressurization since it shuts off the control valve. Pressure switches are common in any industry since all require pressurization of certain components during manufacturing, processing or refining stages. They sell between $200-$2000 depending on the magnitude of the set-point required for protection.

Composition Control: Ratio Control

A ratio controller is used to ensure that two or more process variables such as material flows are kept at the same ratio even if they are changing in value. Control modes can be operated in different types, but mostly feedback PI controller is used for ratio control. In industrial control processes, ratio control is used in the following processes: burner/air ratio, mixing and blending two liquids, injecting modifiers and pigments into resins before molding or extrusion, adjusting heat input in proportion to material flow.

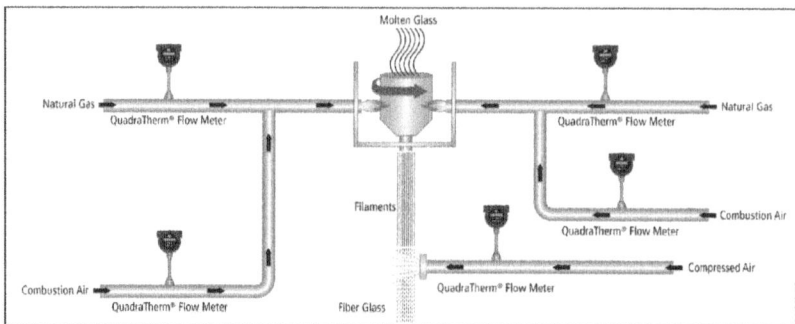

Fig. : A ratio controller.

The prices for industrial ratio controllers start around $500 and increase depending on the sensitivity of the unit to different magnitudes of compositional changes and size of the equipment to be annexed to.

Level Control: Level Switches

Level controls are used to monitor and regulate the liquid level in industrial vessels. There are many various sensors used in level control including ultrasonic,

Fig. : Level Switches.

lasers, and floatation sensors. They all work on the same general principle. A sensor measures the distance from the base of the vessel to the top of the liquid level, mainly by using sonar waves or a laser beam. Based on the time it takes for the wave or for the laser beam to return to the emitting source, the controller sends information to change or maintain the level.

The magnetic float control reads on a sensor located on the wall of the vessel and sends that information to the controller. Some examples of level control in industry are: maintaining the liquid level in a distillation column, protecting from overflow, and measuring the amount of product in storage tanks. Level sensors/controls vary in price based on the type and the accuracy required. Below is a picture of a magnetic float level control

Flow Control: Flow Meters

Flow controls are used to regulate the flow of a liquid or gas through a system. The main form of flow control is a valve. There are many different types of valves, but they all change flow rates by opening or closing based on what amount is needed. A flow sensor reads the flow rate, and a controller will operate the valve to increase or decrease the flow. Some basic types of flow sensors are rotameters and digital mass meters. These controls are used in all forms of industry to control flows including water treatment, product measurement, and fuel flow to furnaces.

Fig. : Flow Control.

The low price end for flow sensors is typically around $500, and depending on the size of the process, and the process material, the price can range into the thousands. Below are two pictures. The first is a manual control valve that can be opened or shut to regulate flow. The second is a valve that can be used in an automatic control system to regulate flow.

TEMPERATURE SENSORS

Temperature sensors are vital to a variety of everyday products. For example, household ovens, refrigerators, and thermostats all rely on temperature maintenance and control in order to function properly. Temperature control also has applications in chemical engineering. Examples of this include maintaining the temperature of a chemical reactor at the ideal set-point, monitoring the temperature of a possible runaway reaction to ensure the safety of employees, and maintaining the temperature of streams released to the environment to minimize harmful environmental impact.

While temperature is generally sensed by humans as "hot", "neutral", or "cold", chemical engineering requires precise, quantitative measurements of temperature in order to accurately control a process. This is achieved through the use of temperature sensors, and temperature regulators which process the signals they receive from sensors.

From a thermodynamics perspective, temperature changes as a function of the average energy of molecular movement. As heat is added to a system, molecular motion increases and the system experiences an increase in temperature. It is difficult, however, to directly measure the energy of molecular movement, so temperature sensors are generally designed to measure a property which changes in response to temperature. The devices are then calibrated to traditional temperature scales using a standard (*i.e.* the boiling point of water at known pressure). The following sections discuss the various types of sensors and regulators.

Temperature Sensors

Temperature sensors are devices used to measure the temperature of a medium. There are 2 kinds on temperature sensors:

1. contact sensors and
2. noncontact sensors.

However, the 3 main types are thermometers, resistance temperature detectors, and thermocouples. All three of these sensors measure a physical property (*i.e.* volume of a liquid, current through a wire), which changes as a function of temperature. In addition to the 3 main types of temperature sensors, there are numerous other temperature sensors available for use.

Contact Sensors

Contact temperature sensors measure the temperature of the object to which the sensor is in contact by assuming or knowing that the two (sensor and the object) are in thermal equilibrium, in other words, there is no heat flow between them.

Examples (further description of each example provide below)
- Thermocouples
- Resistance Temperature Detectors (RTDs)
- Full System Thermometers
- Bimetallic Thermometers

Noncontact Sensors

Most commercial and scientific noncontact temperature sensors measure the thermal radiant power of the Infrared or Optical radiation received from a known or calculated area on its surface or volume within it.

An example of noncontact temperature sensors is a pyrometer, which is described into further detail at the bottom of this section.

Thermometers

Thermometers are the most common temperature sensors encountered in simple, everyday measurements of temperature. Two examples of thermometers are the Filled System and Bimetal thermometers.

Filled System Thermometer

The familiar liquid thermometer consistsof a liquid enclosed in a tube. The volume of the fluid changes as a function of temperature. Increased molecular movement with increasing temperature causes the fluid to expand and move along calibrated markings on the side of the tube. The fluid should have a relatively large thermal expansion coefficient so that small changes in temperature will result in detectable changes in volume.

A common tube material is glass and a common fluid is alcohol. Mercury used to be a more common fluid until its toxicity was realized. Although the filled-system thermometer is the simplest and cheapest way to measure temperature, its accuracy is limited by the calibration marks along the tube length. Because filled system thermometers are read visually and don't produce electrical signals, it is difficult to implement them in process controls that rely heavily on electrical and computerized control.

Bimetal Thermometer

In the bimetal thermometer, two metals (commonly steel and copper) with different thermal expansion coefficients are fixed to one another with rivets or by welding. As the temperature of the strip increases, the metal with the higher thermal expansion coefficients expands to a greater degree, causing stress in the materials and a deflection in the strip.

The amount of this deflection is a function of temperature. The temperature ranges for which these thermometers can be used is limited by the range over which the metals have significantly different thermal expansion coefficients. Bimetallic strips are often wound into coils and placed in thermostats. The moving end of the strip is an electrical contact, which transmits the temperature thermostat.

Resistance Temperature Detectors

A second commonly used temperature sensor is the resistance temperature detector (RTD, also known as resistance thermometer). Unlike filled system thermometers, the RTD provides an electrical means of temperature measurement, thus making it more convenient for use with a computerized system. An RTD utilizes the relationship between electrical resistance and temperature, which may either be linear or nonlinear.

RTDs are traditionally used for their high accuracy and precision. However, at high temperatures (above 700°C) they become very inaccurate due to degradation of the outer sheath, which contains the thermometer. Therefore, RTD usage is preferred at lower temperature ranges, where they are the most accurate.

There are two main types of RTDs, the traditional RTD and the thermistor. Traditional RTDs use metallic sensing elements that result in a linear relationship between temperature and resistance.

As the temperature of the metal increases, increased random molecular movement impedes the flow of electrons. The increased resistance is measured as a reduced current through the metal for a fixed voltage applied. The thermistor uses a semiconductor sensor, which gives a power function relationship between temperature and resistance.

RTD Structure

A schematic diagram of a typical RTD is shown in figure.

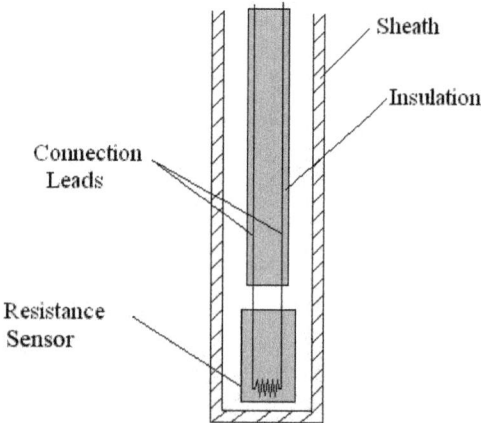

Fig. : Schematic Diagram of Resistance Temperature Structure.

As shown in Figure, the RTD contains an outer sheath to prevent contamination from the surrounding medium. Ideally, this sheath is composed of material that efficiently conducts heat to the resistor, but resists degradation from heat or the surrounding medium.

The resistance sensor itself is responsible for the temperature measurement, as shown in the diagram. Sensors are most commonly composed of metals, such as platinum, nickel, or copper. The material chosen for the sensor determines the range of temperatures in which the RTD could be used. For example, platinum sensors, the most common type of resistor, have a range of approximately -200°C – 800°C. (A sample of the temperature ranges and resistances for the most common resistor metals is shown in Table). Connected to the sensor are two insulated connection leads. These leads continue to complete the resistor circuit.

Table. Common Metal Temperature and Resistance Ranges.

Element Metal	Temperature Range	Base Resistance	TCR($\Omega/\Omega/°C$)
Copper	-100 – 260 °C	10 Ω at 0 °C	0.00427
Nickel	-100 – 260 °C	120 Ω at 0 °C	0.00672
Platinum	-260 – 800 °C	100 Ω at 0 °C	0.003916

There are 4 major categories of RTD sensors. There are carbon resistors, film thermometers, wire-wound thermometers and coil elements.

Carbon resisters are the most commonly used. They are inexpensive and are accurate for low temperatures. They also are not affected by hysteresis or strain gauge effects. They are commonly used by researchers.

Film thermometers have a very thin layer of metal, often platinum, on a plate. This layer is very small, on the micrometer scale. These thermometers have different strain gauge effects based on what the metal and plate are composed of. There are also stability problems that are dependent on the components used.

In wire-wound thermometers the coil gives stability to the measurement. A larger diameter of the coil adds stability, but it also increases the amount the wire can expand which increases strain and drift. They have very good accuracy over a large temperature range.

Coil elements are similar to wire-wound thermometers and have generally replaced them in all industrial applications. The coil is allowed to expand over large temperature ranges while still giving support. This allows for a large temperature range while decreasing the drift.

RTD Operation

Most traditional RTD operation is based upon a linear relationship between resistance and temperature, where the resistance increases with temperature. For this reason, most RTDs are made of platinum, which is linear over a greater range of temperatures and is resistant to corrosion. However, when determining a resistor material, factors such as temperature range, temperature sensitivity, response time, and durability should all be taken into consideration. Different materials have different ranges for each of these characteristics.

The principle behind RTDs is based upon the Callendar – Van Dusen equation shown below, which relates the electrical resistance to the temperature in °C. This equation is merely a generic polynomial that takes form based upon experimental data from the specific RTD. This equation usually takes on a linear form since the coefficients of the higher-order variables (a_2, a_3, etc.) are relatively small.

$$R_T = R_0(1 + a_1T + a_2T^2 + a_3T^3 + a_4T^4 + \cdots + a_nT^n) \quad (1)$$

R_T: Resistance at temperature T, in ohms

R_0: Resistance at temperature = 0°C, in ohms

a_n: Material's resistance constant, in $°C^{n-1}$

Another type of RTD is the thermistor, which operates based upon an exponential relationship between electrical resistance and temperature. Thermistors are primarily composed of semiconductors, and are usually used as fuses, or current-limiting devices. Thermistors have high thermal sensitivity but low temperature measuring ranges and are extremely non-linear. Instead of the Callendar - Van Dusen equation, the thermistor operates based upon the nonlinear equation, equation, shown in degrees K.

$$R_T = R_0 \exp\left(b\left(\frac{1}{T} - \frac{1}{T_0}\right)\right) \quad (2)$$

T_0: Initial temperature, usually set at 298K

b: Material's temperature coefficient of resistance, in K

Errors associated with resistance thermometers will occur due to the individual or collective efforts of: defective insulation, contamination of the resistor, or insecure lead wire connections.

Thermocouples

Another temperature sensor often used in industry is the thermocouple. Among the various temperature sensors available, the thermocouple is the most widely used sensor. Similar to the RTD, the thermocouple provides an electrical measurement of temperature.

Thermocouple Structure

The thermocouple has a long, slender, rod-like shape, which allows it to be conveniently placed in small, tight places that would otherwise be difficult to reach. A schematic diagram of a typical thermocouple is shown in figure.

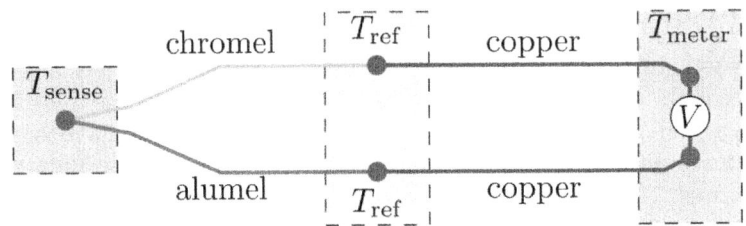

Fig. : Schematic Diagram of Thermocouple Structure.

As illustrated in Figure, the thermocouple contains an outer sheath, or thermowell. The thermowell protects the contents of the thermocouple from mechanical and chemical damage.

Within the thermowell lies two metal wires each consisting of different metals. Various combinations of materials are possible for these metal wires. Three common thermocouple material combinations used for moderate temperature measurements are the Platinum-Rhodium, Iron-Constantan, and Chromel-Alumel metal alloys.

The metal alloys chosen for a thermocouple is based upon the emf value of the alloy pair at a given temperature. Sample emf values for the most common materials at various temperatures are shown in Table. For a given pair of materials, the two wires are connected at one end to form a junction. At the other end, the two wires are connected to a voltage measuring device. These ends of the wires are held at a different reference temperature.

Table. Common Metal Temperature and Emf Values.

Alloy Type	Emf Value at 20 °C	Emf Value at 50 °C	Emf Value at 100 °C
Platinum-Rhodium	0.113 mV	0.299 mV	0.646 mV
Iron-Constantan	1.019 mV	2.585 mV	5.269 mV
Chromel-Alumel	0.798 mV	2.023 mV	4.096 mV

Various methods are used to maintain the reference temperature at a known, constant temperature. One method consists of placement of the reference junction within either an ice bath or oven maintained at a constant temperature. More commonly, the reference temperature is maintained electronically.

Though not as stable as an ice bath, electronically controlled reference temperatures are more convenient for use. Reference temperatures could also be maintained through temperature compensation and zone boxes, which are regions of uniform temperature. The voltage difference across the reference junction is measured and sent to a computer, which then calculates the temperature with this data.

Thermocouple Operation

The main principle upon which the thermocouple function is based on is the difference in the conductivities of the two wire materials that the thermocouple is made of, at a given temperature. This conductivity difference increases at higher temperatures and conversely, the conductivity difference decreases at lower temperatures. This disparity results in the thermocouples being more efficient and useful at higher temperatures. Since the conductivity difference is small at lower temperatures and thus more difficult to detect, they are inefficient and highly unreliable at low temperatures. The conductivity difference between the two wires, along with a temperature difference between the two junctions, creates an electrical current that flows through the thermocouple. The first junction point, which is the point at which the two wires are connected, is placed within the medium whose temperature is being measured.

The second junction point is constantly held at a known reference temperature. When the temperature of the medium differs from the reference temperature, a current flows through the circuit. The strength of this current is based upon the temperature of the medium, the reference temperature, and the materials of the metal wires. Since the reference temperature and materials are known, the temperature of the medium can be determined from the current strength.

Error associated with the thermocouple occurs at lower temperatures due to the difficulty in detecting a difference in conductivities. Therefore, thermocouples are more commonly used at higher temperatures (above -125°C) because it is easier to detect differences in conductivities.

Thermocouples are operable over a wide range of temperatures, from -200°C to 2320°C, which indicates its robustness and vast applications. Thermocouples operate over this wide range of temperatures, without needing a battery as a power source. It should be noted that, the wire insulation might wear out over time by heavy use, thus requiring periodical checks and maintenance to preserve the accuracy of the thermocouple.

To determine the temperature of the medium from the current strength, the emf or voltage values of the current and of the wire materials at the reference temperatures must be known. Often, the measured temperature can be found by using standard thermocouple tables. However, these tables are often referenced at 0°C. To correct for this different reference temperature, equation can be used to calculate the temperature from a given current.

$$\xi_{T_1,T_3} = \xi_{T_1,T_2} + \xi_{T_2,T_3} \tag{3}$$

ξ: emf of an alloy combination generated at two different temperatures

T_1: temperature of the medium whose temperature is to be determined

T_2: reference temperature of the thermocouple

T_3: reference temperature of the standard thermocouple table, which in this case is 0°C

Once the emf between two alloys is calculated relative to a reference temperature when T_3 is 0°C, the standard thermocouple table can be used to determine the temperature T_1 of the medium. This temperature is usually automatically displayed on the thermocouple.

Apart from the common occurrence of the thermocouples being placed in the fluid to measure temperature change, thermocouples can be also embedded in solids with excellent results. This is highly effective while establishing the different thermal properties for a solid. The heat transfer to the thermocouple will now be in the form of conductive heat transfer.

As a result, this setup would be very similar to heat conduction in series, since the thermocouple is almost always made from a different material then the actual solid. Such discrepancies depend on the manner in which the thermocouple is embedded in the solid and should be taken into account when the thermal properties are being calculated and analyzed. One example is shown in the photo below.

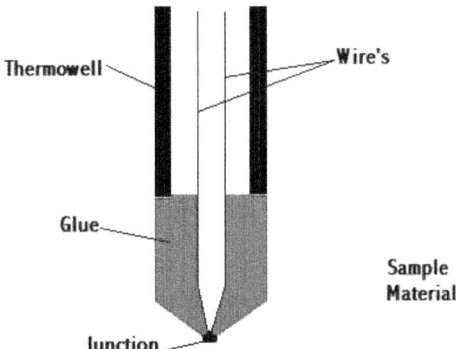

Fig. : Schematic diagram of how the thermocouple function.

Laws for Thermocouples

Law of Homogenous Material

If all the wires and the thermocouple are made of the same material, temperature changes in the wiring do not affect the output voltage. Thus, need different materials to adequately reflect the temperature.

Law of Intermediate Materials

The sum of all the thermoelectric forces in a circuit with a number of dissimilar materials at a uniform temperature is zero. This implies that if a third material is added at the same temperature, no net voltage is generated by the new material.

Law of Successive or Intermediate Temperatures

If two dissimilar homogeneous materials produce thermal emf1 when the junctions are at T1 and T2 and produce thermal emf2 when the junctions are at

T2 and T3, the emf generated when the junctions are at T1 and T3 will be emf1 + emf2.

Application

- Steel industry : Monitor temperature and chemistry throughout the steel making process
- Heating appliance safety: Thermocouples in fail-safe mode are used in ovens and water heaters to detect if pilot flame is burning to prevent fire and health hazard
- Manufacturing : Used for testing prototype electrical and mechanical apparatus
- Process plants :Chemical production plants and refineries use computer programs to view the temperature at various locations. For this situation, a number of thermocouple leads are brought to a common reference block.

Pyrometers

Unlike the thermometer, RTD and the thermocouple, pyrometers (non-contact temperature sensors) measures the amount of heat radiated, rather than the amount of heat conducted and convected to the sensor. Various types of pyrometers, such as total radiation and photoelectric pyrometers, exist. Below is a schematic of an optical pyrometer in figure.

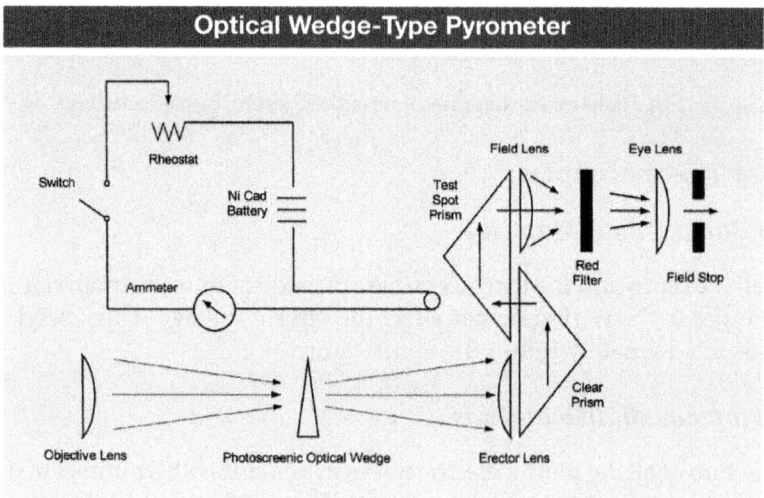

Fig. : Schematic diagram of an optical pyrometer.

These pyrometers differ in the type of radiation they measure. There are many factors that influence the amount of radiated heat detected, thus there are many assumptions that must be made regarding the emissivity, or the measure of the manner in which heat is radiated, of the object.

These assumptions are based upon the manner in which heat is radiated as well as the geometry of the object. Because temperature is dependent on the emissivity of a body, these assumptions regarding the emissivity introduce uncertainties and inaccuracies in the temperature readings. Therefore, because of the error associated with them, pyrometers are not often used in industry.

Table. Summary of Temperature Sensors.

Type		Function	Operating Range (C)	Advantage	Disadvantage
Thermometer	Liquid Filled	Calibrated with markings along the tube to provide a visual reading of the temperature	-200 C to 260 C	Cheap, good for everyday usage	Limited range, can't connect to temperature regulation systems
	Bimetal		-10 C to 110 C	Good for quick checks on temperatures, no batteries, inexpensive	Visual readouts lead to limited accuracy
RTD	Traditional (uses metallic sensing elements)	Provides an electrical means of temperature measurement, making it compatible with a computerized system	-200C to 700C	Easy to recalibrate, provides readings with two significant figures, remain stable for long periods of time	Smaller temperature range than thermocouples, higher start up costs, sensitive to environmental disturbances
	Thermistor (uses a semiconductor sensor)				
Thermocouple		Consists of two metal wires connected to a voltage source to provide an electrical means of	-200C to 2320 C depending on the metals chosen for the two wires	No batteries needed, quick, can be used for wide temperature ranges	Can lose their accuracy after being used for a long time due to wear in the insulation between wires

There are a few different types of pyrometers. There are optical and radiation pyrometers.

How Optical Pyrometers Work:
- Compares the colour of visible light given off by the object with that of a electrically heated wire
- The wire can be preset to a certain temperature
- The wire can be manually adjusted to compare the two objects

How Radiation Pyrometers Work:
- This sensor works by measuring the radiation (infrared or visible light) that an object gives off
- The radiation heats a thermocouple in the pyrometer which in turn induces a current
- The larger the current induced, the higher the temperature is

Pyrometers are usually used at very high temperatures, but can be used at colder temperatures as well. There are lots of industrial applications to pyrometers. Plant operators can use pyrometers to get a sense of what temperature certain processes are running at. The downside to pyrometers is that they are not very accurate as thermocouples or RTD sensors are. This is because they rely on quantifying colours of light.

Temperature Regulators

Temperature regulators, also known as temperature control valves (TCVs), physically control, as well as measure, temperature. Temperature regulators are not capable of directly maintaining a set value; instead, they relate the load (in this case the valve opening) with the control (temperature measurement). These regulators are most useful when temperature is correlated to a flow of a substance. For example, a TCV may be used to control the temperature of an exothermic reaction that requires constant cooling. The TCV measures the temperature of the reaction and, based upon this temperature, either increases or decreases the flowrate of cooling fluid to adjust the temperature of the reaction. Similarly, the regulator could be used to adjust the flow amount of steam, which is typically used to heat a substance. Therefore, by adjusting flowrate, the regulator can indirectly adjust temperature of a given medium.

REGULATOR STRUCTURE

The structure of a typical thermal regulator consists of four main parts, as shown in figure. The temperature detecting element, which in most cases is a temperature sensor, as described above, sends either an electrical or mechanical signal through the connector to the actuator. The actuator then uses this signal to act upon the power source, which determines the position of the valve.

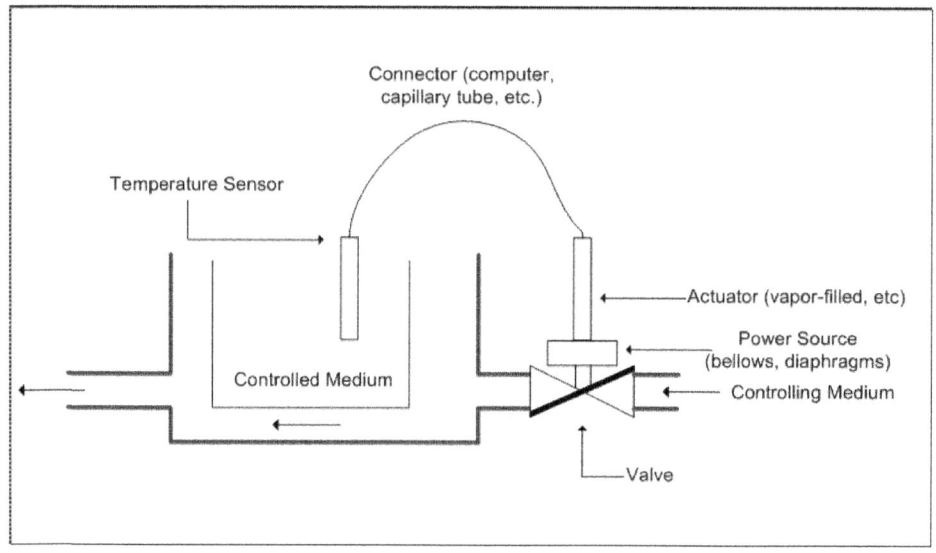

Note: The actuator consists of the capillary tubing, the power source, and the adjustment.

Fig. : Schematic Diagram of Temperature Regulator Structure.

Regulator Operation

The temperature regulator operates based upon a mechanical means of temperature control. As previously mentioned, the bulb of the regulator is typically filled with a heat conducting substance. Due to the thermal expansion properties of this substance, the substance expands as the temperature increases. This expansion causes a change in the pressure of the actuator, which correlates to the temperature of the medium. This pressure change repositions a valve on the regulator, which controls the flowrate of a coolant. The temperature of the medium is then altered by the change in the flowrate of this coolant.

Types of Temperature Regulators

Though all regulators have the same basic build and purpose, they exist in a variety of forms. In particular, these regulators vary in four primary ways: temperature detecting elements, temperature detector placement, actuator type, and valve type.

Temperature Detecting Elements

Most temperature regulation systems use thermocouples or RTDs as temperature sensing devices. For these systems, the connector is a computer. The sensors send an electric signal to the computer, which calculates the temperature. The computer then compares the temperature measured by the sensor to a programmed set-point temperature, thus determining the required pressure in the actuator. The pressure in the actuator changes position of the power source (diaphragm or bellows), which consequently changes the flowrate through the valve.

Some temperature regulation systems use a filled bulb as a temperature sensor. Based on the thermal expansion properties of the material within the bulb, the material expands as the temperature increases. This expansion causes a change in the pressure of the actuator. The pressure change actuator then repositions power source. Again, the change in the power source changes the flowrate through the valve.

Temperature regulation systems using thermocouples or RTDs as temperature sensing devices are much more common than regulation systems using filled bulbs.

Temperature Detector Placement: Internal and Remote Detection

Temperature detection can be done with internal or remote elements. For internal temperature detectors, the thermal actuator and temperature detector are located entirely within the valve. For remote temperature detectors, the primary temperature detecting element is separate from the actuator and valve, and is connected to the actuator with either electrical wiring or capillary tubing, depending on the mechanism of the temperature sensor. Remote temperature detectors are more common, as internal temperature detectors are limited in use. Internal temperature detectors can only measure the temperature of the fluid flowing through the valve and not the temperature of the process.

Actuator Type: Thermal Systems

There are four main categories of thermal actuators used in temperature regulators. Thermal actuators produce power and work, proportional to the measured temperature of the process, on the power source. Actuator types include the vapour-filled system, the liquid-filled system, the hot chamber system, and the fusion-type or wax-filled system. Of all the thermal systems mentioned, liquid-filled systems are the most common, because they relate temperature and pressure change in a linear fashion.

Vapour-Filled Systems

In the vapour-filled system, the thermal actuator is partially filled with a volatile liquid. As the temperature of the sensor increases, the vapour pressure of the liquid also increases. This increases the pressure on the power source, and adjusts the flowrate through the valve.

Liquid-Filled Systems

In liquid-filled systems, the thermal actuator is filled with a chemically stable liquid, such as a hydrocarbon. As the temperature increases, the liquid expands, which produces a force on the power source.

Hot Chamber Systems

In hot chamber systems, the thermal actuator is partially filled with a volatile fluid. An increase in temperature of the system forces some of this fluid into the power unit, where the heat of the unit causes this liquid to turn into a superheated vapour. The pressure increase produces a force on the power source.

Fusion-Type (Wax-Filled) Systems

Of all the systems mentioned, the fusion-type system is the least common. In the fusion-type system, the thermal actuator is filled with special waxes such as hydrocarbons, silicones, and natural waxes. The wax contains large amounts of copper, which increases the heat-transfer quality of the wax. As temperature increases, the wax expands, producing a force that repositions the power source.

Valve Type: Direct and Pilot Actuated

The two main types of valves used in thermal regulators are the direct and pilot actuated valves. In all such thermal regulators, there is a power source (such as bellows and diaphragms) that provides the force required to reposition the valve to control the temperature. These power sources rely on a change in the pressure of the actuator in order to properly regulate temperature. In direct-actuated TCVs, this power unit is directly connected to the valve, which provides the force required to open and close the valve. In pilot-actuated TCVs, the thermal actuator moves a pilot valve, which then transfers energy in the form of pressure to a piston, which then provides the work necessary to reposition the main valve.

Direct-actuated TCVs are often much simpler in structure than pilot-actuated TCVs, and therefore they are also much cheaper. In addition, they respond better to smaller changes in temperature and more accurately reflect the temperature of the medium. Thus, if the exact temperature of the system is essential to ensure correct operation, a direct-actuated TCV should be used. Pilot-actuated TCVs usually have much smaller temperature sensing devices, a faster response time, and the ability withstand much higher pressures through the regulating valve. Therefore, at high pressures or rapid temperature changes, a pilot-actuated TCV should be used.

PRESSURE SENSORS

Pressure must be considered when designing many chemical processes. Pressure is defined as force per unit area and is measured in English units of psi and SI units of Pa. There are three types of pressure measurements:

1. Absolute pressure - atomospheric pressure plus gauge pressure.
2. Gauge Pressure - absolute pressure minus atmospheric pressure.
3. Differential Pressure - pressure difference between two locations.

There are various types of pressure sensors that are available in the market today for use in industry. Each functions best in a certain type of situation.

Sensor Selection Criteria

In order for a pressure controlled system to function properly and cost-effectively, it is important that the pressure sensor used be able to give accurate and precise readings as needed for a long period of time without need for maintenance or replacement while enduring the conditions of the system. Several factors influence the suitability of a particular pressure sensor for a given process: the characteristics of the substances being used or formed during the process, the environmental conditions of the system, the pressure range of the process, and the level of precision and sensitivity required in measurements made.

Process

The pressure sensing element (elastic element) will be exposed to the materials used in the process, therefore materials which might react with the process substances or degrade in corrosive media are unsuitable for use in the sensor. Diaphragms are optimal for very harsh environments.

Environment

The environment (as in the system -- pipes, vibrations, temperature, *etc.*) in which the process is carried out also needs to be considered when choosing a suitable pressure sensor. Corrosive environments, heavy vibrations in the piping units, or extreme temperatures would mean that the sensors would have to have an added level of protection. Sealed, strong casing materials with interior liquid containing glycerine or silicone are often used to encase the internal components of the sensor (not including the sensing element), protecting them from very harsh, corrosive environments or frequent vibrations.

Pressure Range

Most processes operate within a certain pressure range. Because different pressure sensors work optimally in different pressure ranges, there is a need to choose pressure gauges which are able to function well in the range dictated by the process.

Sensitivity

Different processes require different levels of precision and accuracy. In general, the more precise the sensor, the more expensive it is, thus it is economically viable to choose sensors that are able to satisfy the precision desired. There is also a compromise between precision and the ability to detect pressure changes quickly in sensors, hence in processes in which the pressure is highly variable over short periods of time, it is unadvisable to use sensors which take a lot of time to give accurate pressure readings, although they might be able to give precise pressure values.

Pressure Measuring Methods

Several pressure measuring methods have been developed and utilized; these methods include visual inspection of the height of liquid in a column, elastic distortion, and electrical methods.

Height of Liquid in Column

The height of a liquid with known density is used to measure pressure. Using the equation $P = \rho g h$, the gauge pressure can be easily calculated. These types of pressure measuring devices are usually called manometers. Units of length may be used to measure the height of the liquid in the column as well as calibrated pressure units.

Typically water or mercury is used as the liquid within these columns. Water is used when you desire greater sensitivity (its density is much less than liquid mercury, so its height will vary more with a pressure change). Mercury is used when you desire higher pressure measurements and not as great sensitivity.

Elastic Distortion

This pressure measuring method is based on the idea that deformation of an elastic material is directly proportional to the pressure being measured. There are mainly three sensor types that are used in this method of measuring pressure: Bourdon-tubes, diaphragms and bellows.

Electrical Methods

Electrical methods used for measuring pressure utilize the idea that dimensional changes in a wire affect the electrical resistance to the conductor. These devices that use the change in resistance of the wire are called strain gauges. Other electrical sensors include capacitive sensors, inductive pressure transducers, reluctive pressure transducers, potentiometric transducers, piezoresistive transducers and piezoelectric sensors. (refer to "Types of Sensors" Section)

Types of Sensors

There are many different pressure sensors to choose from when considering which is most suitable for a given process, but they can generally be placed into a few categories, namely elastic sensors, electrical transducers, differential pressure cells and vacuum pressure sensors. Listed below each general category are specific internal components, each functioning best in a certain situation.

Elastic Sensors

Most fluid pressure sensors are of the elastic type, where the fluid is enclosed in a small compartment with at least one elastic wall. The pressure reading is thereby determined by measuring the deflection of this elastic wall, resulting in either a direct readout through suitablelinkages, or a transduced electrical signal. Elastic pressure sensors are sensitive; they are commonly fragile and susceptible to vibration, however.

In addition, they tend to be much more expensive than manometers, and are therefore preferentially used for transmitting measured data and measuring pressure differences. A wide variety of flexible elements could conceivably be used for elastic pressure sensors; the majority of devices use some form of a Bourdon tube, bellows, or diaphragm.

Bourdon Tube Gauges

The principle behind all Bourdon tubes is that an increase in pressure on the inside of the tube in comparison to the outside pressure causes the oval or flat shaped cross-section of the tube to try to achieve a circular shape. This phenomenon causes the tube to either straighten itself out in the c-type or spiral cases or to unwind itself for the twisted and helical varieties. This change can then be measured with an analog or digital meter connected to the tube. Tube materials can be changed accordingly to suit the required process conditions. Bourdon tubes can operate under a pressure range from 0.1-700MPa. They are also portable and require little maintenance; however, they can only be used for static measurements and have low accuracy.

Types of Bourdon tubes include C-type, spiral (a more coiled C-type tube), helical and straight tube Bourdon tubes. C-type gauges can be used in pressures approaching 700MPa; they do have a minimum recommended pressure range, though -- 30kPa(*i.e.*, it is not sensitive enough for pressure differences less than 30kPa).

Bellows

Bellows elements are cylindrical in shape and contain many folds. They deform in the axial direction (compression or expansion) with changes in pressure. The pressure that needs to be measured is applied to one side of the bellows (either inside or outside) while atmospheric pressure is on the opposite side. Absolute pressure can be measured by evacuating either the exterior or interior space of the bellows and then measuring the pressure at the opposite side. Bellows can only be connected to an on/off switch or potientiomenter and are used at low pressures, <0.2MPa with a sensitivity of 0.0012MPa.

Diaphragms

Diaphragm elements are made of circular metal discs or flexible elements such as rubber, plastic or leather. The material from which the diaphragm is made depends on whether it takes advantage of the elastic nature of the material, or is opposed by another element (such as a spring). Diaphragms made of metal discs utilize elastic characteristics, while those made of flexible elements are opposed by another elastic element. These diaphragm sensors are very sensitive to rapid pressure changes.

The metal type can measure a maximum pressure of approximately 7MPa, while the elastic type is used for measuring extremely low pressures (0.1kPa - 2.2MPa) when connected to capacitative transducers or differential pressure sensors. Examples of diaphragms include flat, corrugated and capsule diaphragms. As previously noted, diaphragms are very sensitive (0.01MPa). They can measure fractional pressure differences over a very minute range (say, inches of water) (elastic type) or large pressure differences (approaching a maximum range of 207kPa)(metal type).

Diaphragm elements are very versatile -- they are commonly used in very corrosive environments or with extreme over-pressure situations.

Examples of these elastic element pressure sensors are shown here.

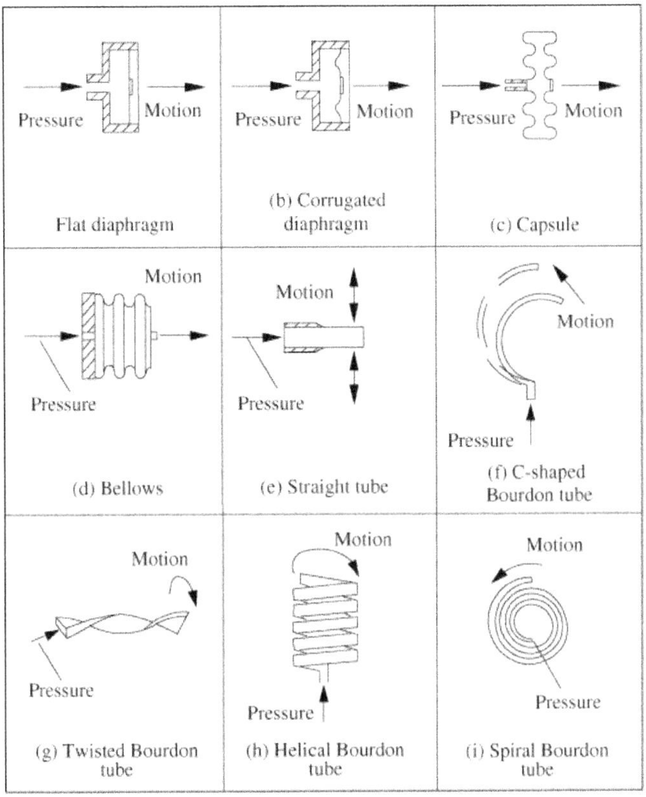

Pressure
Motion
Flat diaphragm
Pressure
Motion
(b) Corrugated diaphragm
Pressure
Motion
(c) Capsule
Motion
Pressure
(d) Bellows
Motion
Pressure
(e) Straight tube
Motion
Pressure
(f) C-shaped Bourdon tube
Motion
Pressure
(g) Twisted Bourdon tube
Motion
Pressure
(h) Helical Bourdon tube
Motion
Pressure
(i) Spiral Bourdon tube

Electric Sensors

Sensors today are not necessarily only connected to a gauge meter needle pointer to indicate pressure, but may also serve to convert the process pressure into an electrical or pneumatic signal, which can be transmitted to a control room from which the pressure reading is determined.

Electric sensors take the given mechanics of an elastic sensor and incorporate an electrical component, thus heightening the sensitivity and increasing the amount of instances in which you could utilize the sensor. The types of pressure transducers are capacitive, inductive, reluctive, piezoelectric, strain gauge, vibrating element, and potentiometric.

Capacitive

A capacitive sensor consists of a parallel plate capacitors coupled with a diaphragm that is usually metal and exposed to the process pressure on one side and the reference pressure on the other side. Electrodes are attached to the diaphragm and are charged by a high frequency oscillator.

The electrodes sense any movement of the diaphragm and this changes the capacitance. The change of the capacitance is detected by an attached circuit which then outputs a voltage according to the pressure change. This type of sensor can be operated in the range of 2.5 Pa - 70MPa with a sensitivity of 0.07MPa.

An example of a capacitive pressure sensor is shown in the figure.

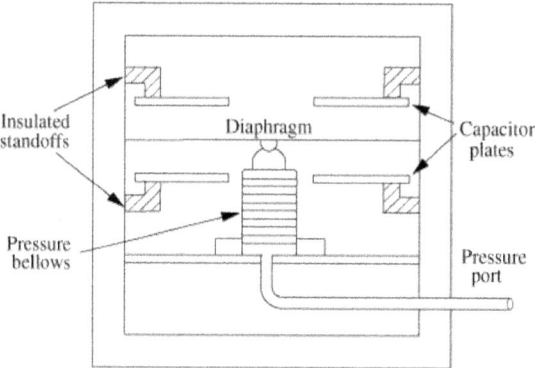

Inductive

Inductive pressure sensors are coupled with a diaphragm or a Bourdon tube. A ferromagnetic core is attached to the elastic element and has a primary and 2 secondary windings. A current is charged to the primary winding. When the core is centered then the same voltage will be induced to the two secondary windings. When the core moves with a pressure change, the voltage ratio between the two secondary windings changes. The difference between the voltages is proportional to the change in pressure.

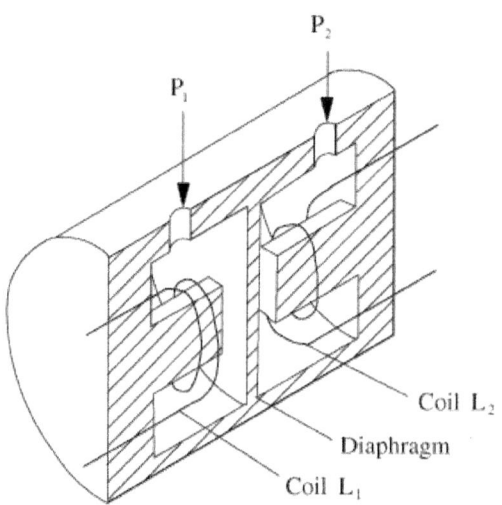

An example of an inductive pressure sensor utilizing a diaphragm is shown below. For this kind of pressure sensor, taking Chamber 1 as the reference chamber with a reference pressure P_1 coming into the chamber and the coil being charged with a reference current. When the pressure in the other chamber changes, the diagphragm moves and induces a current in the other coil, which is measured and gives a measure of the change in pressure.

These may be used with any elastic element (though, it is typically coupled with a diaphragm or a bourdon tube). The pressure reading generated will be determined by voltage calibration. Thus, the range of pressure in which this sensor may be used is determined by an associated elastic element but falls in the range of 250 Pa - 70MPa.

Reluctive

Reluctive pressure sensors also charge a ferromagnetic core. When the pressure changes, the flexible element moves a ferromagnetic plate, leading to a change in the magnetic flux of the circuit which can be measured. The situations in which one would use a reluctive electric element is one in which the inductive sensor does not generate a precise enough measurement. The pressure range is 250 Pa - 70MPa with a sensitivity of 0.35MPa.

An example of a reluctive pressure sensor can be seen in the figure.

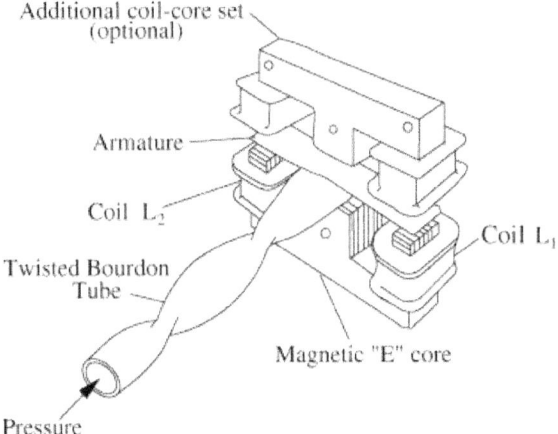

Piezoelectric

Piezoelectric sensors use a crystal sensor. When pressure is applied to the crystal, it deforms and a small electric charge is created. The measurement of the electric charge corresponds to the change in pressure. This type of sensor has a very rapid response time to constant pressure changes. Similar to reluctive electric element, the piezoelectric element is very sensitive, but responds much, much faster. Thus, if time is of the essence, a piezoelectric sensor would be desired. The pressure range is .021 - 100MPa with a sensitivity of 0.1MPa.

Below figue is an example of a piezoelectric pressure sensor.

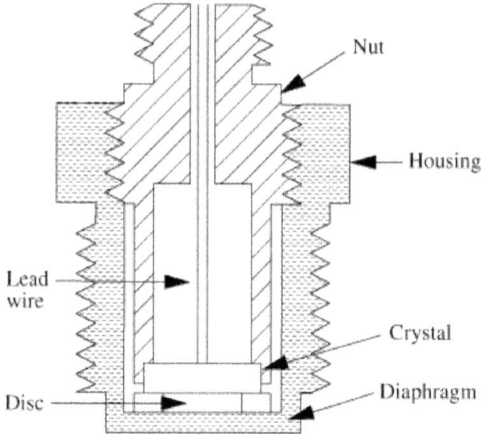

Potentiometric

Potentiometric sensors have an arm mechanically attached to the elastic pressure sensing element. When pressure changes, the elastic element deforms, causing the arm to move backwards or forwards across a potentiometer and a resistance measurement is taken. These sensing elements do posess an optimum working range, but are seemingly limited in their resolution by many factors. As such, they are low end sensors that aren't used for much. With a low sensitivity and working range, they may be best suited as a cheap detector evaluating a coarse process.The pressure range is 0.035 - 70MPa with a sensitivity of 0.07 -0.35MPa.

An example of a potentiometric pressure sensor is shown in the figure.

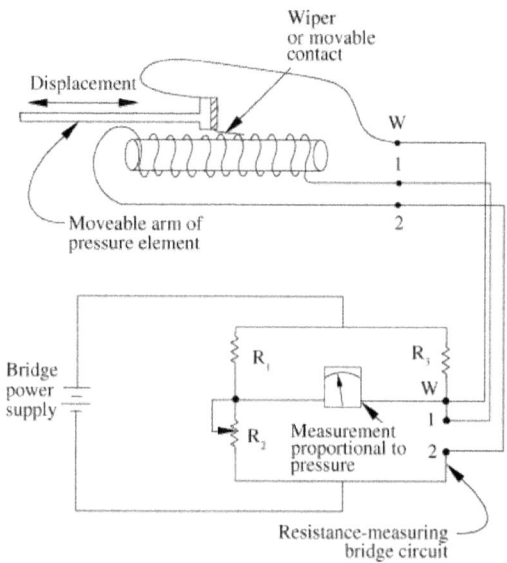

Strain Gauge

The strain gauge detects changes in pressure by measuring the change in resistance of a Wheatstone bridge circuit. In general, this circuit is used to determine an unknown electrical resistance by balancing two sections of a bridge circuit such that the ratio of resistances in one section ($\frac{R_3}{R_2}$) is the same as that in the other section($\frac{R_4}{R_1}$), resulting in a zero reading in the galvanometer in the center branch. One of the sections contains the unknown component of which the resistance is to be determined, while the other section contains a resistor of known resistance that can be varied. The Wheatstone bridge circuit is shown below:

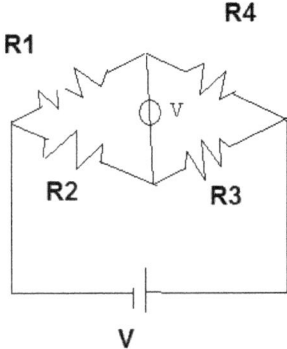

The strain gauge places sensors on each of the resistors and measures the change in resistance of each individual resistor due to a change in pressure. Resistance is governed by the equation $R = \rho \frac{L}{A}$ where ρ = resistivity of the wire, L = length of the wire, and A = cross-sectional area of the wire. A pressure change would either elongate or compress the wire, hence a compression sensor is needed on one resistor and an elongation sensor on the other. To control the effects of temperature (a wire would also either elongate or compress with a change in temperature), a blank sensor would be placed on the remaining two resistors.

These gauges are frequently a type of semiconductor (N-type or P-type). Thus, their sensitivity is much greater than their metal counterparts; however, with greater sensitivity comes a more narrow functional range: the temperature must remain constant to obtain a valid reading. These gauges are affected greatly by variations in temperature (unlike the other types of electrical components). The pressure range is 0 - 1400MPa with a sensitivity of 1.4 - 3.5MPa.

An example of an unbonded strain gauge is shown below. This makes use of strain-sensitive wires one end fixed to an immobile frame and the other end attached to a movable element, which moves with a change in pressure.

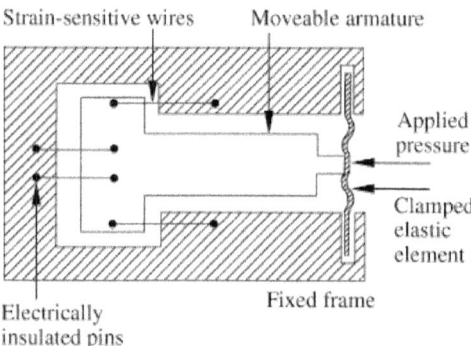

An example of a bonded strain gauge can be seen below. This is placed on top of a diaphragm, which deforms with change in pressure, straining the wires attached to the diaphragm.

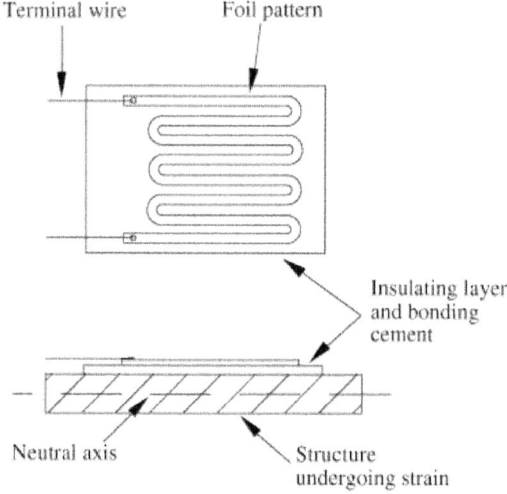

Vibrating Element

Vibrating element pressure sensors function by measuring a change in resonant frequency of a vibrating element. A current is passed through a wire which induces an electromotive force within the wire. The force is then amplified and causes oscillation of the wire. Pressure affects this mechanism by affecting the wire itself: an increase in pressure decreases the tension within the wire and thus lowers the angular frequency of oscillation of the wire. The sensor is housed in a cylinder under vacuum when measuring absolute pressures. These absolute pressure measuring sensors are very efficient: they produce repeatable results and are not affected by temperature greatly. They lack sensitivity in meausurement, though, so they would not be ideal for a process in which minute pressures need monitoring. The pressure range is 0.0035 - 0.3MPa with a sensitivity of 1E-5MPa.

A vibrating wire pressure sensor is shown below.

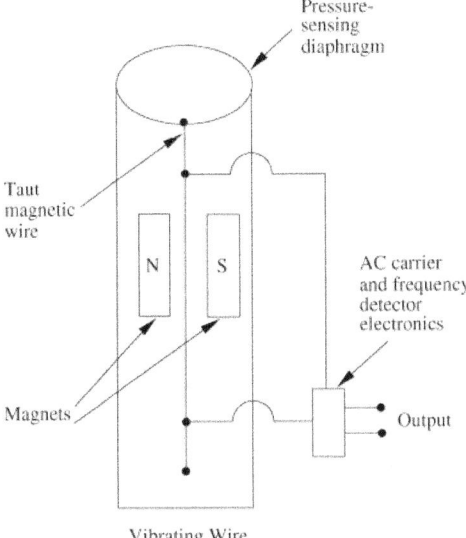

Vibrating Wire

A vibrating cylinder pressure sensor (for absolute pressures) is shown below.

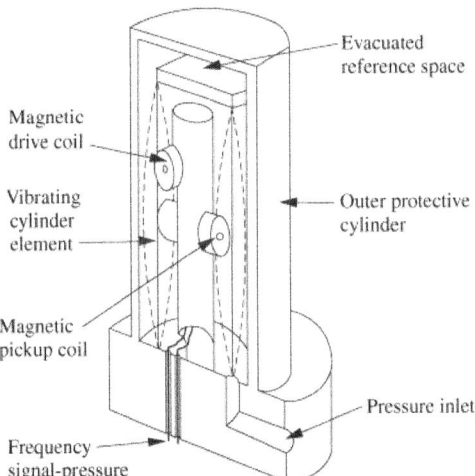

Differential Pressure Cells

Differential pressure cells are used with various kinds of sensors in which measurement of pressure is a result of a pressure differential such as orifice plates, flow nozzles, or venturi meters. The differential pressure cell converts the pressure differential into a transmittable signal. Where the differential pressure (DP) cell is placed depends on the nature of the fluid stream that is being measured.

A typical DP cell is minimally invasive (an external component attached across the point of measurement); it is commonly employed with a capacitive element paired with a diaphragm that allows the capacitive body to separate or move together, generating a signal (*via* change in capacitance) that can be interpreted to a pressure drop. They are often used to detect small differences in large pressure drops. Its placement is similar to connecting a voltmeter in parallel to a resistor to measure its voltage "drop" (analgous to the pressure drop).

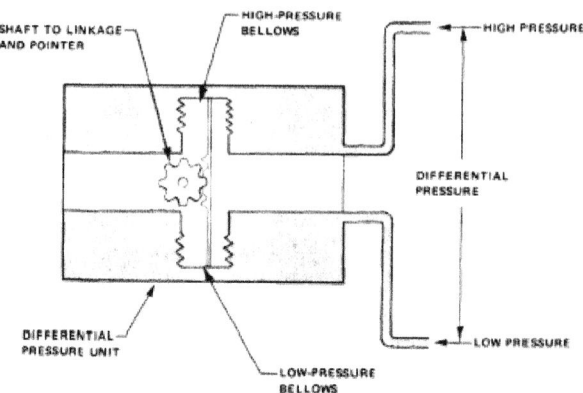

Fig. : An example of a differential pressure cell.

The range of pressure measured and sensitivity of a differential cell depends upon the electric and elastic components used in the cell itself. It is a great sensor to use when measuring a pressure drop; however, for all other applications, it is fairly useless.

Vacuum Sensors

Such sensors are able to measure extremely low pressures or vacuum, referring to pressures below atmospheric pressure. Besides diaphragm and electric sensors designed to measure low pressures, there are also thermal conductivity gauges and ionization sensors.

Thermal Conductivity Gauges

The principle involved here is the change in gas thermal conductivity with pressure. However, due to deviation from ideal gas behaviour in which the relationship between these two properties is linear, these kind of gauges, which are also called Pirani gauges, can only be used at low pressures, in the range of (0.4E-3 to 1.3E-3)MPa. They are amazingly sensitive elements as well (can detect changes of 6E-13MPa).

In these gauges, a coiled wire filament has a current flowing through it, which heats up the coil. A change in pressure changes the rate of heat conduction away from the filament, thereby causing its temperature to vary. These changes in temperature can be detected by thermocouples in the gauge, which are also connected to reference filaments in the gauge as part of a Wheatstone bridge circuit.

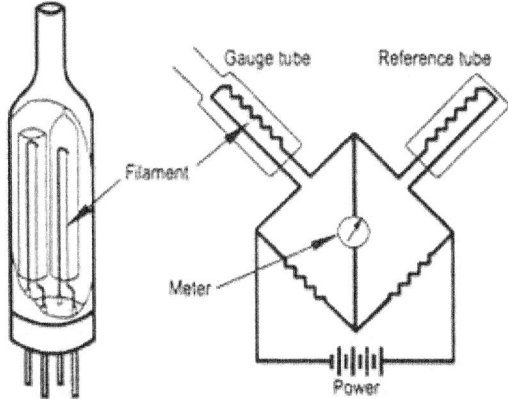

Fig. : An example of a Pirani gauge.

Ionization Gauges

There are two categories for these types of gauges: hot cathode and cold cathode. For hot cathode gauges, electrons are emitted by heated filaments, while for the cold cathode gauges electrons are released from the cathode due to collision of ions. Electrons hit the gas molecules entering the gauge forming positive ions, which are collected and cause an ion current to flow.

The amount of cation formation is related to the gas density and consequently the pressure to be measured, as well as the constant electron current used, hence the ion current flow is a direct measure of the gas pressure. These both are highly sensitive instruments and thus most suited for fractional pressures. The hot cathode gauges are even more sensitive than cold cathode gauges and are able to measure pressures near 10^{-8}Pa. Their sensitivity ranges from (1E-16 to 1E-13)MPa.

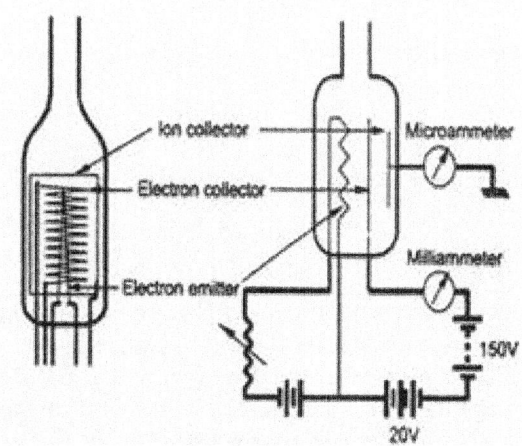

Fig. : An example of an ionization gauge.

Summary

Types of elements	Press. Range	Sensitivity	Advantages	Disadvantages
Bourdon tube	0.1-700 MPa	0.03 MPa	Portable Low-Maintenance	Static measurements Low accuracy
Bellows	< 0.2MPa	0.0012 MPa	Can be used at low press.	Can only be connected to on/off switch or potentiometer
Diaphragms	0.1-2.2 MPa	0.01 MPa	Quick response time Very accurate Generally good linearity Can be exposed to corrosives	Very expensive
Capacitive	2.5 Pa-70 MPa	0.07 MPa	Used for low pressure or vacuum ranges Robust	Electronically complex Capacitor plates can stick together
Inductive	250Pa – 70 MPa	0.35 MPa	High sensitivity	Limited by elastic elements More coarse than reluctive
Reluctive	250Pa – 70 MPa	0.35MPa	High sensitivity	Requires external AC source
Piezoelectric	0.021- 100 MPa	0.1 MPa	Very rapid response time	Affected by high temp. and static forces
Potentiometric	0.035- 70 MPa	0.07 – 0.35 MPa	Can be made very small	Low sensitivity and working range
Strain Gauge	0-1400 MPa	1.4-3.5 MPa	Very high sensitivity Can be used on mobile parts	Extremely slow response time Very small signal output
Differential	Depends on other elements	Depends on other elements	Used for pressure drop	Can only be used for press. drop
Thermal Conductivity	0.4E-3 – 1.3E-3 MPa	6E-13 MPa	Measures vacuum ranges	Has linearity only at low pressures
Ionization	1.3E-13 – 1.3E-8MPa	1E-13 – 1E-16 MPa	Highly sensitive Can measure high and ultra-high vacuum ranges	Limited by photoelectric effect
Vibrating	0.0035MPa - 0.3MPa	1E-5 MPa	Very accurate Not affected by temp. changes	Can't be used at high pressures

LEVEL SENSORS

Level sensors allow for the level control of fluid in a vessel. Examples of where these sensors are installed include reactors, distillation columns, evapourators, mixing tanks, *etc*. Level sensors provide operators with three important data for control:

(1) the amount of materials available for processing,

(2) the amount of products in storage,

(3) the operating condition.

Installing the correct level sensor ensures the safety of the operator and the surrounding environment by preventing materials in vessels from overflowing or running dry.

There are several different types of level monitors, including:

- Visual
- Float
- Valve Controlled
- Electronic
- Radiation

These different types of sensors can also be grouped into categories of process contact and non-process contact. As the name suggests, process contact sensors are within the tank, in physical contact with the material. Non-process contact sensors transmit various types of signals to reflect off of the material and thus measure the level. This sensor design can maintain its integrity within a potentially corrosive material and/or be positioned such that it can monitor the level changes from above the tank.

Visual Level Sensors

Visual level controls were the earliest developed level sensor. These types of monitoring devices can be something as simple as looking into an open container or inserting a marked object such as a dipstick. This type of sensor is the simplest and possibly the most reliable. These devices do not provide a way of connecting to a control device. They require human input with no way of automation; however, electronic issues will not be a possible problem.

Common Uses

Visual indicators can be found in many places and are not limited to chemical engineering applications. Examples of these are: dipsticks found monitoring the oil levels in a car and measuring cups with markings indicating different volumes. In chemical plants, visual indicators are used to measure changes in level as well as for high and low level alarms.

Benefits of Visual Level Sensors

Visual sensors are generally less expensive than other types of sensors. They are more reliable because of the simplicity of design.

Restrictions

Visual indicators do not always accurately measure how much volume is in the tank. They do not allow for any digital connections to process control systems. In the case of the sight tubes, they must be affixed directly to the tank and cannot be read remotely without a camera or other transmission tool. Tanks placed in hard-to-reach places would also cause problems contributing to inaccuracies in measurement, such as not reading markers on eye level or interpolating between marker intervals. Also, the connections between the tube and the tank are subject to residue buildup which could prevent the readings from being accurate.

Weather is also a concern with sight tubes. External temperature changes could affect the fluid in the tube and thus measurement inaccuracy. For example, the fluid in the sensor could freeze, or the sensor could become clogged. This type of indicator requires modification to the vessel wall and would therefore be optimally installed at the time of initial construction. If considered at the time of the initial capital installation of the tank a sight tube will not add a large cost to the project. Yet, later modification to include this in the design and associated tank changes are potentially very costly. Despite these shortfalls, a sight tube is a reliable and appropriate choice in many common level reading applications.

Sight Tube Indicators

Sight tube indicators allow operators to monitor levels with precision while keeping the tank sealed. This type of monitoring device is comprised of a vertical tube equal in height to the actual vessel. This tube is connected in at least two places directly to the vessel so that its contents can flow into the monitoring tube.

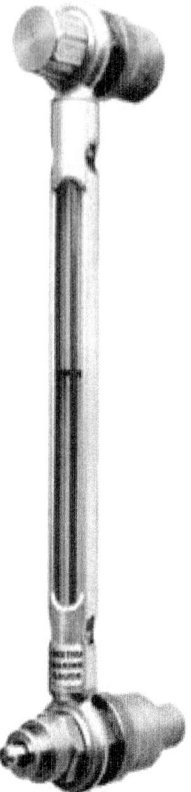

Fig. : A sight tube.

This ensures that the height of liquid in the tank will be equal to the height of liquid in the sight tube. Markings can be affixed to the indicator for both calibration and volume readings. There are many styles of this indicator. Some models have a float ball contained in the tube so that it will float on top of the liquid. Other models of the indicator have many paddles on fixed posts through the length of the tube. These paddles float horizontally when submerged and are vertical when not submerged. This results in a colour change allowing for an easy identification of the tank volume.

Float Type Level Sensors

Float Types of level sensors are based on the principle of buoyancy which is the upward force produced on a submerged object by the displaced fluid. This force is equal to the weight of the displaced fluid. Float sensors take their measurements at the interfaces of materials, where the movement of the float and/or the force on the float are caused by the differing densities of the float and the fluid. There are two broad categories of Float Type level sensors: Buoyancy and Static.

Buoyancy Types

Buoyancy level sensors are less dense than the fluid and thus change position along with the fluid level. The movement of the float transmits the level information through some mechanical linkage to an output such as a valve or operator observation. There are basic three types of mechanical linkage - chain / tape sensors, lever / shaft mechanisms and magnetically coupled devices.

- Chain / Tape sensors – The linkage is by a flexible chain or tape.
- Lever / Shaft mechanisms – The linkage is a rigid shaft.
- Magnetically coupled devices – These devices are similar to the Chain/Tape sensors, except a magnet is attached to the float and another is acted upon by the floating magnet moving a tape like the chain/tape type devices. The moving magnet can be sequestered from the float attached magnet for use in corrosive media.

Schematic of a float type level sensor is shown below. As the fluid level rises or falls, buoyant force is transferred through mechanical linkages to your output device.

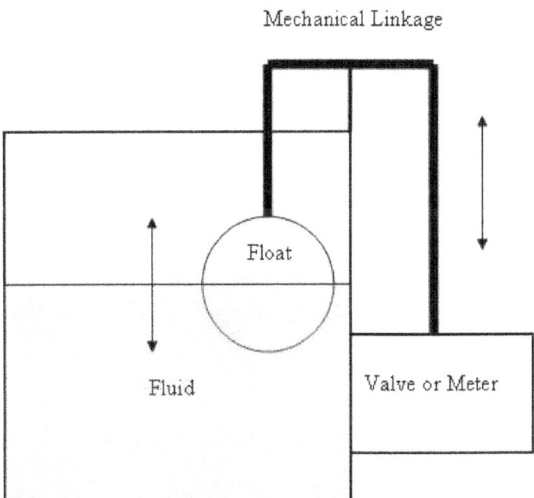

Static Types

Static level sensors are denser than the media being measured and thus do not move. As the level changes, the buoyant force acting on the "float", which is actually a weight, changes. The change in weight is measured by a scale. The level of the tank is calculated by measurements of the weight change in the float, not its actual change in position.

Formulas

Changes in the volumetric rate of material flow, resulting from changes in such things as the pressure of your pump will affect the operation of a float type level sensor. The same movement of the float will have the same effect on the valve, but as valve is moderating a different flow it has a different proportional sensitivity Kc.

$$K_c = \frac{(q)}{k}$$

where q is the flow rate and k is the height the float moves to completely traverse the valves operating range.

This is important because the given change in fluid height used to previously design the system will still change the valve's position in the same manner, but won't affect the same change in flow rate. If your pressure has dropped, your valve will have to be opened wider to match the same flow rate. The amount wider that it has to be opened is called an offset. Manipulating the valve response to the float movement, also known as the gain:

$$Gain = \frac{(valve_{response})}{float_{movement}}$$

can be used to alleviate this problem, but it won't work exactly the same way it did before the change in load.

Operation power is the amount of power the float has available to perform actions to control the system. Changes in the specific gravity of the fluid (or the float) will affect buoyancy of the float and thus will affect the amount of force exert on your system. A temperature is a very common parameter that can change your fluid's specific gravity. Warmer fluids tend to be less dense, and so your float will generate less buoyant force for a given change in fluid level.

Zero buoyancy force level is the level at which the weight of the float is exactly as buoyant as the fluid it replaces. It can be approximated by the following formula:

$$ZeroBuoyantForceLevel = D\frac{(SG_{float})}{SG_{fluid}}$$

where D is the diameter of the float and SG is the specific gravity.

Common Uses

Float Type Level Sensors regulate how much water is in the reservoir of a flush toilet; the float is attached to a lever which has a rotating axis that stops when the flow of water reaches a certain level.

- Chain / Tape sensors - regulating the level on storage tanks at atmospheric pressures.
- Lever / Shaft Mechanisms - regulating the level on vessels under pressure.

Benefits

Float Type level sensors do not require external energy sources to operate. Since they are simple robust machines it is easy to repair. The cost of these units makes them on the economical side, ranging from approximately $20 for the apparatus in a standard toilet to a few hundred dollars for a 3/4" cast iron valve with float.

Prices would increase with higher quality materials and valve size. Floats could be made out of plastics or metals (steel, stainless steel, *etc.*). Material selection will depend on the application; more chemically resistant materials would be used for corrosive mediums.

Restrictions

Float type level sensors should only be used in clean fluids. Fluids that are a suspension of solids or slurries could foul the operation of the machine. Anything that could increase friction on the mechanical linkages could increase the dead band, which is the delay the sensor experiences due to the excess of force required to overcome the static friction of a non-moving system. An excess of force is required to be built up, so a float type level sensor will not respond immediately to changes in level.

Valve-based Level Sensors

Valve-based level sensors not only measure the fluid level, but also cause the fluid level to change accordingly. Two basic types of valves that will be discussed are altitude valves and diverter valves.

Altitude Valves

A simple altitude valve uses a spring that opens and closes different ports and lines when pressure changes due to the changing fluid levels. When fluid levels exceed the setting of the spring, a diaphragm connected to the spring lowers, closes the drain port, and opens the main line pressure. This turns off the main valve and stops fluid from flowing into the tank. When fluid levels decrease, the diaphragm raises, opens the drain port, and closes the main line pressure. This turns on the main valve and more fluid is supplied to the tank.

Common Uses

Altitude valves are on/off controls and can be found in supply lines connected to basins, tanks, and reservoirs. The main job of these valves is to prevent overflow of the fluid and to hold fluid level constant.

Benefits

There is no external power source needed when operating altitude valves because they are controlled by the pressure of the process fluid. More complicated altitude valves can also be used for other purposes other than an on/off control. These features include the following:

- Open the valve when pressure drops to a predetermined point
- Delayed opening so the valve only opens when fluid level drops a certain amount below the set point
- Close the valve slowly to eliminate pressure build-up
- Two-way flow to allow fluid to return when the level has dropped below the set point
- Open the valve while maintaining constant inlet pressure and plant distribution pressure
- Reduce pressure for outlet streams. This may be needed when working with equipment that has high supply pressures, such as aerator basins. Aerator basins are typically used in wastewater treatment because they have the capability of holding large amounts of water.
- Check valves can be installed to allow the valve to close if the pressure reaches a predetermined low point (which could be the cause of equipment failure elsewhere in the plant)

Restrictions

Altitude valves should not be used in pressurized vessels. These valves are constructed from a limited number of materials, such as cast iron. This restricts the number of fluids that can be used with the valves. Therefore, altitude valves are usually used for water service. Since these valves are usually used to operate a large amount of fluid, they are subsequently large with multiple functions that make them very expensive. A 30 inch altitude valve can cost over $50,000. In addition, frequent maintenance is required because of the many components and moving parts. These valves are operated at ambient temperatures and should be checked for freezing of stationary sensing lines.

Diverter Valves

Diverter valves are connected to a dip tube that is submerged in a controlled tank. When the fluid level of the tank drops, the dip tube is exposed to atmospheric pressure. Once air enters the dip tube, it utilizes the Coanda effect. Coanda effects occur when a curved surface exists. Gases tend to follow the nearest curved surface while pushing other fluids in a different direction. In the case of the diverter valve, air from the dip tube follows the curved surface of the control port back to the storage tank and pushes fluid flowing out of the storage tank to the other wall of the valve. This fluid then flows into the controlled tank to adjust the level.

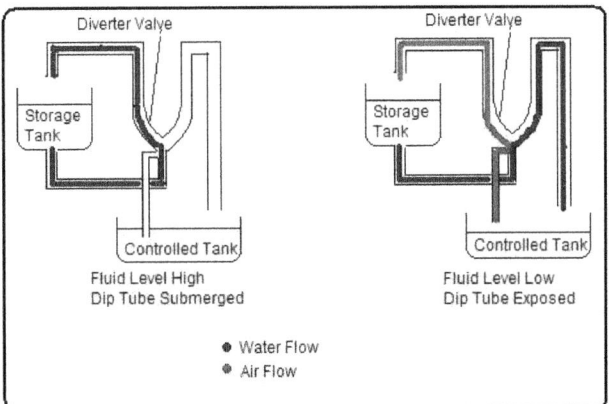

Common Uses

Diverter valves control fluid level by acting as an on/off switch. They can also be modified to be used for throttling. Diverter valves are used as low level indicators so they should only been used when low levels of a tank are a concern.

Benefits

Diverter valves need no external power source for operation. They have no moving parts, plugs, or packing, so they require minimal maintenance. Diverter valves can be operated at any temperature and are not affected by vibration. They can be constructed from many materials such as metals, plastics, and ceramics. Therefore, there is limited restriction to the type of liquid used with the valves. However, when using fluidized solids and slurries, the dip tube must remain open, so build-up does not occur. Any build-up near or in the tube should be removed.

Restrictions

Diverter valves should not be used when operating under conditions other than atmospheric pressure. They are also not for use when dealing with hard to handle processes. Outlet pressure cannot be greater than atmospheric pressure or back pressure will occur and the valve will not operate correctly.

Electrical Level Sensors

Conductive Level Sensing

Conductive level sensors work by applying a low-voltage across two electrodes at different levels in a vessel. When both electrodes are immersed in a conductive liquid, a current flows. This type of electrical conductivity setup is best applicable for point level detection (level detection at a specific point in the material.) They are typically made of titanium, Hastelloy B, or stainless steel.

Example of a conductive level sensor.

VERTICAL INSTALLATION

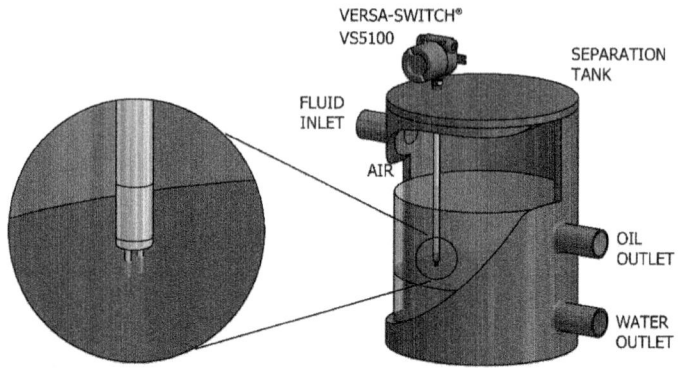

Common Uses

Conductive level sensors are commonly used to measure both conductive and corrosive liquids. A common conductive liquid is water, whereas some common corrosive liquids are nitric acid, ferric chloride, and hydrochloric acid.

Benefits

This method is considered extremely safe due to the low voltages and currents used. Conductive level sensors are also known for their easy installation and use.

Restrictions

The greatest concern with conductive sensors is maintenance. The probe needs to be monitored for buildup on the sensor. Residues from wet or sticky fluids that cause build up can be prevented by coating the sensors with Teflon or polyethylene-based materials.

Capacitance Level Sensing

A capacitor is made up of two conductors (electrodes/plates) that are electrically separated by a nonconductor (dielectric). In the case of level sensing, one of the electrodes is typically a vertically positioned rod while the other is the metallic vessel wall. The dielectric between them is the material being measured in the vessel. The principle of capacitance level sensing is based on the formula below:

Capacitance = Dielectric constant x (Area of the plates ÷ Distance between plates)

If the dielectric is a liquid, the capacitance probe can measure the combined capacitance of both the liquid and gas. When the liquid level rises or falls, the total capacitance value will change. Since the dielectric constant and distance between the plates are constant (the rod and the vessel are stable), the only value changing the capacitance is the area of the plates immersed in the liquid. Total capacitance changes approximately proportionally to the liquid rise or fall in the column. Consequently, the liquid level can be calculated by the change in capacitance.

Common Uses

The list below shows just a few applications of capacitance level sensors found in industry and the materials they sense.
- Chemical/Petrochemical - Oil, clay, soda ash
- Food - Flour, Powered Milk, Sugar
- Charcoal - Wood, Charred sawdust
- Pharmaceuticals - Various powders and liquids
- Mining - Various minerals, metals, stone,

Benefits

Capacitance level sensing is useful in its ability to sense a wide variety of materials such as solids, organic and aqueous solutions, and slurries. For example, materials with dielectric constants as low as 2.1(petroleum oil) and as high as 88(water) or more can be detected with capacitance level sensing. The equipment, typically made of stainless steel, is simple to use, clean and can be specifically designed for certain applications. For example, they can be made to withstand high temperatures and pressures or have built-in splashguards or stilling wells for environments prone to turbulence.

Restrictions

There are limitations to using capacitance level sensors. One major limitation for capacitance probes is found when using tall bins that store bulk solids. Probes are required to extend for the length of the vessel, so in a lengthy vessel the long cable probes can be subject to mechanical tensions and breakage. Another limitation is from build up and discharge of a high-voltage static charge that can result from the rubbing and movement of low dielectric materials, but this danger can be eliminated with proper design and grounding. Also, abrasion, corrosion, and build up of material on the probe can cause variations in the dielectric constant of the material being measured. To reduce this issue, capacitance probes can be coated with Teflon, Kynar, polyethylene or other materials.

Radiation-based Level Sensors

Radiation-based level sensors are based on the principle of a material's ability to absorb or reflect radiation. The common types of radiation used in continuous level gages are ultrasonic, radar / microwave and nuclear.

Ultrasonic (Sonic) Level Sensors

Ultrasonic level sensor transmitters emit high frequency ultrasonic acoustic waves which are reflected back by the medium to the receivers. By measuring the time it takes for the reflected echo to be received, the sensor can calculate the actual distance between the receiver and the fluid level. These sensors can be accurate from a distance of 5mm to 30m.

Common Uses

Ultrasonic level sensors are commonly used for point level detection. They are best used for viscous liquids, slurries and bulk solids.

Benefits

The combination of ultrasonic level sensors' high functionality for relatively low prices makes them a popular choice for non-contacting level sensing.

Restrictions

Since the speed of sound in air can fluctuate in different conditions, ultrasonic sensors are not suitable for use in all applications. Environments that have varying moisture and temperatures will influence the sensors readings. Turbulence, steam, and foam prevent waves from reflecting properly and distort readings. In addition, the level sensor must be mounted properly so that it correctly senses the distance between the transmitter and the fluid level.

Microwave / Radar Level Sensors

Microwave / radar level sensors are similar to ultrasonic level sensors in that they require a transmitter and receiver. In addition to these materials, radar sensors also need an antenna and operator interface to use electromagnetic waves to calculate level distance.

Common Uses

These sensors are frequently used for non-contacting situations that require level sensing in varying temperature and pressure environments.

Benefits

Microwave / radar sensors have an advantage over ultrasonic sensors in that they are able to operate in high pressure and high temperature environments. They can also sense solids with grain sizes larger than 20mm.

Restrictions

When fluid level is under a thick layer of foam or dust, these sensors may not detect the fluid level, but instead detect the level of the dust or foam. Microwave / radar level sensors are also more costly than ultrasonic level sensors.

Nuclear Level Sensors

Nuclear level sensors rely on gamma rays for detection. Although these gamma rays can penetrate even the most solid of mediums, the intensity of the rays will reduce in passage. If the gamma ray emitter and the detector are placed on the top and bottom of a vessel, the thickness of the medium (level) can be calculated by the change in intensity.

Common Uses

These sensors are usually used when the material being measured presents a risk to human life or the environment. These include, for example, materials that are toxic, carcinogenic, or explosive.

Benefits

Nuclear radiation has the ability to detect level even through solid tank walls. Since they appear to "see" through walls, the nuclear gage may be modified and/or installed while the process is running and avoid expensive down time.

Restrictions

Nuclear level sensors are typically the last resort when choosing a level sensor. They not only require a Nuclear Regulatory Commission (NRC) license to install, but are extremely expensive in comparison to other level sensors.

Table. Summary of Benefits and Restrictions.

	Benefits	Restrictions
Visual	• Economical, less expensive • Not subject to much technical/electronical issues	• Require human operator • No connection to process control systems • Cannot be read remotely • Connections between tube and tank subject to clogging, freezing, etc • Require modification to vessel wall
Float	• Do not require external energy source to operate • Economical • Easy to repair	• Should be used in clean fluids
Valve Controlled	• Do not require external energy source to operate • Can be hooked up to process control systems • Allow for two-way flow • Eliminate pressure issues	• Should not be used in pressurized vessels • Require frequent maintenance • Costly

Electronic	Conductive Level Sensing • Considerably safe, low voltage and current • Easy installation	• Maintenance	
	Capacitance Level Sensing • Able to sense a wide variety of materials • Flexible in design	• Probes must extend for full length of vessel, prone to mechanical damage	
Radiation	Ultrasonic Level Sensors • Non-contact level sensing • Low cost	• Speed of sound fluctuates depending on conditions	
	Microwave/Radar Level Sensors • Able to operate in high pressure, high temperature, and vacuum environments • Able to sense solids with grain size >20 mm	• Readings can be influenced by environment • More costly • Cannot work if liquid level is under thick layer of airy foam or dust	
	Nuclear Level Sensors • Able to detect even through solid tank walls • Able to be modified while process is running to avoid expensive down time	• Require licensing • Extremely expensive	

Chapter 4

FLOW CONTROL SENSORS

Flow is defined as the rate (volume or area per unit time) at which a substance travels through a given cross section and is characterized at specific temperatures and pressures. The instruments used to measure flow are termed flow meters. The main components of a flow meter include the sensor, signal processor and transmitter.

Flow sensors use acoustic waves and electromagnetic fields to measure the flow through a given area *via* physical quantities, such as acceleration, frequency, pressure and volume. As a result, many flow meters are named with respect to the physical property that helps to measure the flow.

Flow measurement proves crucial in various industries including petroleum and chemical industries. Consequently, flow measurement becomes a major component in the overall economic success or failure of any given process. Most importantly, accurate flow measurements ensure the safety of the process and for those involved in its success.

Before reading about the intricate details of various flow meters it's a good idea to think about aspects other than the design, governing equations, and the mechanism a flow meter uses in identifying a flow profile. To gain more from this section consider the following questions when learning about each flow meter because when choosing a flow meter the two main things to think about are cost and application:

1. What mechanism does this flow meter use to measure flow?
2. How expensive is it?
3. What impact does it have on the system? (How much power does it withdraw from the system and does the disturbance of flow cause significant problems up or downstream?
4. What are the accuracy limits of the tool? (Increased accuracy and precision results in higher cost especially for those with automated noise filtering and signal amplifying capabilities.

Sometimes you may be given an assignment to purchase a tool for your company's system and given with it a list of qualities to look for in order of importance, sometimes you're going to have to make this list yourself. The following table is a component of the TRIZ method for developing a list of qualities in descending order of importance on the left, perpendicular to the factor it has an impact on, to the right. For a given system you figured out that the cheapest solution for accurately measuring the flow rates of various pipe lines is by placing multiple flow meters in a series of positions.

From a few calculations using propagation of error you find a moderate range of accuracy limits necessary for your system to be considered well monitored. You know the accuracy limits are less important than the total cost of all the flow meters because your boss told you not to waste any money. On the table you can see that **Cost** is at the top of the left column so **Cost** is the main concern. **Accuracy** is in the farthest left column on the top row so **Cost** most dramatically exacerbates the **Accuracy** of the flow meter when it's minimized. In the intersecting box you see that the lowest price flow meters are the ones with the least amount of accuracy, generally speaking.

The next most important quality of your assignment is to get flow meters with the appropriate accuracy, thus **Accuracy** is below **Cost** in the left column. Looking to the top row you see that the **Accuracy** of the flow meter most greatly effects the **Impact on the system**. If you have a low quality flow meter it may be due to the side of the mechanism used to measure flow or the power the meter draws from the system (through mixing, frictional losses, increase in the turbulence of the flow, or buoyant effects caused from heat transfer). Completing the rest of the table you can decide if there are contradictions to what you thought were the most important qualities based on the inputs in the intersecting cells.

Table. Triz model for Cost = dominant factor followed by accuracy, Impact and Maintenance.

	Accuracy	Impact on system	Cost	Maintenance
Cost	Low accuracy/precision meters costs the least	Lower costs will result in larger meters that increase it effects on flow		Lowest cost means the flow meter will need frequent Maint.
Accuracy		Higher accuracy low cost flow meters have more surface area	What is the right price? How much error can I afford?	How long will it be reliable before it needs to be cleaned and recalibrated?
Maintenance	Application dependent, more Maint. Increases reliability	Down time, accumulation of deposits, disturbing flow	Down time, buildup on meter increases drag	
Impact on system	Lower impact may mean lower accuracy		Are there other costs of impact from installing the flow meter	How intricate is the meter, can it be quickly cleaned and recalibrated

An example where this model is important is for a system containing suspended particles in a fluid. If you want to measure the flow rate it might be cheapest to use a pitot tube yet the increased cost of maintaining the flow meter extends into down time of the system, more work from the technicians for a relatively small aspect of the process and lower profits as a result. For this system, maintenance would be the most important factor followed by accuracy, impact on the system and cost. If cost were least important factor you could afford to install a couple doppler meters, gather accurate measurements and the impact of the measurements on the system would be relatively low.

COMMON TYPES OF FLOW METERS

The flow rate as determined by the flow sensor is derived from other physical properties. The relationship between the physical properties and the flow rate is derived from fundamental fluid flow principles, such as Bernoulli's equation.

Differential Pressure

These sensors work according to Bernoulli's principle which states that the pressure drop across the meter is proportional to the square of the flow rate.

$$-\delta p \propto V^2$$

Using the pressure drop across a pipe's cross section is one of the most common manners to determine a flow measurement. As a result, this property has heavy implications for industrial applications. Flow meters characterized by differential pressure come in several different varieties and can be divided into two categories, laminar and turbulent. Differential pressure sensors operate with respect to Bernoulli's principle. Bernoulli's principle states that the pressure drop across the meter is proportional to the square of the flow rate.

Orifice Meter

Orifice plates are installed in flow meters in order to calculate the material balances that will ultimately result in a fluid flow measurement on the sensor. An orifice plate is placed in a pipe containing a fluid flow, which constricts the smooth flow of the fluid inside the pipe. By restricting the flow, the orifice meter causes a pressure drop across the plate.

By measuring the difference between the two pressures across the plate, the orifice meter determines the flow rate through the pipe. The larger the pressure drop, the faster the flow rate would be. There are two types of orifice meters that are usually used in industry, they are the orifice-square edge and the orifice-conic edge.

The orifice-square edge has insignificant friction at the interface between the fluid and the orifice plate. These types of orifice plates are recommended for smooth fluid flows, particularly clean liquids and gases such as steam. Generally, drain holes are incorporated in the design so that liquids and gases do not accumulate inside the pipes. Multi-phase fluids are not recommended for the orifice-squared edge because clogging becomes a significant problem as time progresses.

The orifice-conic edge is similar to the orifice-square edge, the primary difference being that the orifice-conic edge has a gradually expanding cross-sectional area, and this cross-sectional area is circular in shape. A conic-edge design is often a better choice for low velocity, high viscosity flows. Both types operates best under comparable temperature and pressure conditions, pipe sizes and provide similar accuracies.

Orifice meters used in conjunction with DP (Differential Pressure) cells are one of the most common forms of flow measurement. In addition, an orifice meter can be used to measure flows when there is a significant difference in pressure in the pipe, like between the upstream and downstream sides of a partially obstructed pipe, which is exactly what the orifice meter does on its own.

The plate offers a precisely measured obstruction that essentially shrinks the pipe and forces the flowing substance to constrict. A DP cell allows the comparison of pressure on the upstream (unobstructed) side and the downstream (constricted) side. A greater rate of fluid flow would usually result in a larger pressure drop, since the size of the orifice remains constant and the fluid is held longer building potential energy on the upstream side of the orifice.

Some of the other types of orifice plates include concentric, eccentric and segmental plates, each having different shapes and placements for measuring different processes. These plates are available in varied shapes so that the meter has the optimum structure for different applications. Moreover, the density and viscosity of the fluid, and the the shape and width of the pipe also influences the choice of plate shape to be used.

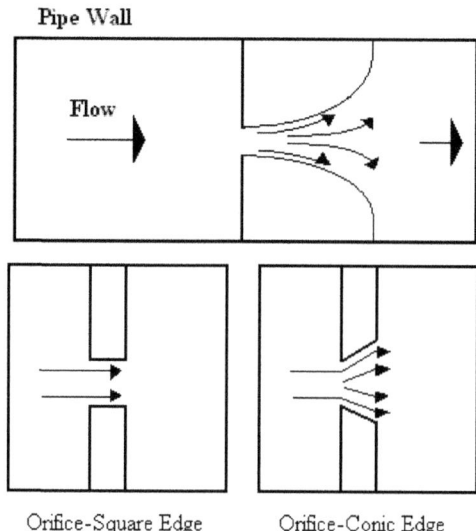

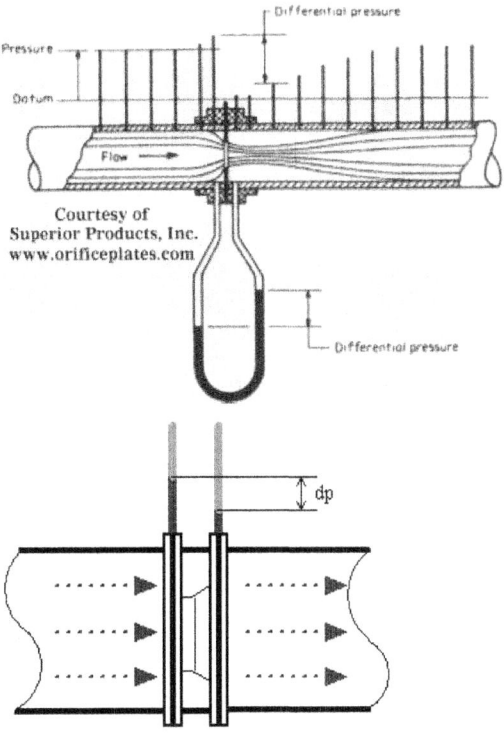

www.EngineeringToolBox.com

Such a pressure drop across the plate is then related to the flow rate using basic fluid mechanics principles that incorporate parameters such as density of the fluid and size of the pipe. The flow rate Q, given by the orifice meter, is usually modeled with the following equation:

$$Q = \frac{C_d A_2}{\sqrt{1 - (A_2/A_1)^2}} \sqrt{\frac{2(p_1 - p_2)}{\rho}}$$

Where $p_1 - p_2$ is the pressure drop across the plate, ρ is the fluid density, A_1 is the pipe cross-sectional area, A_2 is the orifice cross-sectional area, and C_d is the discharge coefficient (usually about 0.6). C_d is used to account for energy losses within the system.

The orifice meter is one of the most commonly used flow meters, since it is inexpensive to install and operate, it is uncomplicated and easy to construct, and it is highly robust and long lasting. Orifice meters are not only simple and cheap, they can also be delivered for almost any application and be made of any material. This simplicity of its design and function is one of its paramount advantages, with the meter essentially consisting of just a modified plate. This not only reduces its initial price but also shrinks its operating costs, maintenance expenses, and spare parts expenditure and availability.

Lower flow rates reduces their accuracy, whereas higher flow rates combined with high quality, unworn orifice plates increases it. The orifice plate is best when a sharp edge is present towards the upstream side of the meter. Wear reduces the accuracy of orifice plates. The turndown rate of orifice plates are generally less than 5:1.

Venturi Meter

Venturi meters can pass 25 – 50% more flow than an orifice meter. In a Venturi meter setup, a short, smaller diameter pipe is substituted into an existing flow line. Because the Venturi meter is insensitive to changes in the velocity profile of the fluid flow, the pipe design does not need to be straight like the orifice meter. Though initially expensive, the Venturi meter has relatively low maintenance and operation costs.

In the Venturi Tube the fluid flowrate is measured by reducing the cross sectional flow area in the flow path, generating a pressure difference. After the constricted area, the fluid is passes through a pressure recovery exit section, where up to 80% of the differential pressure generated at the constricted area, is recovered.

There are two main types of Venturi meters. The first one, known as the classical Herschel Venturi meter, is a very long meter characterized below. Pressure readings at different points in the meter are combined to provide an average pressure reading. Cleaning the classical Venturi meter is limited. The second type of Venturi meter is known as the short form Venturi meter. This differs from its longer counterpart by reduced size and weight.

By Bernoulli's principle the smaller cross-sectional area results in faster flow and therefore lower pressure. The Venturi meter measures the pressure drop between this constricted section of pipe and the non-constricted section.

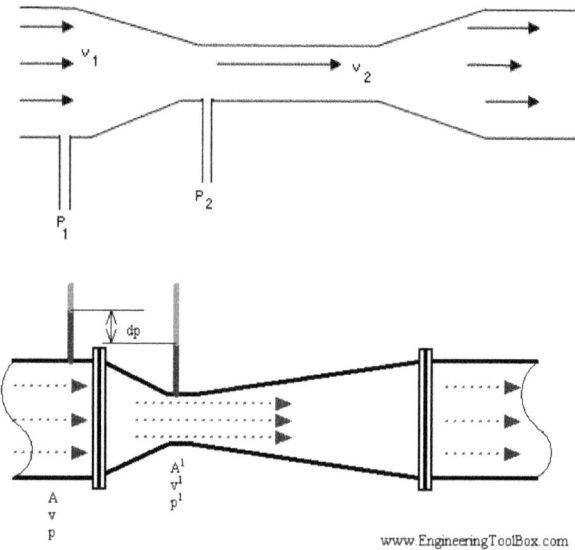

www.EngineeringToolBox.com

The discharge coefficient for the Venturi meter is generally higher than that used for the orifice, usually ranging from 0.94 to 0.99.

The Venturi meter is most commonly used for measuring very large flow rates where power losses could become significant. The Venturi flow meter has a higher start up cost than an orifice, but is balanced by the reduced operating costs.

Due to simplicity and dependability, the Venturi tube flowmeter is often used in applications where higher turndown ratiosor lower pressure drops than orifice plates can provide are necessary. With proper instrumentation and flow calibrating the venturi meter flowrate can be reduced to about 10% of its full scale range with proper accuracy. This provies a turndown ratio of around 10:1.

Flow Nozzle

Another type of differential pressure flowmeter is the flow nozzle. Flow nozzles are often used as measuring elements for air and gas flow in industrial applications. At high velocities, Flow Nozzles can handle approximately 60 percent greater liquid flow than orifice plates having the same pressure drop. For measurements where high temperatures and velocities are present, the flow nozzle may provide a better solution than an orifice plate.

Its construction makes it substantially more rigid in adverse conditions and the flow coefficient data at high Reynolds numbers is better documented than for orifice plates. Liquids with suspended solids can also be metered with flow nozzles. However, the use of the flow nozzles is not recommended for highly viscous liquids or those containing large amounts of sticky solids. The turndown rate of flow nozzles is similar to that of the orifice plate. The flow nozzle is relatively simple and cheap, and available for many applications in many materials.

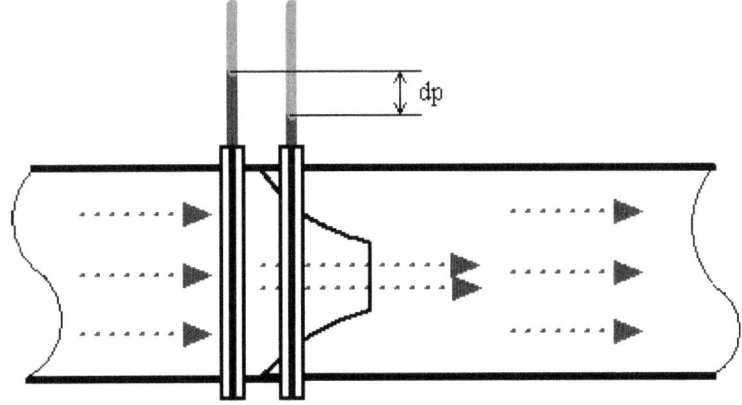

The Sonic Nozzle- Critical(Choked)Flow Nozzle

One type of flow nozzle is the sonic nozzle. The Sonic Nozzle is a converging-diverging flowmeter. It consists of a smooth rounded inlet section converging to a minimum throat area and diverging along a pressure recovery section or exit cone.

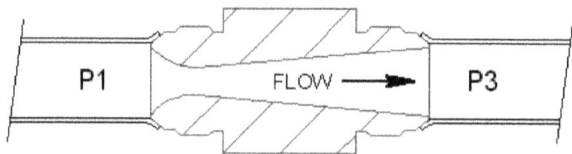

The Sonic Nozzle is operated by either pressurizing the inlet (P1) or evacuating the exit (P3), to achieve a pressure ratio of 1.2 to 1 or greater, inlet to outlet. When a gas accelerates through a nozzle, the velocity increase and the pressure and gas density decrease. The maximum velocity is achieved at the throat, the minimum area, where it breaks Mach 1 or sonic.

At this point it's not possible to increase the flow by lowering the downstream pressure. The flow is choked. Pressure differences within a piping system travel at the speed of sound and generate flow. Downstream differences or disturbances in pressure, traveling at the speed of sound, cannot move upstream past the throat of the Nozzle because the throat velocity is higher and in the opposite direction.

Sonic Nozzles are used in many control systems to maintain fixed, accurate, repeatable gas flow rates unaffected by the downstream pressure. If you have a system with changing or varying gas consumption downstream and you want to feed it a constant or locked flowrate, a Sonic Nozzle is an excellent way to achieve this.

Pitot Tubes

Pitot tubes measure the local velocity due to the pressure difference between points 1 and 2 in the diagrams below. Unlike the other differential flow meters, the pitot tubes only detect fluid flow at one point rather than an overall calculation. The first diagram shows a simple pitot tube configuration while the second shows a compact pitot tube configuration.

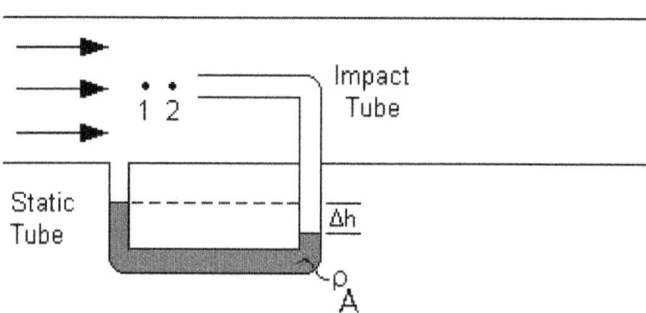

Flow Control Sensors

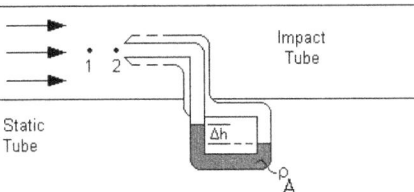

Both tubes work in a similar manner. Each pitot tube has two openings, one perpendicular to the flow and one parallel to the flow. The impact tube has its opening perpendicular to the fluid flow, allowing the fluid to enter the tube at point 2, and build up pressure until the pressure remains constant. This point is known as the stagnation point. The static tube, with openings parallel to the fluid flow gives the static pressure and causes a sealed fluid of known density to shift in the base of the tube. Pressure drop can be calculated using the height change along with the fluid densities and the equation below.

$$\Delta p = \Delta h (\rho_A - \rho) g$$

with Δp as the pressure drop, ρ_A as the known fluid density, ρ as flowing fluid's density, and g as the acceleration due to gravity.

This pressure drop can be related to the velocity after accounting for the losses throughout the piping in the system, given by C_p. This dimensionless coefficient is found through accurate calibration of the pitot tube. The equation below describes this relationship.

$$v = C_p \sqrt{\frac{2(p_1 - p_2)}{\rho}}$$

with v as the fluid velocity, C_p as the loss coefficient, p_1 as the pressure at point 1, p_2 as the pressure at point 2, and ρ as the flowing fluid's density.

By placing the tube at the exact center of the pipe, the maximum velocity can be measured and the average velocity can be calculated *via* the Reynolds number. The governing equation and chart are below.

$$Re = \frac{D v_{max} \rho}{\mu}$$

with Re as the Reynolds number, D as the pipe diameter, v_{max} as the maximum velocity, ρ as the flowing fluid's density, and μ as the flowing fluid's viscosity.

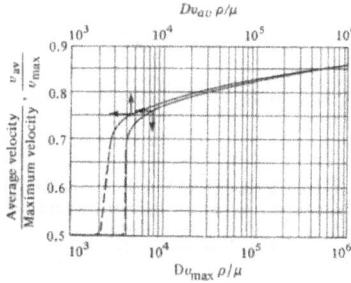

Finally, the flow rate can be found by accounting for the area of the pipe.

$$Q = v_{avg} \frac{\pi}{2} r^2$$

with Q as the volumetric flow rate, v_{avg} as the average velocity, and r as the pipe's radius.

It should be noted that all the equations apply to incompressible fluids only, but these can be used to approximate gas flows at moderate velocities. This flow meter must also be placed at least 100 pipe diameters in distance, downstream of the nearest flow obstruction. This ensures no unwanted pressure fluctuations and accurate pitot tube readings. Furthermore, fluids containing large particles can clog the pitot tube and should be avoided.

DIRECT FORCE

These flow meters are governed by balancing forces within the system.

Rotameter

A rotameter is a vertically installed tube that increases in diameter with increasing height. The meter must be installed vertically so that gravity effects are easily incorporated into the governing equations. Fluid flows in through the bottom of the tube and out through the top. Inside the glass tube there is a float that changes position with the flow rate. When there is no liquid flow, the float rests in the bottom of the meter.

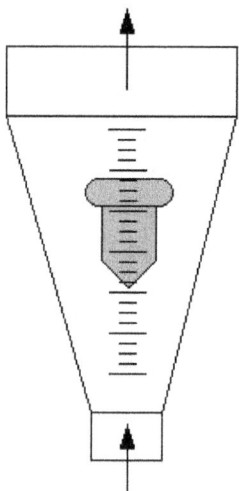

The applied concept for the rotameter is differential area. As the flow rate of the fluid changes, the position of the float changes and annular area change directly, keeping a constant pressure drop across the meter. Changes in float position and annular area are approximately linear with changes in flow rate. Upon achieving a stable flow rate, the vertical forces are balanced and hence the position of the float remains constant. The volumetric flow is given by the following equation:

$$Q = CA_b \sqrt{\frac{2g\left(\frac{V_f(\rho_f-\rho)}{A_f} - \rho h_f\right)}{\rho\left[1 - \left(\frac{A_b}{A_a}\right)^2\right]}}$$

with C being the discharge coefficient, A_b being the cross sectional area of the top of the float, V_f volume of the float, ρ_f the density of the float, ρ the density of the fluid h_f the height of the float, A_a the cross sectional area of the bottom of the float.

Generally, rotameters are inexpensive and simple to use. This allows them to be used in many plant applications.

Turbine Meter

A turbine wheel is placed in a pipe that holds the flowing fluid. As the fluid flows through the turbine, the turbine is forced to rotate at a speed proportional to the fluid flow rate. A magnetic pick-up is mounted to the turbine wheel, and a sensor records the produced voltage pulses. Voltage information can then be translated into the actual flow meter reading.

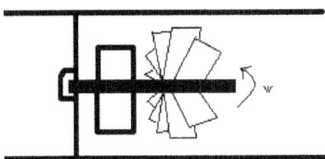

The following is the equation used to model the turbine meter:

$$Q = VA = \frac{\omega\,(\bar{r})^2\,A^2}{\bar{r}Atan\beta - 0.037Re^{-0.2}n\,(R_o + R_i)\,Dsin\beta}$$

with A the pipe area, $\bar{r}$ the root mean squared radius, ω rotational speed, β the angle between the flow direction and the turbine blades, R_o the outer blade radius, R_i the inner radius, and D the distance between blades.

There are two main advantages of the tubine meter over conventional differential head devices

1) The extended are more accurate registatrion of flow in the low flow range of process operation. This results from the registration being proportional to the velocity rather than the velocity square
2) The comparatively low head loss across the meter

Another advantage to using this type of flow meter is reliability. Extensive testing has proven these results. Additionally, the turbine flow meter does not have a high installation cost. However, due to the turbine wheel motion, a low to medium pressure drop can result. Turbine wheel replacement may also be required due to abrasion caused by particles within the fluid.

Propeller Flow Meter

Propeller flow meters have a rotating element similar to the wheel in turbine meters. Again, rotation is caused by fluid flow through the propeller, and voltage pulses are created as the propeller passes a magnetic or optical sensor. Similarly, the frequency of the pulses is proportional to flow rate of the fluid and the voltages can be directly correlated with the fluid flow rate.

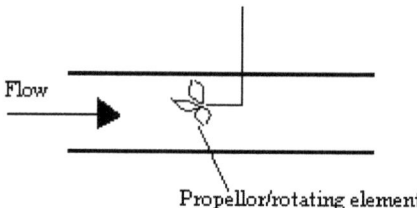

Propellor/rotating element

Propeller flow meters are often used specifically with water, though other fluids may also be used. Low cost coupled with high accuracy make propeller flow meters a common choice in many applications.

Paddle Wheel Sensors

A kind of propeller sensor is the paddle wheel sensor. Paddle wheel sensors are similar to turbine sensors, except for one thing. The shaft of the paddle wheel sensor is perpendicular to the flow of the fluid while the turbine sensor's shaft is parallel to the flow of the fluid. This adds several advantages for the paddle wheel flow sensor. Due to the shaft being perpendicular to the flow, it sustains less axial from the fluid, and thus less friction. Paddle wheel sensors also have a smaller number of blades, so there is less force needed to turn the paddle wheel. This means that a paddle wheel can be accurate at lower flows, have a high efficiency, as well as a longer lifetime.

There are two kinds of paddle wheel sensors, insertion and inline sensors. There is more than one design for an insertion sensor, but one popular design has the bearing built into the rotor and the shaft sliding through it as the center axis of spin. The blade sticks out and is inserted into the pipe with the flowing fluid.

An inline paddle wheel sensor is used for smaller pipes. It contains a rotor assembly with the magnet sealed inside and a main body.

Coriolis Mass Flow Meter

A Coriolis flow meter harnesses the natural phenomenon wherein an object will begin to "drift" as it travels from or toward the center of a rotation occurring in the surrounding environment. A merry-go-round serves as a simple analogy; a person travelling from the outer edge of the circle to its center will find himself deviating from his straight-line path in the direction of the ride's rotation.

Coriolis flow meters generate this effect by diverting the fluid flow through a pair of parallel U-tubes undergoing vibration perpendicular to the flow. This vibration simulates a rotation of the pipe, and the resulting Coriolis "drift" in the fluid will cause the U-tubes to twist and deviate from their parallel alignment. This Coriolis force producing this deviation is ultimately proportional to the mass flow rate through the U-tubes.

$$MassFlow = \frac{F_c}{2wx}$$

where Fc is the Coriolis force observed, w is the angular velocity resulting from rotation, and x is the length of tubing in the flow meter.

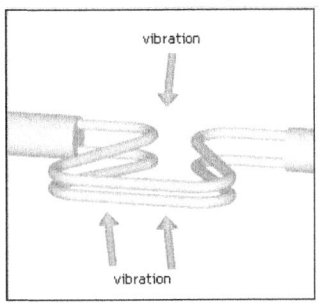

Coriolis flow meter undergoing no flow.

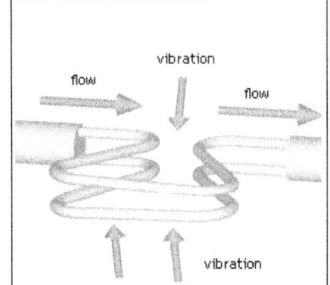

Coriolis flow meter exhibiting deflection as a result of mass flow

Adapted from Wikipedia's entry on Coriolis Mass Flow Meters.
http://en.wikipedia.org/wiki/Mass_flow_meter

Because the Coriolis flow meter measures the mass flow rate of the fluid, the reading will not be affected by fluctuations in the fluid density. Furthermore, the absence of direct obstructions to flow makes the Coriolis flow meter a suitable choice for measuring the flow of corrosive fluids. Its limitations include a significant pressure drop and diminished accuracy in the presence of low-flow gases.

To understand how the mass flow rate is measured with this device, refer to the following attachment.

Frequency

These flow meters use frequency and electronic signals to calculate the flow rate.

Vortex Shedding Flow Meter

A blunt, non-streamline body is placed in the stream of the flow through a pipe. When the flow stream hits the body, a series of alternating vortices are produced, which causes the fluid to swirl as it flows downstream. The number of vortices formed is directly proportional to the flow velocity and hence the flow rate.

The vortices are detected downstream from the blunt body using an ultrasonic beam that is transmitted perpendicular to the direction of flow. As the vortices cross the beam, they alter the carrier wave as the signal is processed electronically, using a frequency-to-voltage circuit. The following diagram shows the basic principle of the vortex-shedding flow meter:

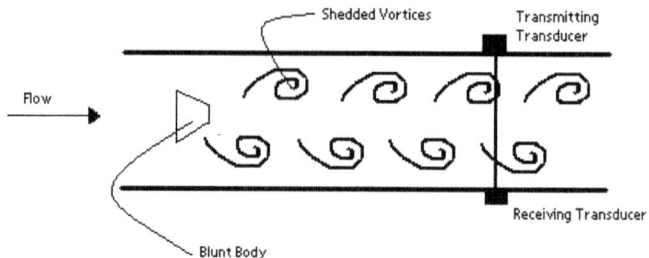

Vortex-shedding flow meters are best used in turbulent flow with a Reynolds number greater than 10,000. One advantage of using this type of flow meter is its insensitivity from temperature, pressure, and viscosity. The major disadvantage to using this method is the pressure drop caused by the flow obstruction.

ULTRASONIC FLOW METERS

There are two types of Ultrasonic meters, the transit time/time of flight and Doppler models, both of which have unique equations representing the principles behind them. The basis for these meters is monitoring ultrasonic waves in fluid passing through a pre-configured acoustic field. These meters are based on the technique of sound waves that change.

Transit Time/Time of Flight Flow Meters

Transit time meters have two opposing transducers outside of the pipe to measure the time of a signal sent from a transducer upstream to a transducer downstream and vice versa.

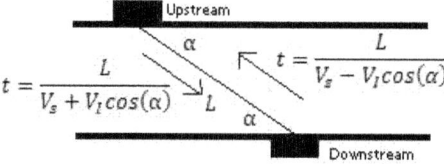

This allows the average velocity and hence the flow rate, Q, to be determined.

$$Q = K \left(\frac{\pi d^3 \tan\alpha}{8}\right) \left(\frac{1}{t_{UD}} - \frac{1}{t_{DU}}\right)$$

where d is the diameter of the pipe, α is the angle between direction of the flow and the pipe, t_{UD} is the time for the signal to reach downstream transducer from the upstream transducer, and t_{DU} is the time for signal to reach upstream transducer from the downstream transducer.

Flow Control Sensors

With the Time of Flight Ultrasonic Flowmeter the time for the sound to travel between a transmitter and a receiver is measured. This method is not dependable on the particles in the fluid.

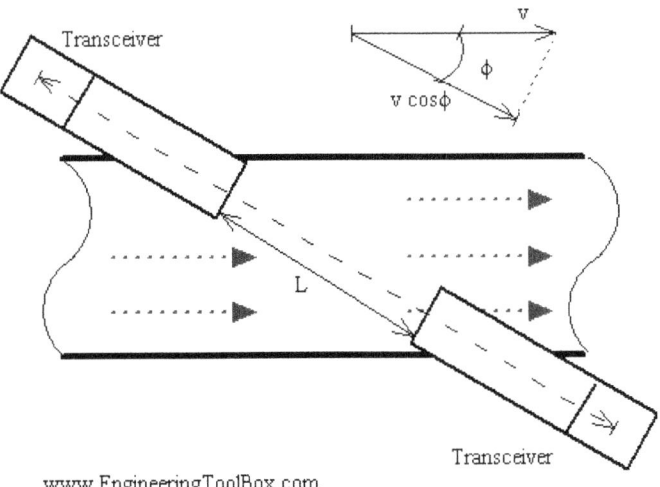

www.EngineeringToolBox.com

Two transmitters / receivers (transceivers) are located on each side of the pipe. The transmitters sends pulsating ultrasonic waves in a predefined frequency from one side to the other. The difference in frequency is proportional to the average fluid velocity.

Downstream pulse transmit time can be expressed as

$$t_d = L / (c + v \cos\Phi)$$

where t_d = downstream pulse transmission time L = distance between transceivers

Downstream pulse transmit time can be expressed as

$$t_u = L / (c - v \cos\Phi)$$

where t_u = upstream pulse transmission time

Since the sound travels faster downstream than upstream, the difference can be expressed as

$t = t_d - t_u$ $t = 2 v L \cos\Phi / (c^2 - v^2 \cos 2\Phi)$ $t = 2 v L \cos\Phi / c^2 (4)$ (since v is very small compared to c)

Doppler Meters

Doppler meters use the frequency shift of an ultrasonic signal when it is reflected by suspended particles or gas bubbles (discontinuities) in motion. The Doppler Effect Ultrasonic Flowmeter uses reflected ultrasonic sound to measure the fluid velocity. By measuring the frequency shift between the ultrasonic frequency source, the receiver, and the fluid carrier, the relative motion are measured. The resulting frequency shift is named the *Doppler Effect*.

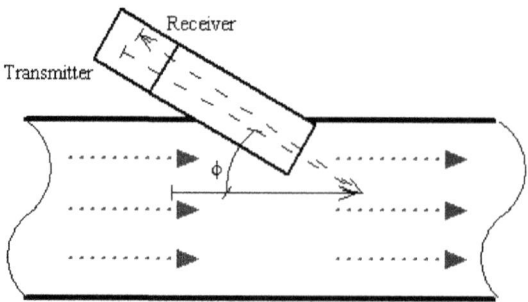

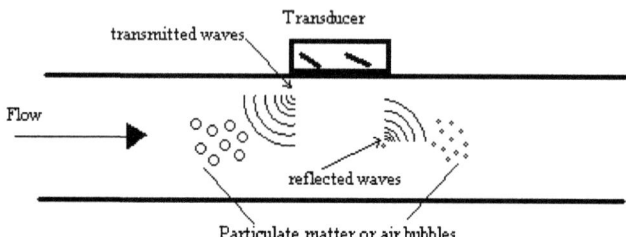

These reflected signals travel at the velocity of light.

$$V_i = \frac{V_s \Delta f}{2 f_{actual} \cos \alpha}$$

where f is the actual frequency and Δf is the change in frequency or frequency shift.

The fluid velocity can be expressed as

v = c (fr - ft) / 2ftcosΦ

where

fr = received frequency

ft = transmission frequency

v = fluid flow velocity

Φ = the relative angle between the transmitted ultrasonic beam and the fluid flow

c = the velocity of sound in the fluid

This method requires that there are some reflecting particles in the fluid. The method is not suitable for clear liquids.

Advantages with the Doppler Effect Ultrasonic Flowmeter

Doppler meters may be used where other meters don't work. This could be in liquid slurries, aerated liquids or liquids with some small or large amount on suspended solids. The advantages can be summarized to:

- Obstruct less flow

- Can be installed outside the pipes
- The pressure drop is equal to the equivalent length of a straight pipe
- Low flow cut off
- Corrosion resistant
- Relative low power consumption

Limitations with Doppler Effect Ultrasonic Flowmeters

- Doppler flowmetersperformance are highly dependent on physical properties of the fluid, such as the sonic conductivity, particle density, and flow profile.
- Non uniformity of particle distribution in the pipe cross section may result in a incorrectly computed mean velocity. The flowmeter accuracy is sensitive to velocity profile variations and to the distribution of acoustic reflectors in the measurement section.
- Unlike other acoustic flowmeters, Doppler meters are affected by changes in the liquid's sonic velocity. As a result, the meter is also sensitive to changes in density and temperature. These problems make Doppler flowmeters unsuitable for highly accurate measurement applications.

Benefits with Ultrasonic Flowmeters as a Whole

- Obstruction less flow
- Pressure drop equal to an equivalent length of straight pipe
- Unaffected by changes in temperature, density or viscosity
- Bi-directional flow capability
- Low flow cutoff
- Corrosion-resistant
- Accuracy about 1% of flow rate
- Relative low power consumption

Both meters are effective in measuring open channels and partially filled pipes but are very sensitive to flow conditions and hence should be calibrated with care. Also, there is no pressure drop since there are no obstructions in the flow path.

Limitations with Ultrasonic Flowmeters as a Whole

- The operating principle for the ultrasonic flowmeter requires reliability high frequency sound transmitted across the pipe. Liquid slurries with excess solids or with entrained gases may block the ultrasonic pulses.
- Ultrasonic flowmeters are not recommended for primary sludge, mixed liquor, aerobically digested sludge, dissolved air flotation thickened sludge and its liquid phase, septic sludge and activated carbon sludge.

- Liquids with entrained gases cannot be measured reliably.

The following link will help to show how both types of the Ultrasonic meter works and how the above equations are derived.

OTHER TYPES

Magnetic Flow Meter

One magnetic model flow meter positions electric coils around the pipe of the flow to be measured. A pair of electrodes is set up across the pipe wall. The fluid flowing has a minimum value of electrical conductivity, the movement of the fluid through the pipe acts as a conductor moving across the magnetic field. There is an induced change in voltage between the electrodes, which is proportional to the flow velocity.

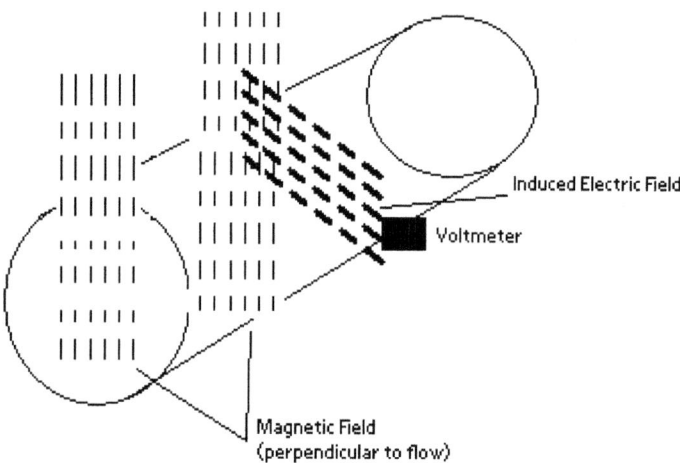

The flow velocity is found by measuring the changes of induced voltage of the conductive fluid passing through a controlled magnetic field at right angles. According to Faraday's Law, the magnitude of the voltage induced is directly proportional to the product of the magnetic flux, distance between probes and the velocity of the medium (fluid).

$$E = -N\frac{d\phi}{dt} = -NB\frac{dA}{dt} = -NBD\frac{dz}{dt} = -NBDv$$

where E is the voltage of induced current, N is the number of turns, B is the external magnetic field, ϕ is the magnetic flux, D is the distance between electrodes and v is the velocity of the fluid.

Some of the advantages are minimum pressure drop because of minimum obstructions in flow path; low maintenance cost because of no moving parts. One of the disadvantages is that it usually requires a fluid with an electrical conductivity of higher than 3 µS/cm.

Calorimetric Flow Meter

This type of flow meter is suitable for monitoring the flow rates of liquid and gaseous mediums. The calorimetric principle is based on two temperature sensors in close quarters of one another but thermally insulated from one another.

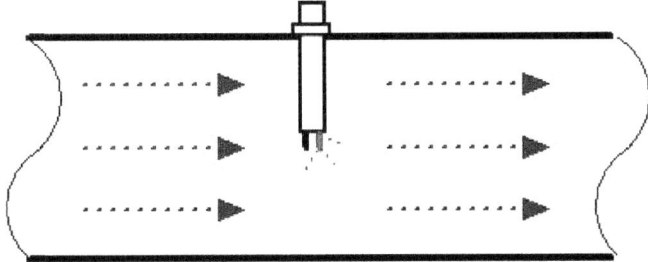

www.EngineeringToolBox.com

One of the two temperature sensors is constantly heated and the cooling effect of the flowing fluid is used to monitor the flow rate. In a stationary phase fluid condition there is a constant temperature difference between the two temperature sensors. When the fluid flow increases, heat energy is extracted from the heated sensor and the temperature difference between the sensors are reduced. The reduction is proportional to the flow rate of the fluid. The calorimetric flow meter can achieve relatively high accuracy at low flow rates.

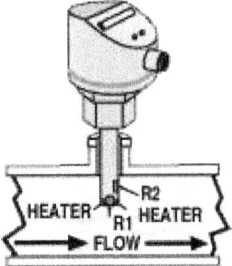

Common applications: air compression, argon compression, nitrogen compression, carbon dioxide compression and flow detection of all fluids (liquids and gases)

Gear Flow Meter

This type of flow meter has oval shaped gears with fitting teeth which control the amount of fluid passing through. The flow rate is calculated by number of times the gears are filled and emptied. These meters have high accuracy and are used for measuring low flow and for high viscosity fluids. It is very easy to install these types of meters because it requires no pipe.

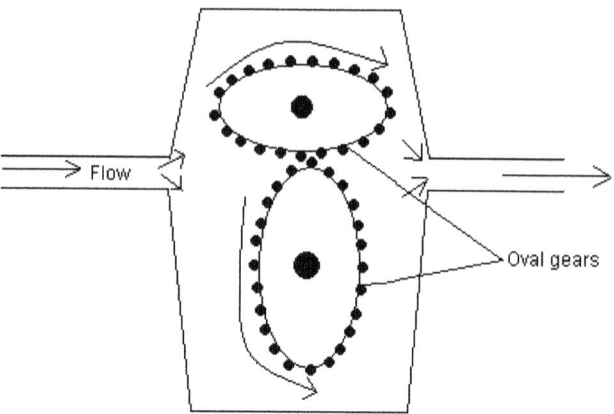

Thermal Flow Meters

These flow meters take advantage of the thermal properties of the fluid to measure the flow of the fluid in a pipe. In most thermal flow meters, a measured amount of heat is applied to the heater of the sensor.

Portions of this heat is lost to the fluid as it flows. Accordingly, as the fluid flow increases more heat is lost. The amount of heat lost is tracked using a temperature measurement instrument in the sensor. Then an electronic transmitter uses the heat input and temperature measurements to calculate the fluid flow, taking into account the thermal properties of the fluid.

Common applications of thermal flow meters are measuring the flow rate of clean gases like air, nitrogen, hydrogen, helium, ammonia, and argon. Most thermal flow meters are used to measure pure gases that would be used in laboratory experiments or semiconductor production. Mixtures like flue stack flow can also be measured but the mixture compositions must be known to use the appropriate thermal properties of each component in the mixture. The key advantage of this technology is its dependence on the thermal properties which are mostly independent of the gas density.

However caution should be taken if this type of flow meter is used to measure a fluid of unknown or varying composition. Additionally, thermal flow meters should not be used to measure abrasive fluids because they may damage the sensor. Some fluids can coat the sensor causing it to be inoperable and must be regularly cleaned to be useful.

Table of Flow Meters

The following table outlines specifics for most of the flow meters. Accuracy is given in terms of FSD (full scale deflection) which is the deflection of meter's pointer to the farthest point on the scale. This implies the highest measurement.

Flow Control Sensors

Type of Flowmeter	Operating Temperature of Sensor (K)	Pressure Change (Comparative)	Pipe Size (mm)	Accuracy	Measurables	Advantages	Disadvantages
Orifice-Square Edge	<800	High	>40	±1-2% FSD	Clean gases, steam, clean liquids, most dirty liquids and most corrosive liquids	Negligible friction at the boundary between the fluid and the orifice plate	Not suited for multi-phase fluids because clogging becomes a significant problem as time progresses
Orifice-Conic Edge	<800	High	>40	±2% FSD	Clean liquids, viscous liquids, mostly dirty liquids, and most corrosive liquids	Has a gradually expanding cross-sectional area and is good for low velocity, high viscosity flows	Not recommended for high velocity, low viscous flows
Venturi	<800	80-90% of Energy Recovered	>50	±1-2% FSD	Clean gases, most dirty gases, steam, clean liquids, most viscous liquids, most dirty liquids, most corrosive liquids, and most fibrous, abrasive slurries	Pipe design does not need to be straight, relatively low maintenance and operation costs	Insensitive to changes in the velocity profile of the fluid flow, relatively expensive to orifice flow meters
Pitot Tube	<800	Low	>10	±1-2% FSD	Clean gases, most dirty gases, steam, clean liquids, most viscous liquids, most dirty liquids, most corrosive liquids, and most fibrous, abrasive slurries	Detects fluid flow at one point from pressure drop	Clogs occur from fluids containing large particles
Electromagnetic	230-450	Low	Jun-00	±0.2 flow rate to 1.0% FSD	Clean liquids, viscous liquids, dirty liquids, corrosive liquids, fibrous slurries and abrasive slurries	Minimum P-drop because of few obstructions in flow path; low maintenance cost because of no moving parts	Requires a fluid with an electrical conductivity > 3 μS/cm
Turbine	10-530	Medium	6-600	±0.5 flow rate	Clean gases, clean liquids, and most corrosive liquids	Relatively high-tested reliability and low installation cost	May cause low to medium P-drop from turbine wheel motion, may require turbine wheel replacement due to abrasion
Ultrasonic-Transit Time	100-530	Low	>12	±1 flow rate to ±2% FSD	Clean liquids, most viscous liquids, and corrosive liquids	Effective in measuring open channels and partially filled pipes	Very sensitive to flow conditions and thus need to be calibrated with care
Doppler	100-400	Low	>12	±1 flow rate to ±2% FSD	Most viscous liquids, dirty liquids, corrosive liquids, fibrous, and abrasive slurries	Effective in measuring open channels and partially filled pipes	Very sensitive to flow conditions and thus need to be calibrated with care
Rotameter	Glass < 400 Metal < 800	Low	<=75	±0.5 flow rate to ±1% FSD	Clean gases, clean liquids, viscous liquids, and most corrosive liquids	Relatively inexpensive and simple to use	Typically fragile, must specifically be installed vertically with fluid flowing upwards
Vortex	80-680	Low	15-400	±0.5 flow rate to ±1 flow rate	Clean gases, most dirty gases, steam, clean liquids, and most corrosive liquids	Insensitive to temperature, pressure, and viscosity	Causes pressure drop due to flow obstruction
Coriolis	Insensitive to Temperature	Low	10-100	±1% FSD	All. Exception – Both clean and dirty gases can be measured only at high pressures	Effectively measures flow rates of high viscous liquids at a low pressure drop, insensitive to temperature, pressure, corrosive liquids, and fluctuating densities	Significant pressure drop and diminished accuracy in the presence of low-flow gases

Flow Profile Distortion

In the real world, the flow profile is not always symmetrical. Pipe fittings such as elbows, tee-pieces, and reducers can change the flow profile. One example of a fitting that alternates the flow profile is a sharp elbow which causes pure swirls throughout the fluid. Some flow meters are more sensitive to particular types of flow distortion. More complex flow conditions produce better velocity profile but there is a trade off since they are more expensive and give higher pressure drops.

Turndown Ratio

The turndown ratio is a term used to describe the range of accurate operability of a specific flow meter. This rangeability is critical when selecting flow meters for specific applications. Typically, in a plant setting gas flow may not be constant, but accurate measurement of gas flow is needed ranging from no flow to full flow.

For example if nitrogen gas is being used in a plant conducting multiple batch reactions, sometimes the little nitrogen (100 m^3/min) will be needed and other times full nitrogen flow (1000 m^3/min) will be needed. For this system, the turndown ratio is 10:1. Accordingly, a flow meter must have a turndown ratio of at least 10:1.

For each type of flow meter, the turndown ratio is limited by the theoretical and physical constraints. For example, for an orifice meter the accuracy may be compromised near the limits of the rangeability. Orifice meters create a pressure drop in the measured fluid which is proportional to the velocity squared. If the range of differential pressures becomes too large the overall accuracy of the flow meter at its range limits may be inconsistent.

Turndown ratio can be expressed as:

TR = qmax / qmin(1)

where

TR = Turndown Ratio

qmax = maximum flow

qmin = minimum flow

The table below shows a list of typical turndown ratios for different flow meters.

Type	Turndown Ratio
Orifice Meter	3:1
Turbine Meter	10:1
Rotary Displacement Meter	10:1-80:1
Diaphragm Meter	80:1
Ultrasonic Meter	50:1
Thermal Meter	10:1-100:1

Example - Turndown Ratio for an Orifice Meter

The turndown ratio - TR - for an orifice meter with maximum flow of 12 kg/s and a minimum flow of 3 kg/s can be calculated as:

TR = (12 kg/s) / (3 kg/s) = 4 - normally expressed as turndown ratio of 4:1

This is a typical turndown ratio for a orifice plate. In general a orifice plates has turndown ratio between 3:1 and 5:1.

Turndown Ratio and Measured Signal

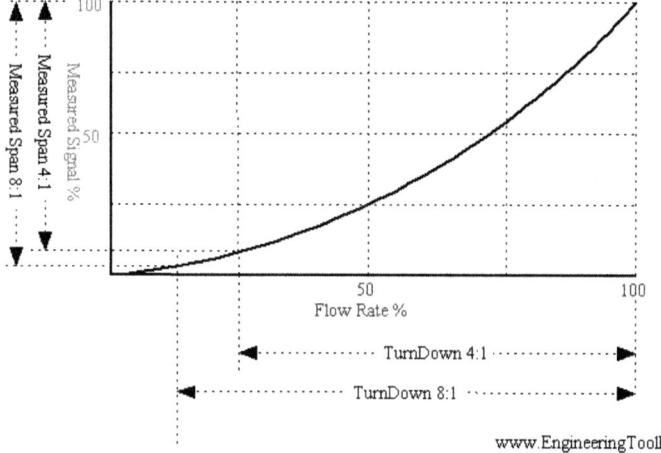

www.EngineeringToolBox.com

The graph above shows how the turndown ratio effects the measured signal % and flowmeter %. With an increased turndown ratio there is a larger range that the flowmeters can operate within. In a flow meter based on the orifice or venturi principle, the differential pressure upstream and downstream of an obstruction in the flow is used to indicate the flow. According the Bernoulli Equation the differential pressure increases with the square of flow velocity. A large turndown ratio will cramp the measurement signal at low flow rate and this is why venturi and orifice meters are not accurate at low flowrates.

COMPOSITION SENSORS

There are many useful analytical tools, such as photometric analysis, electrometric analysis, chromatography, mass spectrometry, thermal conductivity, and various physical property measurements (density and specific gravity), which can be used to determine the composition of mixtures.

A wide array of methods to measure composition are available, so it is important to choose the best method given a set of conditions. The first step is to decide between measuring physical or chemical properties to determine the composition.

If you choose to measure a physical property, make sure that it is unique to the desired component of the mixture and will accurately allow you to determine the composition. The goal of this chapter is to explain the various analytical methods and tools used to determine the composition of a given sample. After reading this, you should be able to determine which method of composition measurement is most appropriate for a given circumstance.

Types of Testing: On-line vs. Off-line

On-line analysis is the continuous monitoring of the composition of a sample, which is under the influence of a control system and directed by an actuator which can respond and regulate the operating conditions in real time such that the desired set points are maintained.

On-line testing can either be performed in-line or by slip stream testing. In on-line testing, the sensor is attached directly to the line and provides feedback *via* a transmitter. On the other hand, in slip stream testing, a side stream of the process runs alongside the main line. Such an apparatus closely resembles the set up of a bypass.

The slip stream process conditions can be continuously manipulated to make the measurements easier to obtain. Similar to in-line testing, the sensor is directly attached to the slip stream and provides feedback through a transmitter. Advantages of on-line analysis include an immediate and continuous feedback responding to changes in process conditions.

The main disadvantage of on-line testing is that it is usually much more complicated and more expensive than off-line testing. Also, off-line testing is more robust and has more varied applications, while on-line testing may not work in every situation. On-line testing prevents the continued production of undesired product, with an immediate response and correction of the flawed material.

Off-line analysis involves the extraction of a sample from the process or reaction, and its subsequent testing in a machine that may be situated at a location far away from the process line in a lab. In this case, a sample is manually removed which is later sent to the composition analyzer. The results of the analysis are examined and then they are sent to the control system or actuator to make the appropriate adjustments.

On-line analysis is comparable to off-line testing, the primary difference being that in on-line testing, the samples are analyzed on a machine that is next to the process line. This greatly reduces time lost in transporting the sample, though it still permits the prospect of introducing contaminants into the sample. In off-line analysis, some of the disadvantages include sample dead-time, which is the time lost during transportation, variability of sample testing locations, and lag time for adjustments to be made to the process.

In off-line analysis, the lag time between removing the sample from the system and receiving the results of the test could cause significant losses to the company, since the defective product is produced and the process is not corrected until the results are received. This prolonged defective processing could result in the accumulation of losses worth thousands of dollars, which could have been avoided if an on-line analysis system was set up in place.

While both methods are widely used in industry, there is a push towards more on-line testing. A paper published in 2003 urging drug companies to adopt more in-process analytical testing such as on-line testing, endorsed by the Food and Drug Administration reflects this trend. This has been a part of a broader effort by the FDA to encourage companies to move towards better manufacturing practices.

While on-line testing maybe comparatively expensive to install at first, the savings from the process would not only recoup the costs associated with installation, but also prevent significant losses. Many companies have adopted lean manufacturing techniques, and one of its hallmarks is to correct a defect as early in the production process as possible, not allowing defective products to move on to the next processing step. Allowing defective products to perpetuate through the system may even contaminate other non-defective materials in the production process.

Standards and Calibration Curves

Before discussing the different ways to measure the composition of a sample, one needs to understand that composition sensors use standards or calibration curves to measure an unknown composition against a known one(s). The standard can either be an internal standard against which all other measurements are compared, such as a reference cell or a series of standards used to create a calibration curve.

With a calibration curve, a series of known standards are measured with the sensor and the signal produced by the sensor is graphed on the y-axis of a plot; the known information about each sample is graphed on the x-axis. From this information, a relationship can be developed between the signal output from the sensor and the known quantity that you are trying to measure.

The data measured for the unknown sample can then be compared to the calibration curve graphically or using the equation written to describe the curve.

Using this information, the measurement made by the sensor can be interpreted as a useful composition measurement of weight percent, volume percent, mole percent, *etc.*

Photometric Analysis

Photometric analysis is the measurement of the intensity of visible light and other electromagnetic (EM) waves. Through the measurement of these values, the composition of samples or flows can be determined in various different ways. A wide range of photometric devices based on many differing principles are used in the chemical engineering industry for this purpose. A few of the most common instruments are covered in this chapter.

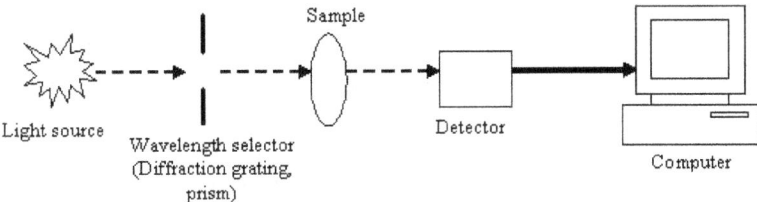

Types of Photometers

A photometer is any type of instrument used to measure characteristics of light. There are two broad categories of photometers: dispersive and non-dispersive.

Dispersive Photometers

In dispersive photometers, the light from the source is dispersed and a narrow spectral band is selectively directed to the sample and detector. A monochromator is usually the tool that performs this function. The dispersing element of the monochromator can be a diffraction grating, prism, and interference filters.

Diffraction gratings are the most common; for each position of the grating a narrow band of dispersed light passes through an exit slit. Dispersive devices can scan across a spectrum and make measurements at several wavelengths. This capability allows them to be used for the analysis of multiple components.

Non-dispersive Photometers

Non-dispersive photometers use a narrow-band-pass filter to block out a large amount of undesired radiation. They make measurements at selected discrete wavelengths. The filter passes radiation at the selected reference and measurement wavelengths. The reference wavelength filter selects a band where none of the components present in a process stream absorb radiation; the measurement wavelength filter is selected to match the absorption band of the component being analyzed. The ratio of the transmitted light at the reference and measurement wavelengths is measured by the photometer. Non-dispersive photometers are usually used to measure a single component in the process stream.

An example of such a device is the Non-Dispersive Infrared (NDIR) analyzer that is frequently used to measure the concentrations of certain gases in a mixed gas flow. The NDIR analyzer uses a reference non-absorbent gas such as O_2, Cl_2 and N_2 and compares that to a sample gas like CO, CO_2, SO_2, CH_4, etc. Two beams of infrared radiation are used: a reference beam for the non-absorbent gas and an analyzing beam for the sample.

Spectrophotometers

Spectrophotometers are widely used as composition sensors. Spectrophotometers are photometers that measure the intensity of light at specific wavelengths. Many different types of spectrophotometers exist, and they are typically categorized based on the ranges of wavelengths they monitor: ultraviolet (280-380 nm), visible light (400-700nm), infrared (0.7-1000μm), or microwave (1-300mm). Other important distinctions are the measurement technique, the method of obtaining the spectrum, and the source of intensity variation.

The emission spectrum recorded by the spectrophotometer typically consists of a series of narrow peaks of varying heights, also known as spectral lines. The height or amplitude of those lines can be related to the concentration or abundance of the specific material through a calibration curve and their position on the spectrum distinguishes the components of the material. While there are numerous methods for obtaining a spectrum, the most common application of a spectrophotometer is for the measurement of light absorption.

Absorption

The degree of absorption of electromagnetic waves exhibited by the components in the sample is a distinctive feature that can be used to reveal the composition of the sample. Absorption is related to the inverse of transmittance by:

$$A = log\left(\frac{1}{T}\right)$$

A = Absorbance (Absorbance unit, AU)

T = Transmittance (as percentage)

Spectrophotometers based on absorption have a radiation source that will emit a specified range of EM waves to be absorbed by the sample by measuring the intensity of light that passes through the sample. A useful relationship used to quantitatively determine the concentration of an absorbing species in solution is the *Beer-Lambert Law*:

$$A = -log\left(\frac{I}{I_0}\right) = \epsilon * c * L$$

I = Intensity of incident light at a particular wavelength

I_0 = Intensity of transmitted light

epsilon = molar absorptivity or extinction coefficient (1 / M * cm)

c = concentration of the absorbing species

L = path length through the sample

According to the *Beer-Lambert Law*, the absorbance of a solution is directly proportional to its concentration. The constant *epsilon* is a fundamental molecular property for a given solvent. It can be found from references tables or determined through the construction of a calibration curve.

The *Beer-Lambert Law* is useful for characterizing many compounds, but does not hold for all species. For example, very large, complex molecules such as organic dyes often exhibit a 2nd order polynomial relationship between absorbance and concentration.

Two widely-used devices that measure the absorbance of samples are *Ultraviolet* and *Infrared* spectrophotometers.

Ultraviolet (UV) Spectrophotometers

UV spectrophotometry is a useful method for detecting and quantifying substances in solution. Certain compounds in the gaseous or liquid state absorb ultraviolet radiation, resulting in absorption peaks that can be correlated with specific functional groups.

These species include aromatic and carbonyl compounds in organic substances and transition metal ions in solution. This property is useful for categorizing and measuring the composition of these types of samples, however this property cannot be used as a specific test for a given compound.

The nature of the solvent, the pH of the solution, temperature, high electrolyte concentrations, and the presence of interfering substances can influence the absorption spectra of compounds, as can variations in slit width in the spectrophotometer. Many organic solvents have significant UV absorption and therefore will add peaks to the species dissolved in them.

Solvent polarity, high electrolyte concentration, and pH can effect the absorption spectrum of an organic compound by causing protonation or deprotonation of the compound. Temperature can also cause complex molecules to change in conformation and can also alter the properties of the solvent used. Variations in the slit width will alter the effective bandwidth, distorting wavelength measurements.

The basic components of the overall analyzer are the UV light source, sample holder, optical filters, and detector (typically a phototube). Optical filters (typically a monochromator or diffraction grating) separate the wavelengths of light going through the sample and ensure that only the desired wavelengths reach the detector.

Most UV analyzers are dispersive photometers that also function in the visible light spectrum and are called UV-Vis spectrophotometers. Since these instruments are generally found in laboratories, they are generally used for offline analysis. However, there are inline UV machines and sensors available for real-time inline measurements of process flow streams. UV spectrophotometers used throughout many industries for various applications.

Infrared (IR) Spectrophotometers

Infrared spectroscopy is based on the fact that there are specific frequencies at which molecular bonds rotate or vibrate when exposed to discrete energy levels of radiation. A beam of infrared light is passed through the sample, either one wavelength at a time with a monochromatic beam or using a Fourier transform instrument that measures all wavelengths at once. Inspection of the transmitted light reveals how much energy was absorbed at each wavelength, this data is then plotted.

This plot allows the frequency of the vibrations to be associated with a particular bond type. The characteristic IR absorption spectra obtained for a sample can be used to identify the compound, while the amplitude of absorbance is proportional to the analyte concentration. IR analyzers are used to analyze gaseous or liquid compounds that can absorb IR waves. The system components are generally similar to those of an UV spectrophotometer, with the exception of the IR light source.

IR instruments have been used regularly in a fixed laboratory setting for decades now. Both dispersive and non-dispersive IR techniques exist. More recently, manufacturers of instruments have significantly reduced their size and power requirements. In addition to inline IR sensors for process streams, portable instrumentation is available for field analysis such as the monitoring of stack gases.

Another relatively modern technology, the Fourier Transform Infrared (FTIR) Spectrometer, is widely used in industry today because of its high sensitivity and broad range of application. For example, FTIR analysis is used in the quantification of harmful organic peroxides in lubricants and fuels.

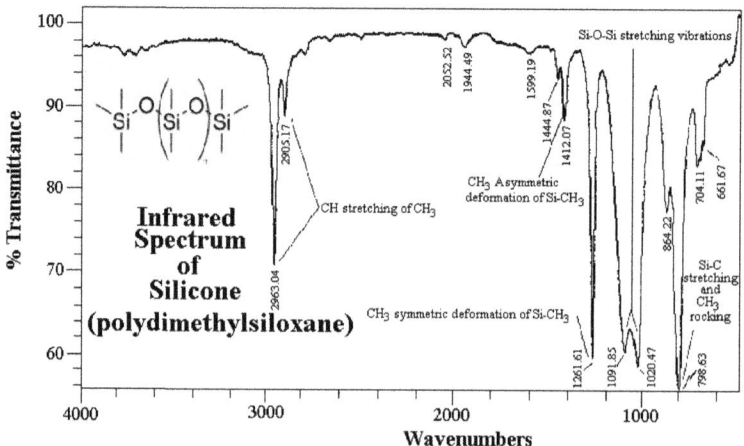

Radiation

Instead of measuring the absorption of light, many spectrophotometers measure the radiation of EM waves. Samples may radiate EM waves in many forms such as heat, luminescence, fluorescence, and phosphorescence. The measurement of such radiation is often employed in biological and pharmaceutical areas because of its high sensitivity and specificity.

For example, fluorescence spectrophotometers are used to assay the presence and concentration of target biological compounds. Fluorescent probe molecules are introduced to the sample that will attach selectively to the biological molecule of interest and the device will then detect and measure the amount of fluorescent radiation. In these experiments, the appropriate standards must be measured to properly calibrate the radiation measured with sample concentration. Additionally, contamination of the sample can affect the radiation patterns.

Examples of other techniques utilizing the radiation EM waves include flame emission spectroscopy, particle-induced X-ray emission, and X-ray fluorescence. These methods are strictly used for laboratory analysis.

Photometry Using Visible Light

The manner in which certain materials interact with visible light can be used to analyze the composition of the sample material. Refractometers, turbidimeters, and opacity meters are all analyzers which utilize this principle.

Refractometry

The theory behind refractometry is that visible light bends (or refracts) when it passes two opaque media of differing densities. Snell's Law relates the angle of incidence α to the refractive index η by:

$$\frac{n_1}{n_2} = \frac{\alpha_2}{\alpha_1}$$

α_1, α_2 = angles of incidence

η_1, η_2 = refractive indices of the two media (unitless)

Alternatively, the critical angle may be used to analyze the composition. The critical angle is defined as the angle of incidence that produces a 90° refraction with respect to normal of the interface *i.e.* $\alpha_2 = 0°$. Thus, Snell's Law becomes:

$$\frac{n_1}{n_2} = \frac{1}{\sin \alpha_c}$$

α_c = critical angle

Please refer to the diagram below.

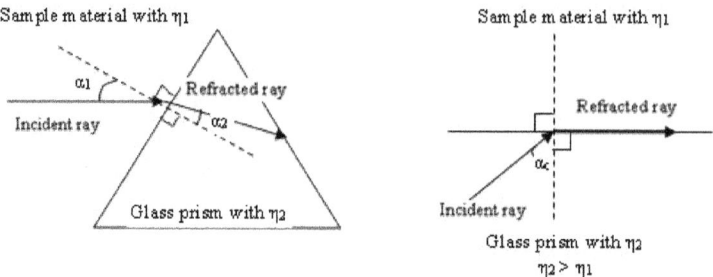

Fig. : Light refraction at a glass-medium interface

Different materials will have different refractive indices; the experimental refractive index measured can be compared to a list of refractive index values at a data bank. Continuous process or online refractometers generally measure the changes in the critical angle at a glass-sample interface caused by the variations in composition concentrations. Refractometers are used in numerous industries such as food, chemical, metalworking industry because of its high accuracy and repeatability, low maintenance and easy installation. An example is the use of a refractometer to monitor the amount of juice concentrate in juice.

Turbidimeters

Turbidimeters measure the turbidity, or sample clarity, of fluids. A light beam passes through the analyzer and if the water, or other chemical that is to be tested, is "pure," then the light would travel undisturbed. However if the sample contains suspended solids or particles, the light would interact with the particles causing the particles to absorb the light and reflect it into different directions. An example is smoke in air that gives the air a hazy look.

Most modern turbidimeters are nephelometers. Nephelometers or nephelometricturbidimeters measure the *scattered* light reflected off the suspended particles in the sample. This instrument contains a light source, some sort of container to hold the sample, and photodetectors to sense the scattered light. A tungsten filament lamp is most commonly used as a light source. If the suspended particles, typically measured in nanometers, are smaller than the wavelength than the incident light, the reflected light is scattered equally in all directions. If the particles are larger than the wavelength of the incident light, the pattern of scattering will be focused away from the incident light.

Nephelometers are usually used when solids are present in the sample in small concentrations. When there is high turbidity in a sample, multiple scattering occurs and the turbidimeter may no longer be able to analyze the fluid. Disadvantages of nephelometers include being prone to fouling and incapability of analyzing fluids that have high concentrations of solid particles.

Turbidimeters are important to industry, specifically for chemical plants, oil refineries, and waste water treatment plants. For example, nephelometers are used in detecting the pollutants in water in water quality monitoring.

Opacity Monitors

Opacity monitors measure the *attenuation* of light due to *scattering and absorption* by the sample. Attenuation is defined as the decrease in intensity and amplitude of the measured light beam when compared to the emitted light beam. Opacity monitors differ from turbidimeters in that they analyze samples by measuring the percentage of transmission of the light source that passes through the sample and NOT the scattering of the light source. A common application is the measurement of smoke density in chimney stacks. The density of the pollutant or particle is expressed as percent opacity, percent transmittance, or optical density.

$$\% \text{ opacity} = 100 \times \text{opacity} = 100 - \% \text{ transmittance}$$

ELECTROMETRIC ANALYSIS

Electrometric analysis uses the principles of electrochemistry to analyze the composition of different samples.

Conductivity Cells

Conductivity cells measure a liquid's ability to conduct an electric current. Since liquids conduct current due to the presence of ions, the composition can be determined by measuring the concentration of the ions present. Conductivity cells determine concentration from the definition of molar conductivity (lambda). Molar conductivity is a property that can be influenced by temperature. It also varies with concentration if the chemical being measured is a weak electrolyte. It is defined by the following equation:

$$\Lambda = \frac{\kappa}{C}$$

Λ = molar conductivity [Sm^2/kmol]

κ = conductivity [S]

C = concentration [kmol/m^3]

Since it is difficult to directly measure conductivity, conductivity cells measure the resistance (R) of a sample liquid between two electrodes when a small current is applied. Resistance is related to conductivity by the following equation:

$$\kappa = \left(\frac{A}{x}\right)\left(\frac{1}{R}\right)$$

κ = conductivity [S]

A = area of electrode [m^2]

x = distance between electrodes [m]

R = resistance [ohm, Ω]

Note that the electrode area, A, and distance between the two electrodes, x, are fixed for a particular sensor and are often expressed as a cell constant (k_c).

The following flow chart shows how composition is determined in terms of concentration:

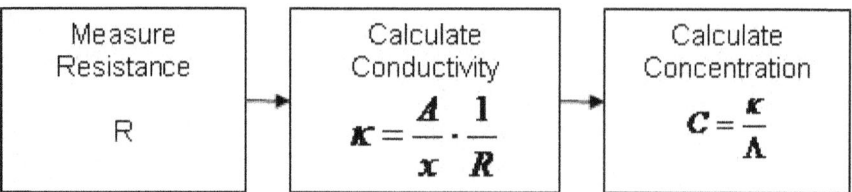

Conductivity cells can be used in flow system and installed directly into the system of pipelines. Conductivity cells can come in three types: two electrode probe, multiple electrode probe, or a non-contact, electrode-less probe (toroidal). The two electrode probe is one of the most common types and typically uses carbon or stainless steel as the electrodes. The multiple electrode probed has either two, four, or six electrodes.

This probe has the advantage of reducing errors due to polarization (electrochemical reaction on the electrode surface) and fouling. Finally, the toroidal probe has no electrodes and instead measures the AC current in a closed loop of solution using two embedded toroids(a magnetic core with wire wrapped around it to make an inductor).

The electrode-less probe is good for solutions that tend to have problems with polarization or oily fouling. In-line conductivity cells are suitable for temperatures up to 480K and pressures up to 1700kN/m^2.

Since concentration is found from molar conductivity, which is a property of the fluid that is strongly dependent on temperature, a temperature sensor is often paired with a conductivity cell so temperature effects can be accounted for in concentration calculations.

Conductance cells are best used in applications where a single type of ion is present in a dilute solution. Higher concentrations lead to complications that cannot be described by the simplistic linear relation between concentration and conductivity, since molar conductivity starts to vary with higher concentrations of weak electrolytes.

Ion Selective Electrodes

pH Electrode

The pH electrode is commonly used in both the lab and industrial setting. This electrode is sensitive to the concentration of the H^+ ions. The pH electrode basically measures the potential difference between a reference solution and the sample.

Other Ion Selective Electrodes

Electrodes can also be specific to other ions such as: Na^+, K^+, Ag^+, ClO_4, BF_4, SO_2, and CO_2. An electrode's ion sensitivity depends mainly on the method of separating ions in the solution. The biggest problem with ion selective electrodes is the problem of not enough selectivity for the desired ion, and subsequent interference by other ions with similar chemical properties to the desired ion. Most electrodes use a membrane to separate a specific ion, though other techniques exist.

Selective ion electrodes can be divided into the following three categories based on the mode of separation: solid membrane (glass or crystalline), liquid membrane (classical ion exchanger with a neutral or charged carrier), and membrane in a special electrode (gas sensing or enzyme electrode). Examples of ion selective electrodes used in industry include nitrate sensors for sewage treatment or river quality monitoring, hydrazine and sodium sensors for in boiler water, and choride and fluorine sensors for in water treatment plants.

Oxidation-Reduction Potential (ORP) Sensors

ORP sensors, or also called Redox sensors, can determine concentration of ions that exist in two different oxidation states in solution. For example, this type of electrode can be used to analyze the composition of a solution with Fe^{2+} and Fe^{3+} ions. This type of electrode measures the potential caused by a solution's reduction or oxidation ability.

This type of sensor directly measures electric potential which indicates the reducing or oxidizing strength of a solution. The reducing or oxidizing power of a solution depends on whether the higher or lower oxidation state is preferred by the ion.

When an ion favors the higher oxidation state (or has reducing properties), the ions tend to give up electrons and the electrode becomes more negatively charged.

When an ion favors the lower oxidation state (or has oxidizing properties), the ions tend to gain electrons and the electrode becomes more positively charged.

When purchasing an ORP sensor from a vendor, many models come already packaged with a pH sensor, since these two measurements are often both monitored to keep a process running at the right conditions. For instance, in the pool and spa industry, its important to monitor both the concentration of sanitizers in the water with an ORP sensor, as well as measure the pH of the water.

Polarographic Sensors

Polagraphic sensors work by performing electrolysis with two types of electrodes: one which is polarizable and one which is not. They operate very similarly to voltameters, but with voltameters the potential is changed in a controlled manner and the voltage measured. The oldest system of polarographic measurement used dropping mercury electrodes.

Here the principle is that that the surface tension of mercury changes with how much charge is applied between the top and bottom of the electrode, and this surface tension change can be seen in how the weight of the mercury drops varies with time.

More modern types of polarography use two different types of metals which are connected in a solution with electrolytes, and electrons will move towards the more positively charged metal. This movement of electrons induces a current, which eventually stops once the charges are balanced (polarized). If all the charges are balanced within the system, no current will flow. Therefore a small potential is applied to a polarographic sensor so that a current exists for the sensor to measure.

The sample being analyzed needs to have the ability to depolarize the system, or tip the balance of charges. The ion of interest crosses a selective membrane and reacts by an oxidation/reduction reaction with a metal surface (often called the measuring electrode). The reaction between the ion and electrode causes an uneven distribution of charges. Once a system is depolarized, the potential of the system increases since a more positive and a more negative region exists separately.

The increase in potential causes current to flow which is the signal measured by the sensor. This creates a graph showing the relationship between applied current versus measured potential. A disadvantage to this type of measurement tool is that this graph requires a calibration curve in order to interpret the data, and so analysis of composition using polarography can't be done in-line. The advantage of this type of method is that it is highly accurate and very reproducible.

Below is a basic schematic of a polarographic sensor.

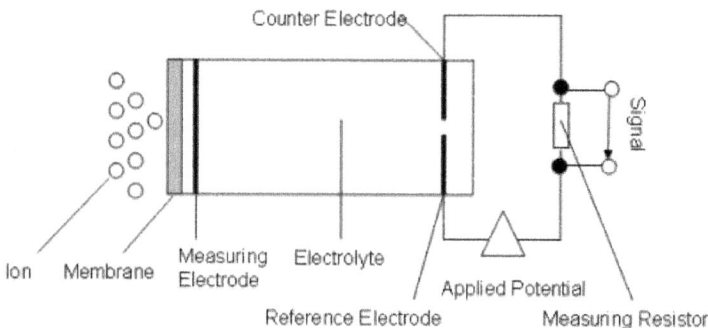

Polagraphic sensors are most commonly used assessing pollutants in air such as Cl_2, HCl, HBr and HF. It an also be used to measure O_2 and other inert gases in flue.

High Temperature Ceramic Sensors

These types of sensors have a heated section of zirconium oxide (ZrO_2), also known as zirconia, that is slightly doped with yttria (Y_2O_3). This stable lattice structure is able to measure oxygen ion conduction at high temperatures. This sensor is exclusively used for measuring O_2 and is often used to measure O_2 composition in flue gases.

CHROMATOGRAPHY

Chromatography is a set of physical methods used to separate and thus analyze complex mixtures. Chromatography consists of a mobile phase, usually a gas or a liquid, and a stationary phase, typically a solid or a liquid. These two phases are combined in a column of the stationary phase based on the type of chromatography one wishes to complete (*i.e.* Gas-liquid, Gas-solid, Liquid) in order to separate the mixture.

The retention time of a component is the time before the component elutes from the column. The retention time and elution order are both based on the relative interaction between each solute and each phase. The stronger the solute's interaction with the mobile phase, relative to the stationary phase, the sooner the solute will elute from the column. Solute-phase interaction is a function of the charges of the molecules, the absorption of the stationary phase, and the relative solubility.

Chromatography by itself does not detect substances, but it is routinely coupled with a detection unit. Each substance is detected as leaves the chromatograph. The retention time for each substance can then be compared to a standard to determine what substances were in the mixture. For unknown substances, the elution order can determine some properties of the substance. For example, a substance that initially elutes after ethanol in a nonpolar column is less polar than ethanol. However, other composition detectors are far better suited for analysis of unknown substances.

Chromatography is the most widely used composition detecting process for on-line analysis of samples within a complex mixture. Due to its precision, chromatography is commonly used to separate or purify samples within the chemical or biochemical industries. Depending on the type of chromatography used, even delicate mixtures like proteins can be analyzed.

However, chromatography is primarily useful when there are some expectations for the components in a mixture; chromatography is not best for a completely unknown sample, but it is useful for detecting the composition of a process stream where most components are known and a suitable standard exists.

Types of Chromatography

Gas-Liquid Chromatography (GLC)

GLC is one of the more common types of chromatography. It is commonly referred to as GC (gas chromatography). In a typical run, the liquid mixture to be analyzed is added to the system and vapourized before entering the column. A carrier gas, often Helium, carries the vapourised sample through a column. The column can either be packed with stationary phase, or the column can have a very small diameter in which case it is lined with stationary phase (capillary tube).

The gas mobile phase runs through the stationary phase within the column at various rates determined by relative volatility and affinity to the stationary phase. The longer a sample molecule spends in the gas phase, the faster it will elute from the column. For this reason, temperature and the chemical identity of the coating on the stationary phase are two important variables which may need to be manipulated to receive clear separation of the components in the mixture.

In general, lower temperature will result in better separation but longer elution times and sample spreading due to flow effects. Efficient use of this apparatus requires a balance between these competing effects.

GC is relatively simple for a technician to use; a calibrated gas chromatograph can run several samples without a large amount of sample preparation. Gas chromatography works well on substances that vapourize below 300°C and are free of ions; however, it does not work well on substances that decompose below 300°C.

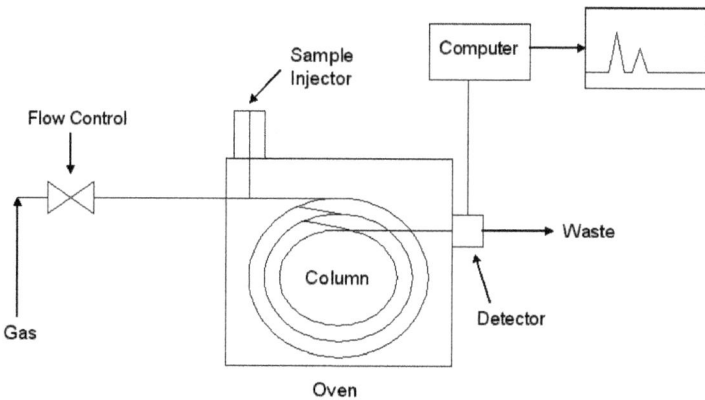

Fig. : Diagram of Gas Chromatography.

Gas-solid Chromatography (GSC)

GSC uses a gas mobile phase which runs through an absorbent solid (*e.g.* silica gel or alumina) stationary phase within the glass column as determined by absorption. This method operates along the same principles as gas-liquid chromatography. It is most commonly used with gases (as opposed to a vapourized substance) and with low molecular weight hydrocarbons.

Liquid-Chromatography (LC)

LC analyzes low volatility liquids within a solid stationary phase. Separation occurs from adsorption/desorption due to the solutes having different degrees of attraction to the stationary phase, typically due to the stationary phase having different polarity than the mobile phase. It is common to use a mixture of solvents to accurately separate mixtures. Liquid chromatography is used in several industries, including the pharmaceutical and petrochemical industries.

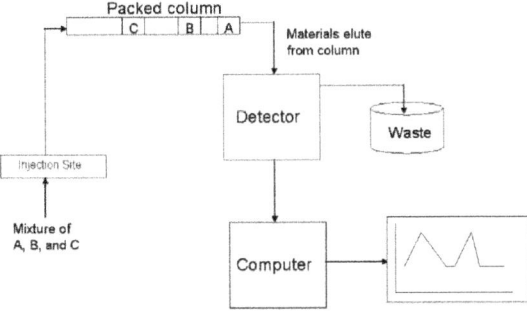

Fig. : Example of Liquid Chromatography.

Mass Spectrometry

Mass spectrometry is among the most precise compositional sensing tools, and, accordingly, one of the most expensive. The basic idea is illustrated below:

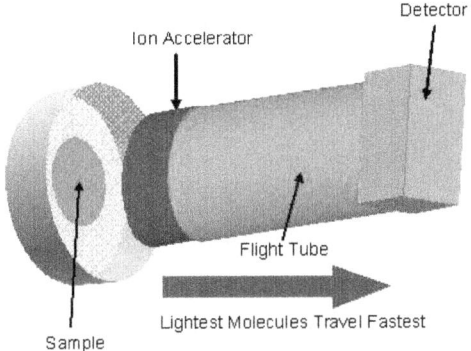

A sample is ionized – this may be performed by a number of techniques – the ions are subjected to an electrical force and accelerated through a tube. Because the electrical force applied to each molecule is the same, a molecules acceleration through the tube depends on its mass (F=ma). The acceleration is also dependent on any charge on the molecule due to magnetic attractive or repulsive forces. At the end of the tube, a detector calculates the time of flight for each of the molecules.

Mass spectrometry is essentially a tool to create a spectrum of distinct mass/charge ratios. It is very often used after chromatography separation techniques to serve as a molecule identification technique.

MS Components

There are 3 fundamental components for mass spectrometry – an ion source, a mass analyzer, and a detector.

The ion source is the component responsible for ionizing the sample. There are many different methods including chemical, electrical, or laser ionization. Selection of the proper ion source depends on the characteristics of the sample (phase, biologically active, *etc.*).

The mass analyzer is the technique and tool used to differentiate the ions in a spectrum based on the mass/charge ratio. The technique described earlier is the Time-of-Flight technique. Another common method is a sector field mass analyzer where the molecules are accelerated through a tube and a magnetic force is applied perpendicular to the direction of flight of the molecules. The molecules will deflect at different magnitudes depending on their size.

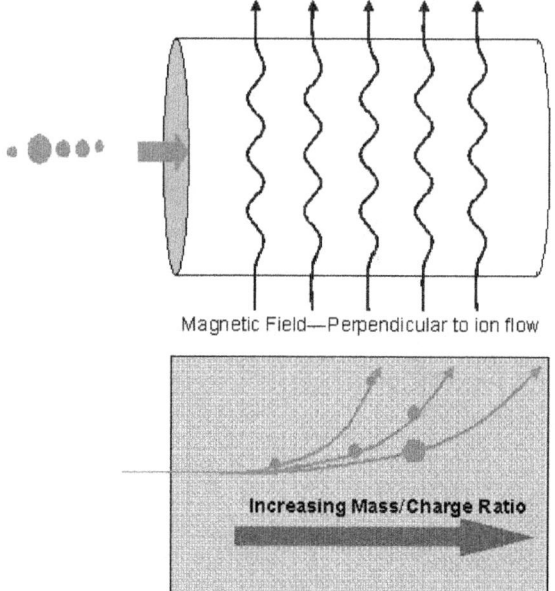

The detector generally amplifies changes in an electrical current that occur when molecules hit the detector.

Thermal Conductivity for Gases

Changes in thermal conductivity may be used to detect variations in a mixture of gases. When a heat source is present, gases and vapour will conduct heat. The change in heat between the source and gases will stabilize to a common temperature. The temperature is mainly dependent on the thermal conductivity and therefore the composition of the gas.

Typical equipment for thermal conductivity analyzers includes: a reference cell, a sample cell, a combined heat source (wire filaments or thermistors), and detector. The sample cell and reference cell are usually placed in a holder where the detector may be mounted. The reference is an identical cell of the sample cell, through which a known gas will flow.

The reference-detector resistance will be constant and the sample-detector resistance will vary depending on the composition. Resistance is a function of temperature and the output from the detector bridge will be a function of sample composition.

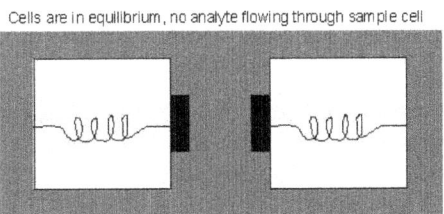
Cells are in equilibrium, no analyte flowing through sample cell
Reference Cell Sample Cell

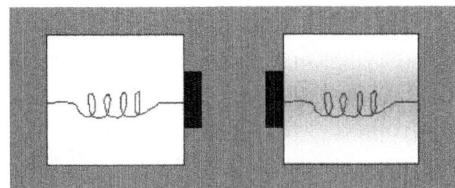
Analyte flowing through sample cell. The change in thermal conductivity from wire filament to surroundings results in an altered resistance. Comparison of this resistance with the reference cell serves as the composition identification mechanism

Thermal conductivity, like mass spectrometry, is often used in combination with gas chromatography

Physical Property Measurements

Physical properties can be used to indirectly measure the composition, but this technique is not as versatile as those described above, since it only applies to a limited number of circumstances. The advantage to using physical properties to indirectly measure composition is that these types of measures are often cheaper than the types of sensors described above.

Density and Specific Gravity

Density and specific gravity measurements require a binary mixture with at least one fluid phase to make accurate measurements. Given the temperature and pressure, the density of a gas can be found using the ideal gas law or some other relationship, and since density is a function of composition, the composition can be measured indirectly. Specific gravity is simply the ratio of density of water over the density of the non-compressible component at the same physical conditions, which means that it is simply a dimensionless measurement of density.

Four different measuring devices for finding density are described below.

Liquid Column

In a liquid column, the pressure measurements in a column are used to determine the density. The column may be open to the atmosphere, where the gauge pressure at the bottom of the fixed-height column is measured. Or, the column may be closed, in which case the differential pressure measurement is made between the bottom of the fixed-height column and at the top of the column immediately below the liquid surface.

Displacement – Hydrometer

A hydrometer is usually made of glass and consists of a cylindrical stem and a bulb weighted with mercury or lead shot to make it float upright. The liquid to be tested is poured into a tall jar, and the hydrometer is gently lowered into the liquid until it floats freely. The point at which the surface of the liquid touches the stem of the hydrometer is noted. Hydrometers usually contain a paper scale inside the stem, so that the specific gravity can be read directly.

The operation of the hydrometer is based on the Archimedes principle that a solid suspended in a liquid will be buoyed up by a force equal to the weight of the liquid displaced. Thus, the lower the density of the substance, the lower the hydrometer will sink.

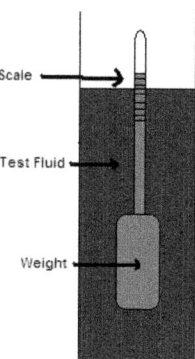

Direct Mass Measurement

With this technique, fluid is continuously passing through a U-shaped tube, where the tube is oscillating at its natural frequency. An electrochemical device strikes the tube periodically at the curved portion of the tube. The change in frequency between the electrochemical strike and the natural frequency is proportional to the changes in density which correlates to the composition of the substance.

Radiation-Density Gauges

Gamma radiation inside a pipe/vessel is used to determine the density of the liquid. Below is a diagram of the setup.

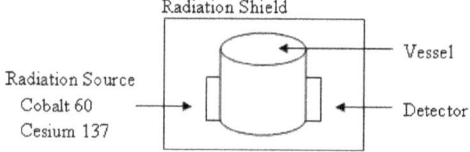

If the path-length for the radiation source is under 610 mm then cesium 137 is used. Above 610mm, cobalt 60 is used. The detector is usually an ionization gauge. The density is a function of the absorption of the radiation being detected. Decay of the vessel must be taken into consideration when calibrating

Comprehensive Charts

For simple, easy to read charts summarizing the advantages and disadvantages of each type of sensor.

Table. Photometric Analysis.

Sensor	Advantages	Disadvantages	Example
UV spectrophotometers	+ Available in-line or off-line + Reliable	- Must be in solution (gas or liquid)	Presence of aromatics in organic compounds
IR	+ Available in-line or off-line + Reliable	- Must be in solution (gas or liquid)	Quantifying organic peroxides in lubricants using FTIR
Radiation	+ Highly selective + Very specific	- Lab use only - Too specific	Fluoresence spectrophometers to detect specific proteins in cells
Turbidimeters	+ Ideal for solids in small concentration	- Prone to foul - Solids must be present	Detect pollutants in water (water quality)
Opacity	+ More accurate than turbidimeters	- Lack stability	Measures smoke density in chimney stacks

Table. Electroanalytical Measuringtechniques.

Sensor	Advantages	Disadvantages	Example
Conductivity Cells	+ Can be installed in-line	- Doesn't work with higher concentrations - Not suitable for solutions with more than one ion	Measure electrolytes in de-ionized water
pH	+ Highly accurate if properly calibrated	- Can't distinguish between multiple bases or acids	Monitor pH of pool water
Ion Specific	+ Highly selective	- Only good for one specific ion - Can have interference from other ions	Measure fluorine in water treatment plants
ORP	+ Highly selective	- Only works for multiple oxidation state ions	Measure levels of Fe_{+2} and Fe_{+3}
Polarographic	+ Highly accurate + Reproducible	- Can't be done in-line - Requires a calibration curve	Measure O_2 in flue gases

Table. Other types of Measuring Techniques

Sensor	Advantages	Disadvantages	Example
Chromatography	+ Precise	- Off-line typically (lag time) - Expensive	Separate Proteins
Mass Spectrometry	+ Accurate	- Expensive - Off-line	Detect iron content in cereals
Thermal Conductivity	+ None	- Complicated calculations - Low accuracy due to so many variables	Detect polymers

For complete summary charts including more sensors not discussed in the article, please follow the links here for a PDF version of these summary charts.

Table. Physical Property Measurements.

	Density & Specific Gravity					Thermal Conductivity
Phase	Binary Mix (L &/or g) OR Sol'n (s or g) in solvent					Gases
Relationship	r = f(composition) @ T & P S.G. = r/ref(r) @ T & P					R = 6.1)nf(s.m(composition)
Equipment	Liquid Column	Pressure	Displacement (Hydrometer) Degree of immersion	Direct Mass Measurement Frequency of fluid	Radiation-Density Gauges Gamma radiation inside pipe/vessel	Thermal Conductivity Analyzer Resistance
Operating Conditions	*Column open to atmosphere	*Closed column	*Constant weight, variable immersion device	*Fluid continuously passing through U-shaped tube	*Detector & source mounted on opposite sides of pipe	Detector mounted to reference cell and sample cell to determine composition
Comments	*Measure gauge pressure at base	*Differential pressure measurement btwn bottom of column & vapor over column	*Degree of immersion is a measure of density when weight of hydrometer equals weight of displaced liquid	*Tube is at natural frequency *Electrochemical dive strikes periodically at curved portion Δf α Δr	*Cesium-137 source for path lengths under 610mm *Cobalt-60 source used above 610mm *Absorption = f(r)	
Examples	*Manometers to measure pressure or fluid flow		*Alcoholometer determines alcohol strength *Saccharometer measures amount of sugar in solution	*Guitar		*Not ideal for composition detection *Polymers
Advantages	*Easy to set up *Low cost *Able to detect pressure down to 1 militorr		*Accessible operation		*Precise (NMR)	*No inline advantages
Disadvantages	*Not very durable *Portable for rough field applications		*Simple solutions	*Relies on change *Not as precise	*Safety	*Complicated calculations *Low accuracy - calculations depend on many variables (structure and temperature)

Table. Chemical Composition Analyzers.

	Chromatography			Spectrophotometers		Photometry		Visible Light Methods		Spectrometry
Methods	Liquid	Gas-Liquid	Gas-Solid	Radiation	Absorption	Refractometry	Turbidimeters		Opacity Monitors	Mass
Working Principle	Separation occurs in the column based on varying affinities of components to bind to column packing. A detector measures some physical property that relates to concentration			*Heat, luminescence, fluorescence *Monitors wavelengths. EM spectrum. The degree of absorption/radiation reveals molecular structure & concentration of sample	*Ultraviolet *Infrared	*Measures the refractive index using Shell's law	*Measures turbidity in sample clarity		*Measures attenuation of light *absorption or scattering of light)	Detects molecular weight of composition in mixture
Comments	*Widely used to separate and measure volatile compounds			*Often used in biological or pharmaceutical industries	*Gas/liquid *Nondispersive	*For optically transparent materials only	*Fluids with suspended solid particles		*Sample must transmit light	*Mass to charge ratio *Must liquify sample
Examples	*Chemical or biochemical processes *Separate proteins			*Fluorescence spectrophotometers to detect specific proteins in cells	*Presence of aromatics in organic compounds	*Quantifying organic pesticides in lubricants using FT-IR analysis	*Monitoring juice concentrate in orange juice		*Detect pollutants in water (turbidity check)	Detecting iron content in cereal
Advantages	*Precise			*High sensitivity *Very specific	*Reliable *Large data bank for reference	*Relatively simple set-up *Simple calculations	*Ideal for small concentration of solids in fluids		*Better than turbidimeters, accounts for scattering and absorption of light	*Accurate
Disadvantages	*Offline *Lag time *Expensive			*Too specific	*Applicable to certain EM range, gas/liquid compounds *Durability/mobility	*Prone to fouling	*Fouling *Stability			*Expensive *Offline

Table. Electroanalytical Instruments.

	Electrodes				Conductivity Cells
Methods	pH	Specific Ion	Redox	Polarographic	
Working Principle	*Measures concentration of H^+ ions	*Separates & measures conc. of specific ions	*Measures potential caused by solutions' oxidation/reduction ability	*Measures current caused by addition of depolarizing agent in ionic solution	*Measures a liquids ability to conduct an electric current - compositions analyzed in terms of conc.
Comments	*Uses a reference solution	*Uses a membrane to separate several types	*Applied for ions that have multiple oxidation states	*Samples is depolarizing agent	*Molar conductivity is strongly dependent on temperature (temp sensors used)
Examples	*Using pH electrodes to measure extent of hydrolysis rxn of acetic anhydride	*Calcium electrodes used in blood-gas analysis	*Analyzing composition & solution with Fe^{2+} & Fe^{3+} ions	*Measure O_2 in flue gases	*Measuring concentration of electrolyes (ie. Dionize H_2O)
Advantages	*High accuracy if calibrated properly	*Highly selective	*Highly selective	*Highly selective	*Can be installed directly into pipelines suitable for a max T of 480K, Pressure of 1700KN/m²
Disadvantages	*If multiple acids or alkalines present, cannot show difference	*Useful for only selected ion	*Limited to ions with multiple oxidation states	*Limited approach	*Not suitable for solutions with multiple type of ions or non-dilute solutions *Multiple species of electrons that compete

Applied Example: Mars Pathfinder Spectrometer

The Mars Pathfinder uses an Alpha Proton X-Ray spectrometer to identify the chemicals that comprise the Martian surface. The incredible tool shoots X-Rays at the surface of the planet and uses a backscatter detector to collect the alpha particles, protons, and X-rays characteristic of specific elements. These atom specific scattering events allow NASA scientists to determine the surface composition.

Chapter 5

PH AND VISCOSITY CONTROL SENSORS

INTRODUCTION TO PH

pH is a measure of concentration of H^+ ions ($[H^+]$) in solution, and can be defined as follows:

$$pH = -\log([H^+])$$

Typically, pH values range between 0(strongly acidic) and 14(strongly basic, or alkaline). However, that is not to say that 0 and 14 are the limits of pH values. A solution that has a H^+ concentration of 10 M, such as very strong HCl, would have a pH of -1. Pure water, which has a pH of 7, is said to be neutral because its H^+ concentration is equal to its OH^- concentration. An acid solution is said to have a greater H^+ concentration than its OH^- concentration. Conversely, a base, or alkaline solution, is said to have a greater OH^- concentration than its H^+ concentration. See for reference:

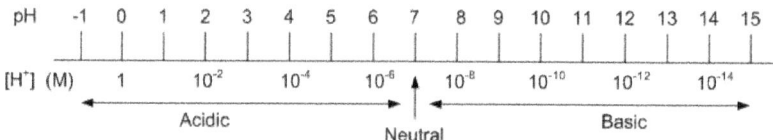

For weak acids and bases, the above equation is appropriate to use when calculating pH. However, many solutions are not weak acids and bases, which means that the other ions in solution play a large role in acidity and alkalinity, and thus a more accurate method is needed to calculate the pH. pH sensors are used to determine the acidity or alkalinity of a particular process solution. When dealing with a strong solutions, the sensors measure the effective concentration, or activity, of H^+ ions in solution, and do not measure the actual concentration. Hence, for the purposes of pH sensors, pH is defined as:

$$pH = -\log(a_{H^+})$$

$$a_{H+} = (\gamma_{H+})\,[H^+]$$

γ_{H+} = activity coefficient

The activity coefficient accounts for the interactions between all the ions present. These interactions may prevent some of the H⁺ ions from contributing to the acidity of the solution, and thus decrease the effective concentration of H⁺ ions. This effective concentration is the concentration used when calculating pH.

Why is pH Relevant for Chemical Engineering?

pH sensors, while being relatively simple pieces of equipment, have numerous uses. Knowing the pH of a solution is valuable to an analytical chemist attempting to determine the contents of an unknown solution, as well as a farmer trying to determine the appropriate applications to his fields and many people in between. In industry pH sensors can be used as a simple way to monitor reactions involving acids or bases as either a reactant or product of a chemical process.

They can also be used to monitor the conditions of a reactor to ensure optimal conditions. This is especially important in biological systems where very minor changes in pH can result in significantly lowered production levels or even a complete stoppage of production due to the death of the organisms on which the reaction is dependent.

An example of pH measurement's importance in industry is seen in food processing during the manufacture of fruit jelly. The gel formation brought about by pectin occurs over a small pH range, which is further complicated by the concentration of sugar. Too high of a pH value results in an unacceptably runny liquid. Too low of a pH causes the mixture to stiffen prematurely, resulting in a nearly-solid product. Continuous pH adjustment using edible acids such as citric acid yields the optimal gel consistency.

It is also important to keep in mind that application requirements should be carefully considered a pH sensor is chosen. In the case above, it is important that the pH sensor be made of a material that will not interfere with the quality of the product. For instance, the sensor should not be made of glass in order to avoid broken glass entering the product.

Accurate pH measurement and the precise control that it allows for helps to optimize a process and can result in increased product quality and consistency. Continuous pH monitoring also controls chemical usage (such as the edible acids above), resulting in minimized system maintenance (*i.e.* less cleaning of hardened gel from containers, less corrosion of containers, *etc.*).

Typical pH Sensor Construction

Because it is difficult to directly measure the H+ concentration of a solution, pH is typically measured by comparing the potential of a solution with a known [H⁺](this is known as a reference solution) with the potential of a solution with an unknown [H⁺]. The potentials of these solutions are measured using a reference half-cell and a sensing half-cell, respectively.

Reference Half-Cell

The reference half-cell generally consists of a chamber with a conductor submerged in a reference electrolyte. The conductor, or external reference electrode wire, is typically silver coated with silver chloride (Ag/AgCl(s)) or mercury coated with mercurous chloride (Hg/Hg$_2$Cl(s)). The reference electrolyte is a standard solution such as KCl.

The last component of the reference half-cell is a porous plug, which serves as a liquid-liquid interface between the standard solution and the process solution being analyzed. This plug allows standard solution to travel from the chamber out into the process solution, but does not allow process solution into the chamber. The purpose of this interface is to establish the electrical connection, which provides the reference potential.

Sensing Half-Cell

The sensing half-cell is of similar construction to the reference half-cell. A glass chamber contains an electrode (Ag/AgCl(s)), known as the internal reference electrode, which is submerged in a standard solution of constant pH. However, instead of a porous plug acting as the liquid-liquid interface between the standard and process solutions, there exists a glass membrane.

This membrane is coated on the inside and outside with a hydrated gel. When the sensing half-cell is placed in solution, metal ions (Na⁺) in the gel on the outside of the membrane diffuse out into the sample being analyzed and are replaced by H⁺ ions from the sample.

The change in energy associated with this substitution creates the change in electric potential that is detected by the electrode. The combination of the hydrated gel and the glass membrane are what make the sensor specific only to H⁺ ions. The solution of constant pH ensures that the surface of the glass membrane and the electrode are at the same pH before the cell is exposed to the process solution being analyzed.

The electrode wires from the reference and sensing half-cells are each connected to the voltage measuring device.

The diagram below shows the setup of a typical pH measurement system:

While this figure shows the essential components of a pH sensor, it is not an accurate depiction of what pH sensors actually look like. The reference and sensing cells in most pH sensors are generally encased together in a sturdy housing (usually made of some sort of plastic). This does not affect the operation of the sensor, but allows for ease of use, as there is only one major piece of equipment that the user needs to carry around.

pH and Viscosity Control Sensors

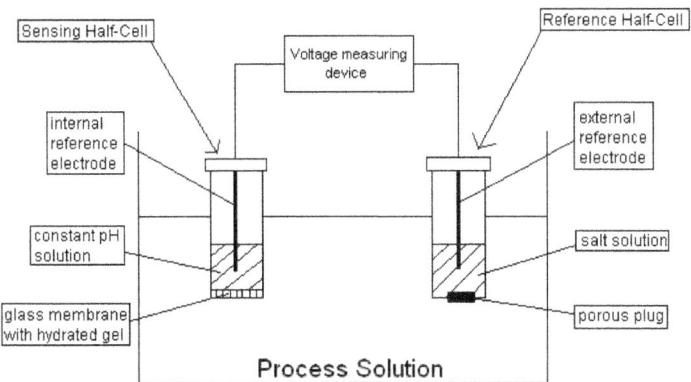

pH sensors can be used both online and offline. If the sensor is being used to carefully control the pH of a system or to provide an accurate record of the pH of a solution it is often desirable to have an online pH sensor. In order to have an online pH sensor determine accurate readings it must be in a position to take a representative sample.

This is most often done by diverting a side stream from the main process in which the pH sensor can be inserted. Often times continuous pH measurements are not necessary, but rather infrequent measurements to make sure that the system is running correctly are of interest. In these cases it is much more efficient to use an offline pH sensor.

How do pH Sensors Calculate pH?

pH is calculated using the potential drop between the reference and sensing electrodes. The potential across the overall pH measurement system can be expressed as:

$$E = E_{internal\,reference} + E_{membrane} + E_{porous\,plug} - E_{external\,reference}$$

The potential of the porous junction is usually negligible, and the potentials of the internal and external references can be lumped into a constant value, E_0. Thus, we get:

$$E = E_0 + E_{membrane}$$

Using the Nernst equation, the potential of the membrane in a half-cell reaction can be expressed as:

$$E = E_0 + \frac{RT}{nF}\ln(a)$$

R = Universal Gas Constant

$F = 9.6490 * 10^7$ *coulombs/kmol* (Faraday's Constant)

n = net number of negative charges transferred in the reaction

a = activity of the ions involved in the reaction

This equation is the standard equation used in electrochemistry to find the potential with respect to a reference electrode. It can be viewed as a calculation of the change in free energy at an electrode, due to changing ion concentrations at that electrode. Assuming that the electrode is completely selective to H^+ ions (a = 1) and that there is only one positive ion carrying one elementary charge in each half-cell reaction (n = 1) the pH of the unknown solution can be expressed as:

$$pH = \frac{F(E_0 - E)}{2.303\ RT}$$

T = temperature of solution being analyzed

Temperature Compensation

As one can see from the Nernst equation, the pH as determined using the potential difference between the reference and sensing half-cells varies linearly with changes in temperature. That is, the voltage drop/pH unit across the cells increases with increasing temperature. Thus, pH measuring system needs to have some way to compensate for temperature changes in the process solution.

Many pH control kits come equipped with automatic temperature compensation (ATC), which continuously measures and incorporates temperature changes into their pH calculations. Other older pH control kits require manual temperature compensation, meaning that the operator or user must manually input the process solution temperature into the control system.

Alternative Methods for Determining pH

While glass membrane pH sensors are an effective method of measuring pH they are not always the best choice. Certain properties of the system of interest make using a glass membrane pH sensor impractical. Glass membrane pH sensors can become clogged by viscous solutions or solutions containing suspended solids. Harsh conditions can also scratch or break the glass membrane. There are also cases where a glass membrane pH sensor would work but using one would be an unnecessary cost.

In these situations there are several alternatives that can be used to determine pH. One method is using pH paper. This paper changes colour when submerged in a solution. The colour of the paper can be compared to a standard scale included with the paper to determine an approximate pH.pH paper is an excellent choice when all that is needed is a "ballpark" pH value.

Another method of determining pH is the use of indicators. These are substances that change colour depending on the pH of the solution. They can be very useful if the primary interest is determining if the pH of the system is above or below a particular point. Numerous indicators exist that change colours at different pH values.

Other Ion-Specific Sensors

The same concepts that allow pH to be measured using a reference and measuring electrode can also be applied to measure the concentration of other ions. The differences between the pH sensors and ion specific sensors are the composition of the measuring electrode the membrane used in the measuring electrode. For some ions of interest glass membranes can still be used, however for other ions solid state membranes or liquid ion exchange may be necessary. The choice of membrane depends on both the size and the charge of the ion of interest.

While ion specific electrodes can be fairly accurate a number of problems can compromise their accuracy. Ion specific electrodes can suffer from interference. This can be especially problematic in complex solutions, especially those containing many other ions similar in charge and size to the ion of interest. The measurements reported by ion specific sensors are also functions of the ionic strength of the solution. This can also result in significantly decreased accuracy of the sensor if the effect of the ionic strength of the solution is not accounted for.

As with pH the property that is actually being measured with an ion specific sensor is the activity of the ion, not its concentration. The activity can no longer be considered negligable in solutions with high ionic strength. In order to obtain proper readings the user must consider what effect the ionic strength of the solution has on the activity coefficient and adjust the measurement accordingly.

Despite these limitations ion specific sensors can still provide accurate and valuable information, provided an appropriate measuring electrode membrane is used and the ionic strength of the solution is accounted for, and are widely used.

Problems with pH Sensors

Several problems can occur with pH sensors that can negatively affect the accuracy of these devices. The majority of these problems involve the reference electrode. One such problem is a partial clogging of the porous plug. Another error associated with the plug is due to a junction potential which can vary depending on the electrode setup.

This junction potential adds error to the pH measurement when the ionic composition of the solution being tested is significantly different than the ionic composition of the standard buffer used to calibrate the electrode.

In general, a junction potential develops at the interface of two solutions with different ionic compositions because the various ions in each solution have different mobilities, and as oppositely charged ions diffuse at different rates a charge separation develops at the interface between the solutions.

A different type of error can occur in measurements of solutions with very low hydrogen ion concentration and very high sodium ion concentration; pH electrodes can "mistake" the sodium for hydrogen, leading to a lower pH reading than the true value.

Other problems include the contamination of the electrolyte in the reference electrode of a complete depletion of the electrolyte. Any of these problems will result in an incorrect potential in the reference electrode and ultimately inaccurate pH measurements.

Problems can also occur with the measuring electrode. Any damage to the glass membrane including cracks, etching, or abrasion, as well as coating of the glass membrane will result in poor pH measurements. In addition, if the electrode is dry it will not measure pH correctly; it must be soaked in some aqueous solution for a few hours.

Problems with pH sensors can be fairly easily detected. Any problems such as difficulty calibrating the sensor, unsteady pH readings, or significant drift in the readings usually indicate damage to the pH sensor. It is now possible to test some forms of pH sensors online to quickly and easily determine which electrode is the cause of the problems and potentially the source of these problems. In some cases these problems can be fixed by a careful cleaning of the sensor, however if problems persist it may be necessary to replace the pH sensor.

INTRODUCTION TO VISCOSITY

Viscosity quantifies the systematic deformation of the surface of a fluid upon an applied shear stress. Fluid, while in motion, typically travels at varying velocities due to the geometry of contact surfaces. It may be characterized either as a Newtonian fluid or a Non-Newtonian fluid. A Newtonian fluid flows continuously in a uniform manner irrespective of the forces acting upon it (a common example, water). Conversely, the viscosity of a non-Newtonian fluid changes upon differing flow rates (common examples, table ketchup or mud).

Also, the effect of temperature on fluid viscosity is imperative. From a thermodynamic perspectice, temperature measures the random movement of molecules of a given substance, and as movement increases internal energy increases, and temperature also increases. For most traditional substances, having higher internal energy/temperature also implies a lower viscosity, as the substance will be less resistant to movement (*i.e.* will deform sooner to an applied shear stress). For example, consider honey: at room temperature it oozes out of the squeeze bottle, but after a minute in the microwave it may be easily poured out, similar to a syrup.

Nearly all modern viscometers have built in controls to address temperature issues, and a competent operator or engineer may input temperature settings to that the appropriate calculations are carried out correctly.

Viscosity may be further subdivided into two distinct forms: dynamic viscosity and kinematic viscosity. Dynamic viscosity (μ), a figure representing shear stress as proportional to the strain rate, has the SI unit Pa*s. Kinematic viscosity, ($v = \mu/\rho$), describes shifts in momentum and has SI units $m^2 s^{-1}$, but is also commonly reprented by the Stoke, $cm^2 s^{-1}$.

pH and Viscosity Control Sensors

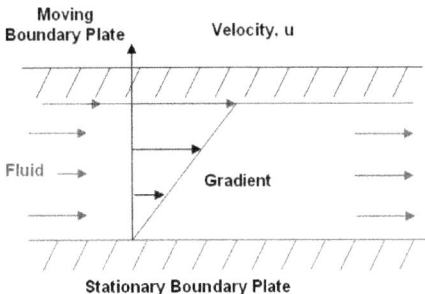

As noticed above, the typical Newtonian fluid experiences friction upon contacting a surface and resultantly develops a characteristic velocity profile. This profile may be described by examining the known properties of the fluid and the surrounding structure through which it travels, commonly piping for plant chemical engineering.

In the above diagram we notice a non-cylindrical surface featuring two boundary plates, one stationary to the observer, and one mobile. Fluid contained within the two boundaries may behave according to the constraining conditions of the boundary plates, and subsequently characterized by fluid mechanics:

$$\tau = \mu * \frac{\partial u}{\partial y}$$

Assuming μ as the coefficient of viscosity, the above equation describes the relationship between the shear (parallel) stress and velocity of fluid travel per unit height above the stationary boundary plate. Fluids having a relatively high viscosity resultantly also require greater force to induce motion—resulting in a larger shear stress.

Why is Viscosity Relevant for Chemical Engineering?

For engineers a thorough knowledge of the physical and chemical properties of products is essential to the successful implementation of any design. Viscosity, similar to volatility, density, or any other physical property, offers significant insight into the potential behaviour of a substance, whether classified formally as a solid, liquid or gas. By taking viscosity into account, engineers may correctly select and place instrumentation in an environmentally friendly, sustainable method.

Many professional assignments necessitate a genuine understanding of viscosity and its effects on process engineering. He speaks at length regarding the cost- effective selection of specific types of blades used in mixing processes.

As an example, a key parameter in food process monitoring is viscosity. The viscosity is directly related to the flow characteristics of the product, which impact pumpability, pourability, and spreadability. In the food-processing environment there are a number of challenges to viscosity measurement, such as harsh process conditions and the complex rheological properties (relationships between deformations and stresses of materials).

Offline measurements are often cumbersome, labor-intensive, and prone to operator error. Online viscometers must be able to deliver continuous measurement day after day with minimal maintenance. Therefore, considerations when choosing viscosity sensors for food processing should include ease of cleaning, minimal risk of fouling, and whether or not they meet sanitary requirements.

Off-line Instruments

For the measurement of viscosity in controlled settings (typically laboratory) where the majority of variables are maintained as constant, engineers and scientists use off-line viscometers. Typical examples include: Capillary, Couette, Falling Ball, Cone and Plate and Oscillating Cylinder. It is very important to keep these instruments in a regulated environment with a stable temperature (such as in a water bath) because of viscosity's sensitivity to changes in temperature.

Capillary

The Ostwald U- Tube viscometer functions by measuing the amount of time a specified quantity of fluid takes to move from its initial position to a final position.

Typically, a suction device holds the bottom of the fluid meniscus at the start position (indicated by the solid horizontal red line), and upon its release the fluid drops an approximate distance of 5 mm (to reach the solid horizontal green line). The time for this drop is recorded so that the acquired data may then be used in Poiseuille's Law to determine the outcome of kinematic viscosity.

Both the dynamic and kinematic viscosities may be found by appling these formulae (where K_0 is the non-dimensionless viscometer constant):

Dynamic Viscosity: $\mu = K_0 \rho t$

Kinematic Viscosity: $v = K_0 t$

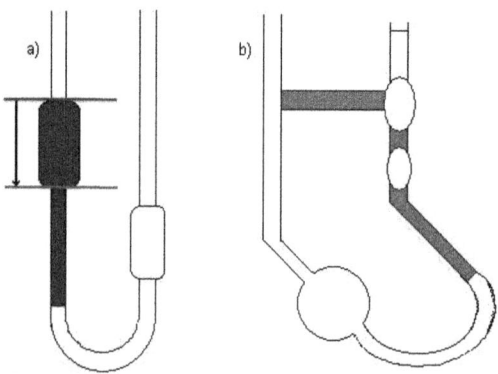

Examples of Capillary Viscometers:

a) Simple Ostwald Type
b) Cannon- Fenske Type

Content for Diagram Adapted from:

Coulson and Richardson's Chemical Enginering, Volume 3, Third Edition, Chemical, Biochemical Reactors, and Process Control; contributing editors: J.F. Richardson & D.G. Peacock,

Couette

This device may be used in both off-line and on-line applications with appropriate modifications in the output settings.

A couette type viscometer measures viscosity by spinning a cylinder encapsulated in fluid. This is accomplished through the synchronization (identical frequency, no phase difference) of a motor with magnetic coupling to rotate a magnet which in turns forces the inner cylinder to revolve within the fluid.

The torque reaction of the motor is resultant of the viscous drag on the rotating cylinder. This torque on the motor is effectively counteracted by the torsion bar (a thin rod connecting the control/ measuring bandto the linear variable displacement transformer).

The deflection of the torsion bar, a function of the fluid viscosity, is then subsequently converted into local signal available for laboratory analysis.

Engineers use viscometers of the couette type in in-line or in-tank applications. Also, the meters are appropriate for both Newtonian and non-Newtonian fluids. The acceptable range of viscosity spans from 10^{-3} to $5 * 10^3 Ns / m^2$.

Information for this device has been adapted from Richardson's Chemical Engineering.

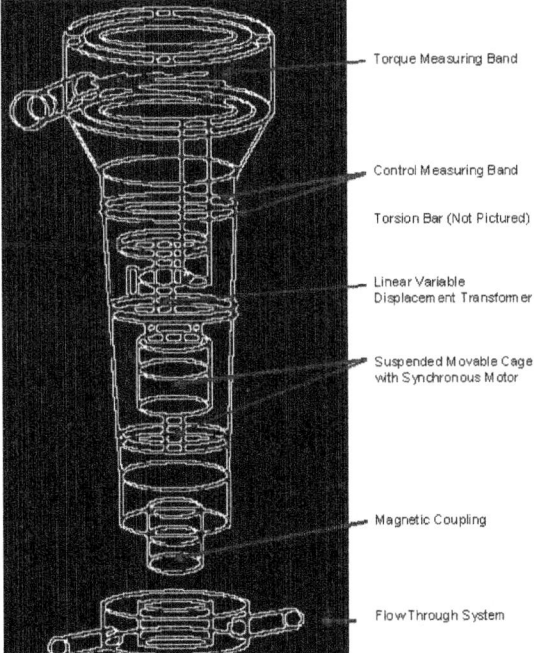

Couette- Type Viscometer for a Continuous Process

Content for Diagram Adapted from:

Coulson and Richardson's Chemical Engineering, Volume 3, Third Edition, Chemical, Biochemical Reactors, and Process Control; contributing editors: J.F. Richardson & D.G. Peacock,

Falling Ball

A substantially more direct approach to viscosity measurement, testers allow a ball to fall through a column containing liquid of unknown viscosity and then measure the amount of time necessary for the ball to reach the bottom of the column.

A commonly used method involves a stainless steel ball which sinks through the liquid under scrutiny; researchers measure the time necessary for the ball to drop from one preselected level to another. To attain the terminal velocity of the ball in the liquid, u_0, timers write the initial timing mark a minimum of six ball diameters below the ball release point.

Also important to note, the ratio of ball diameter to column tube diameter should not exceed 1:10 to account for the drag effects of the column tube walls on the terminal velocity of the falling steel ball.

A simple viscosity derivation from buoyancy principles is presented in Richardson's text, and is shown again here:

$$v = \frac{K_v(\rho_s - \rho)}{\rho}$$

K_v = Dimensional viscometer constant
ρ_s = Density of the ball
ρ = Density of the liquid
t = Time required for the ball to sink the measured distance

Cone and Plate

Another popular method used employs a cone placed in a manner so that the tip touches the center of a stationary plate. Fluid encapsulates the cone and researchers take a measurement of the amount of torque required to keep the plate stationary.

This process allows the dynamic viscosity to be quantitatively described as:

$$v = \frac{3}{\pi \gamma r^3} T$$

T = Torque required to keep the top plate in place
γ = Rate of shear

The rate of shear is the ratio of the angular velocity of the cone to angle between the cone and the plate (note: this angle must be small enough so that the sin of the angle approximately equals the angle itself).

Important Additional Notes:

Edge effects of the cone or plate are neglected. The equation may also be used for Non-Newtonian fluids provided that γ is relatively constant.

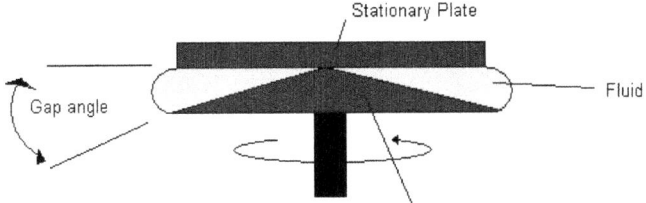

Example of a Cone and Plate Viscometer

Parallel Plates

Parallel plates or discs viscometer is similar to the cone and plate method. The cone in the case is replaced with another plate. In this viscometer, the fluid flows in a gap between two parallel discs. One of the discs rotates with an angular velocity ω which creates the shear. Torque is applied to the other plate so it stays stationary. A normal force, F, is created by fluid elasticity and acts as to separate the two plates. No slip at boundaries is assumed.

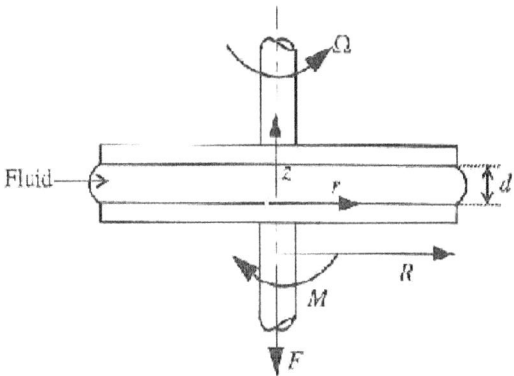

Parallel Plates viscometer

ω, T and F are all experimental parameters. Shear rate is given by:

$$\dot{\gamma} = \frac{wr}{d}$$

$\dot{\gamma}$ = Shear rate
r = The radius of the disc
d = The distance between the two plates

Shear stress, τ, is given by :

$$\tau = \frac{T}{2\pi r^3}\left(3 + \frac{d\ln T}{d\ln \dot{\gamma}}\right)$$

Viscosity is, therefore, the ratio of shear stress to shear rate. Some limitations associated with parallel plates include:

1. Sufficient data of T vs w and F vs w must be available
2. Uniform temperature at different points in the gap between the two plates is required
3. Error might be caused from edge fracture, wall slip, misalignment of the plates and viscous heating.

Oscillating Cylinder

This instrument involves an arrangement requiring a rotating cylinder placed in a quantity of a viscous fluid. The level of disturbance noticed in the fluid resulting from the rotation of the cylinder is then noted, and may be used if the following equation to determine viscosity:

$$v = \frac{(r_2^2 - r_1^2)}{4\pi w_v r_2^2 r_1^2} T'$$

r_2 = Outer cylinder radius

r_1 = Inner cylinder radius

ω_v = Angular velocity

The outer cylinder of radius r_2 rotates with an angular velocity of ω_v while the inner cylinder of r_1 remains stationary thus allowing for the torque (alternatively reffered to as 'viscous drag'), T', to be measured at predetermined values of ω_v.

On-line Instruments

On-line instruments are those that are capable of giving a precise measurement of viscosity under plant conditions. This means that they will be able to withstand the wide range of viscosities and the variable temperature, pressure, and flow rate that occur within a process plant while still obtaining the accuracy that is required. In all on-line instruments automatic temperature compensation is always required because of viscosity's temperature dependence.

Capillary

This type of viscometer is derived from Poiseuille's law:

$$\frac{dv}{dt} = v_s * \pi * r^2$$

v_s = Fluid velocity

r = Internal radius

The instrument has many temperature and pressure sensors in the intricate system because there must be a precisely controlled pressure drop and temperature in the stream. Response time is minimized by adding a fast loop sampling system. A capillary viscometer would be useful when using Newtonian fluids

pH and Viscosity Control Sensors

such as lubricating oils or fuel oils. It can endure viscosities ranging from 2×10^{-3} to $4 Ns/m^2$.

Couette

Identical to the description provided earlier, except now with a change in the output data processing.

The deflection of the torsion bar, a function of the fluid viscosity, is now converted into a 4-20 mA output signal and is readily available for additional plant data mangement/processing.

Vibrating Rod or Cylinder

This type of viscometer can also be used in-line or in-tank. It is suitable for Newtonian, non-Newtonian fluids, and mineral slurries. The vibrating rod or cylinder works by using an electromechanical system as the driving force, or by a pulse of current through a specific alloy.

The lowest viscosity that it can handle is $10^{-4} Ns/m^2$ and it can extend to $2 \times 10^3 Ns/m^2$. Normally on lower viscosity ranges the cylinder is used, while with higher ranges the rod is used. Given a plants' natural vibration care must be taken when deciding on the location of an oscillation viscometer because of their sensitivity.

Table of Viscometer Uses

This table found in Richardson's text offers some guidance on the device selection process.

Common Viscometer Ranges (Content from Richardson's Chemical Engineering)

Viscometer Type	Lowest Viscosity (Ns/m^2)	Highest Viscosity (Ns/m^2)	Shear Rate Range (s^-1)
Offline (Laboratory Instruments)			
Capillary	2*10^-4	100	1 - 1.5*10^4
Falling Ball	1	200	indeterminate
Couette	5*10^-4	4*10^6	10^-2 - 10^4
Cone and Plate	10^-4	10^9	10^-3 - 10^4
Oscillating Cylinder	10^-4	5*10^6	7*10^-3 - 9*10^3
Online Instruments			Applications
Capillary	2*10^-3	4	Newtonian Fluids
Couette	10^-3	5*10^3	In-line or in-tank *
Vibrating Rod or Cylinder	10^-4	2*10^3	In-line or in-tank * #

* Also, Newtonian and Non Newtonian Fluids.
\# May be used also for mineral slurries.

One may notice that the viscosity ranges of some of the viscometers overlap. In a situation where this occurs and it is necessary to choose between multiple viscometers, one needs to take into account process conditions such as changes in temperature, pressure, and solution flow rate. The appropriate viscometer should be able to handle these conditions without disrupting the process.

pH/Viscosity Sensor Key Learnings

pH and viscosity measurements do not have much in common with each other. However, there is a common thread between these two types of sensors that needs to be acknowledged, and that is the method that needs to be taken when deciding what type of sensor to use. While there is only one single type of pH sensor available, before it is to be used the user needs to evaluate whether there is an easier and cheaper way (*e.g.*, using pH paper or a pH activated dye) to accomplish the task at hand. With viscometers, there are multiple types of viscometers to use over various ranges of viscosity. The user needs to evaluate process conditions such as temperature, pressure, and flow rate, in order to decide which type of viscometer is optimal for the application.

The math associated with both types of sensors is good to know for background purposes and to gain a fundamental understanding of how a sensor produces a measurement, however it is not information that greatly helps the user decided what type of pH indicator or viscometer to use.

MISCELLANEOUS SENSORS

Humidity Sensors

"Humidity sensors are used to control the amount of water vapour present in many industrial processes. Textile, wood, and chemical processing is very sensitive to humidity".

Psychrometers

Psychrometers use latent heat of vapourization to determine relative humidity in the system. This can be done by using a dry bulb thermometer with a wet bulb thermometer. The two temperatures recorded can be used with a psychrometric chart to obtain the relative humidity, water vapour pressure, heat content, and weight of water vapour in the air.

Hygrometers

Hygrometers are devices that sense the change in either their physical or electrical properties. Some materials such as hair or thin strips of wood change length depending on water content. The change in length is directly related to the humidity.

Dew Point Measuring Devices

These devices measure humidity by cooling the air until water starts to condense on the object. The amount that the air needed to be cooled before water started to form on the object can be used to determine the relative humidity.

Other Humidity Sensors

Microwave absorption by water vapour can be used to measure the humidity in a material

Infrared absorption can be used by hitting the object with infrared radiation and measuring the energy of the reflectance.

Sound Sensors

Sound sensors are important because they can be used in industrial applications such as detection of flaws in solids and location and linear distance measurements. Sound pressure waves can also cause vibration and failure.

Microphones

Microphones are pressure transducers, and are used to convert sound pressures into electrical signals. There are six types of microphones: electromagnetic, capacitance, ribbon, crystal, carbon, and piezoelectric.

Smoke Sensors

Smoke sensors are useful for not only safety for workers, but also environmental concerns and purity issues of processes.

Infrared Sensors

These sensors detect changes a signal received from an LED due to smoke or other objects in the light path from the LED.

Ionization Chambers

These devices can detect the difference in current between two plates that have a voltage between them. The difference is due to carbon particles from smoke that provide a conductive path between the two plates.

Taguchi-Type

Taguchi-type sensors are used to detect hydrocarbon gases, such as carbon monoxide and carbon dioxide. The sensor is coated with an oxidized element that when combined with a hydrocarbon creates a change in the electrical resistance of the sensor.

VALVE TYPES SELECTION

Control valves are imperative elements in any system where fluid flow must be monitored and manipulated. Selection of the proper valve involves a thorough knowledge of the process for which it will be used. Involved in selecting the proper valve is not only which type of valve to use, but the material of which it is made and the size it must be to perform its designated task.

The basic valve is used to permit or restrain the flow of fluid and/or adjust the pressure in a system. A complete control valve is made of the valve itself, an actuator, and, if necessary, a valve control device. The actuator is what provides the required force to cause the closing part of the valve to move. Valve control devices keep the valves in the proper operating conditions; they can ensure appropriate position, interpret signals, and manipulate responses.

When implementing a valve into a process, one must consider the possible adverse occurrences in the system. This can include noise due to the movement of the valve, which can ultimately produce shock waves and damage the construction of the system. Cavitation and flashing, which involve the rapid expansion and collapse of vapour bubbles inside the pipe, can also damage the system and may corrode the valve material and reduce the fluid flow.

There are four general types of valves.
1. Electronic, or electrical valves. The movement of the ball or flap that controls flow is controlled electronically through circuits or digitally. These types of valves have very precise control but can also be very expensive.
2. Non-Return valves. These valves allow flow in only one direction, and are common in various industries. When pressure in the opposite direction is applied, the valve closes.
3. Electromechanical valves. These valves have electro magnets controlling whether the valve is open or closed. These valves can only be fully open or fully closed.
4. Mechanical Valves. These valves use mechanical energy in the process of opening and closing the actual valve. Larger valves can be opened and closed using mechanical processes such as levers and pulleys, whereas smaller mechanical valves can be opened or closed *via* a turning wheel or pulling a level by hand.

There are four major valve types: ball valves, butterfly valves, globe valves, and plug valves. There is also an array of many other types of valves specific to certain processes. Selecting which type of valve to use depends on what task the valve must carry out and in what space said valve can fit to carry out the task.

Some general features that one can take into consideration when selecting a valve are the following:
1. Pressure rating
2. Size and flow capacity
3. Desired flow condition
4. Temperature limits
5. Shutoff response to leakage
6. Equipment and pipes connected
7. Material compatibility and durability
8. Cost

Valve Types

There is a vast abundance of valve types available for implementation into systems. The valves most commonly used in processes are ball valves, butterfly valves, globe valves, and plug valves. A summary of these four valve types and their relevant applications is in the table below.

Valve Type	Application	Other information
Ball	Flow is on or off	Easy to clean
Butterfly	Good flow control at high capacities	Economical
Globe	Good flow control	Difficult to clean
Plug	Extreme on/off situations	More rugged, costly than ball valve

Following is a detailed description of the four main valve types.

Ball Valves

A ball valve is a valve with a spherical disc, the part of the valve which controls the flow through it. The sphere has a hole, or port, through the middle so that when the port is in line with both ends of the valve, flow will occur. When the valve is closed, the hole is perpendicular to the ends of the valve, and flow is blocked. There are four types of ball valves.

A full port ball valve has an over sized ball so that the hole in the ball is the same size as the pipeline resulting in lower friction loss. Flow is unrestricted, but the valve is larger. This is not required for general industrial applications as all types of valves used in industry like gate valves, plug valves, butterfly valves, *etc.* have restrictions across the flow and does not permit full flow. This leads to excessive costs for full bore ball valves and is generally an unnecessary cost.

In reduced port ball valves, flow through the valve is one pipe size smaller than the valve's pipe size resulting in flow area becoming lesser than pipe. But the flow discharge remains constant as it is a multiplier factor of flow discharge (Q) is equal to area of flow (A) into velocity (V).

$$A_1V_1 = A_2V_2;$$

the velocity increases with reduced area of flow and decreases with increased area of flow.

A V port ball valve has either a 'v' shaped ball or a 'v' shaped seat. This allows the orifice to be opened and closed in a more controlled manner with a closer to linear flow characteristic. When the valve is in the closed position and opening is commenced the small end of the 'v' is opened first allowing stable flow control during this stage. This type of design requires a generally more robust construction due to higher velocities of the fluids, which would quickly damage a standard valve.

A trunnion ball valve has a mechanical means of anchoring the ball at the top and the bottom, this design is usually applied on larger and higher pressure valves (say, above 10 cm and 40 bars).

Ball valves are good for on/off situations. A common use for a ball valve is the emergency shut off for a sink.

Butterfly Valves

Butterfly valves consist of a disc attached to a shaft with bearings used to facilitate rotation. The characteristics of the flow can be controlled by changing the design of the disk being used. For example, there are designs that can be used in order to reduce the noise caused by a fluid as it flows through. Butterfly valves are good for situations with straight flow and where a small pressure drop is desired. There are also high performance butterfly valves.

They have the added benefit of reduced torque issues, tight shutoff, and excellent throttling. It is necessary to consider the torque that will act on the valve. It will have water moving on both sides and when being used to throttle the flow through the valve it becomes a big factor. These valves are good in situations with high desired pressure drops.

They are desirable due to their small size, which makes them a low cost control instrument. Some kind of seal is necessary in order for the valve to provide a leak free seal. A common example would be the air intake on older model automobiles.

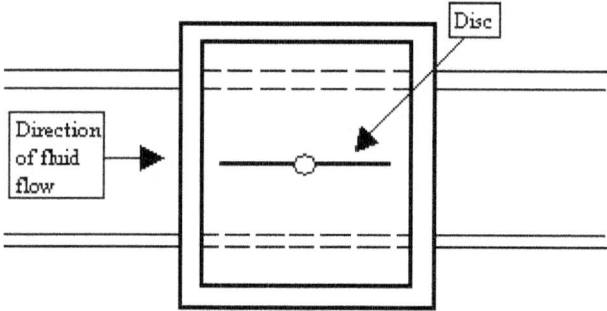

Fig. : Top view, open configuration.

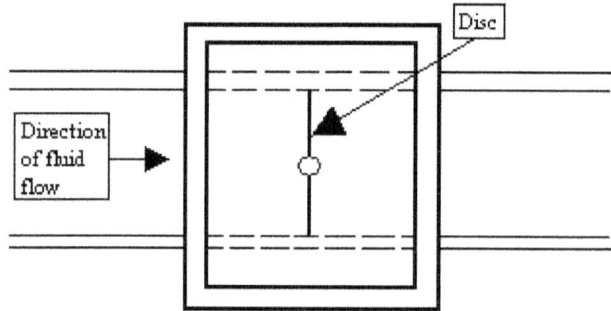

Fig. : Top view, closed configuration.

Globe Valves

A globe valve is a type of valve used for regulating flow in a pipeline, consisting of a movable disk-type element and a stationary ring seat in a generally spherical body. The valve can have a stem or a cage, similar to ball valves, that moves the plug into and out of the globe. The fluid's flow characteristics can be controlled by the design of the plug being used in the valve.

A seal is used to stop leakage through the valve. Globe valves are designed to be easily maintained. They usually have a top that can be easily removed, exposing the plug and seal. Globe valves are good for on, off, and accurate throttling purposes but especially for situations when noise and caviatation are factors. A common example would be the valves that control the hot and cold water for a kitchen or bathroom sink.

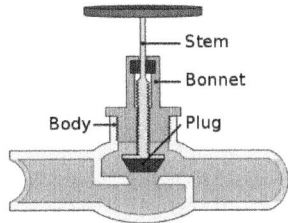

Plug Valves

Plug valves are valves with cylindrical or conically-tapered "plugs" which can be rotated inside the valve body to control flow through the valve. The plugs in plug valves have one or more hollow passageways going sideways through the plug, so that fluid can flow through the plug when the valve is open. Plug valves are simple and often economical. There are two types of plug valves.

One has a port through a cylindrical plug that is perpendicular to the pipe and rotates to allow the fluid to proceed through the valve if in an open configuration. In the closed configuration, the cylinder rotates about its axis so that its port is no longer open to the flow of fluid. An advantage of these types of valves is that they are excellent for quick shutoff. The high friction resulting from the design, however, limits their use for accurate modulating/throttling. Schematics of this type of plug valve are below.

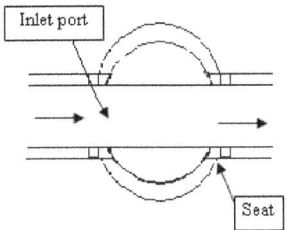

Fig. : Top view, open configuration.

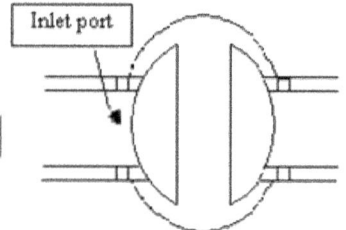

Fig. : Top view, closed configuration.

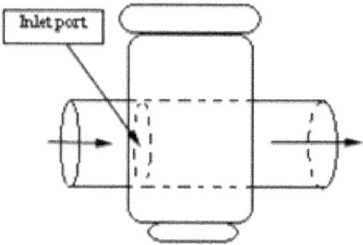

Fig. : Side view.

The other type of plug valve is the eccentric plug valve. In this design, the plug rotates about a shaft in a fashion similar to a ball valve. To permit fluid flow, the plug can rotate so that it is out of the way of the seat. To block fluid flow, it rotates to the closed position where it impedes fluid flow and rests in the seat. A schematic of this valve is below.

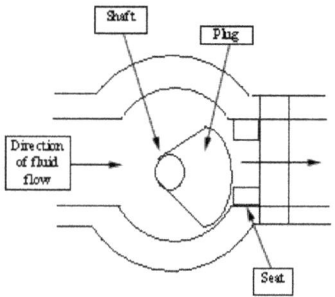

Fig. : Top view, near-closed configuration.

A common example would be a spray nozzle at the end of a garden hose.

Safety Valves

Pressure Relief Valves

Pressure relief valves are used as a safety device to protect equipment from over-pressure occurrences in any fluid process. Loss of heating and cooling, mechanical failure of valves, and poor draining and venting are some of the common causes of overpressure. The relieving system depends on the process at hand; pressure relief valves either bypass a fluid to an auxiliary passage or open a port to relieve the pressure to atmosphere.

Some areas of common usage include reaction vessels and storage tanks. In the Petroleum Refining Industry, for example, the Fluidized Catalytic Cracker (FCC) reactor has several pressure relief valves to follow saftey codes and procedures on such a high pressure/high temperature process. Each of the pressure relief valves have different levels of pressure ratings to release different amounts of material to atmosphere in order to minimize environmental impact.

Here are three examples of pressure relief valves:

Conventional Spring Loaded Safety Valve

As the pressure rises, this causes a force to be put on the valve disc. This force opposes the spring force until at the set pressure the forces are balanced and the disc will start to lift. As the pressure continues to rise, the spring compresses more, further lifting the disc and alleviating the higher pressure. As the pressure inside the vessel decreases, the disc returns to its normal closed state.

Advantages:
- Most reliable type
- Versatile

Disadvantages:
- Pressure relief is affected by back pressure
- Susceptible to chatter

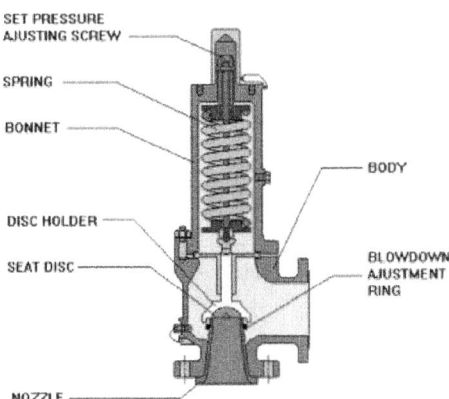

Bellows Spring Loaded Safety Relief Valve

The bellows spring loaded safety relief valve has the same principle as the conventional spring valve, with the exception of a vent located on the side of the valve. This vent lets releases the contents of the valve out to the surrounding environment.

Advantages:
- Pressure relief is not affected by back pressure
- Can handle higher built-up back pressure
- Spring is protected from corrosion

Disadvantages:
- Bellows can be susceptible to fatigue
- Not environmentally friendly (can release of toxics into atmosphere)
- Requires a venting system

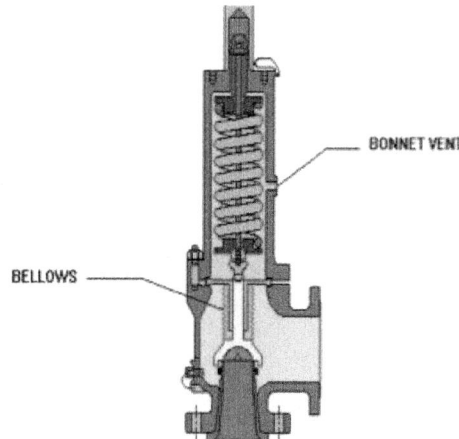

Pilot Assisted Safety Relief Valve

The pilot operated safety relief valve is also similar to the conventional safety relief valve except a pneumatic diaphragm or piston is attached to the top. This piston can apply forces on the valve when the valve is closed to balance the spring force and applies additional sealing pressure to prevent leakage.

Advantages:
- Pressure relief is not affected by back pressure
- Can operate at 98% of set pressure
- Less susceptible to chatter

Disadvantages:
- Pilot is susceptible to plugging
- Has limited chemical use
- Condensation can cause problems with the valve
- Potential for back flow

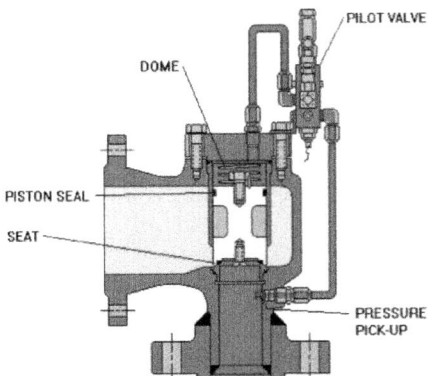

Steam Traps

Steam traps are devices that exist in low lying places within a pressurized steam line to release condensate and non-condensable gases from the system. Steam lines in industry are used to open/close control valves, heat trace pipelines to prevent freezing, *etc.* These steam traps are used in industry to save money on the prevention of corrosion and loss of steam. When these traps fail, it can mean a lot of money for the industry.

There can be several hundred to several thousand in one process unit, therefore it is important to maintain and check the condition of each trap annually. The checks can be done by visual, thermal, or acoustic techniques. Many suppliers have equipment to read the flow within the pipeline to see if it is: blocked, cold, leaking, working.

There are many types of steam traps that differ in the properties they operate on including Mechanical (density), Temperature (temperature), and Thermodynamic (pressure).

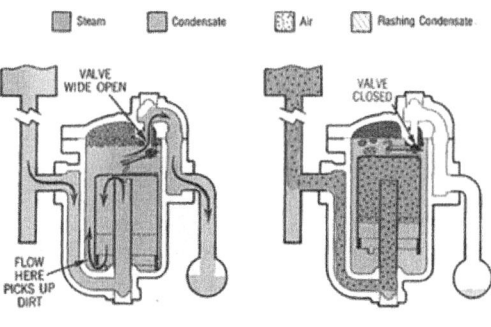

1. Steam trap is installed in drain line between steam heated unit and condensate return header. At this point, bucket is down and valve is wide open. As initial flood of condensate enters the trap and flows under bottom edge of bucket, it fills trap body and completely submerges bucket. Condensate then discharges through wide open valve to return header.

2. Steam also enters trap under bottom edge of bucket, where it rises and collects at top, imparting buoyancy. Bucket then rises and lifts valve toward its seat until valve is snapped tightly shut. Air and carbon dioxide continually pass through bucket vent and collect at top of trap. Any steam passing through vent is condensed by radiation from trap.

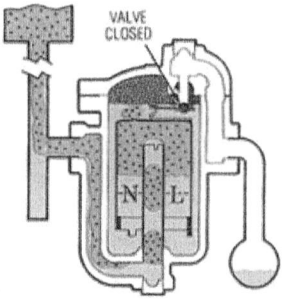

3. When entering condensate brings the condensate level slightly above the neutral line the bucket exerts a slight pull on the lever. The valve does not open, however, until the condensate level rises to the opening line for the existing pressure differential between the steam and the condensate return header.

4. When the condensate level reaches opening line the weight of the bucket, times leverage, exceeds the pressure holding valve to its seat. Bucket then sinks and opens trap valve. Any accumulated air is discharged first followed by condensate. Discharge continues until more steam floats bucket at which time cycle begins to repeat.

Other Safety Valves

- *Rupture Disc* - A rupture disc (also called a "bursting disc" or "safety disc") is a thin membrane of material (usually metal) that acts as a one-time use pressure relief device. At a critical pressure, the disc will fail and rupture allowing flow and the release of pressure. Often, rupture discs are used as a back-up to a conventional spring-controlled safety valve. Given primary safety valve failure (either no or inadequate pressure relief) the rupture disc will burst open and relieve pressure. Rupture discs are commonly used in chemical process plants and large air-cooled steam condensers.
- *Pressure Vacuum Valves* - Pressure vacuum (P-V) valves protect against both over-pressure and under-pressure conditions. They are commonly used in atmospheric storage tanks, to prevent the build up of excessive pressure or vacuum which can dangerously unbalance the system or damage the storage vessel.

Other Valves

In addition to the four main types of control valves, there are numerous other valves that may be necessary to manipulate fluid flow in a certain system. Some valves important to chemical engineering processes are stated below with a brief description of their design and application.

- *Angle valves* include inlet and outlet ports that are oriented at a 90 degree angle. The fluid leaves at right angles to the direction in which it enters the valve.
- *Bleed valves* vent signal line pressure to atmosphere before removal of an instrument or to assist in calibration of control devices. Common bleed valves include ball and plug bleed valves.

- *Check valves* are one way valves. Check valves only allow fluid in one way and out the other. They are often placed on individual fluid streams when mixing fluids so as to prevent the mixture from flowing back into the original streams. Also, the speed with which the valve closes is significant to prevent reverse-flow velocity. There are different types of check valves which include lift check, swing check, tilting disk and diaphragm.
    o *Lift check valves* are quick closing. This valve should only be used for low viscosity fluids because they can be slowed by viscous fluids.
    o *Swing check valve* has a disc like closing method from a hinge which may or may not be spring loaded.
    o *Tilting disk check valve* is spring loaded for quick response. These are often more expensive and harder to fix when broken.
- *Diaphragm valves* have excellent shut-off characteristics and are used when there is particulate matter in the fluids. Not a good choice for controlling flow. A diaphragm valve has both a flexible and a rigid section. One advantage is that there are no crevices that affect the flow of the fluid when open. Used mostly in the pharmaceutical and food industry for their ability to be sanitary.
- *Piston valves* have a closure member shaped like a piston. When the valve opens, no flow is observed until the piston is fully moved from the seat bore, and when the valve closes, the piston removes any solids that might be deposited on the seat. Therefore, piston valves are used with fluids that have solid particles in suspension.
- *Gate valves* work by raising a gate or barrier out of the way of the flowing fluid. The water spigot outside a house works as a gate valve. Have the positive quality that, when fully open, the fluid flow is totally unobstructed. Two major types of gate valves are used: parallel and wedge gate valves. The wedge gate valve, in which the closure member has a wedge shape, provides the advantage of sealing against low pressure, as well as high pressure, flow. Gate valves have the ability to open very quickly. Advantages of gate valves are that they have a high capacity, have good seals, relatively inexpensive, and don't have very much resistance to flow. Some disadvantages of gate valves are that they sometimes can have poor control, they can cavitate at lower pressures, and they cannot be used for throttling.
- *Needle valves* are similar to gate valves; however, they are usually applied to a much smaller orifice. Needle valves are excellent for precise control of fluid flow, typically at low flow rates.
- *Flush Bottom valves* are usually at the lowest point of a tank or reactor and used to drain out contents. Unique because it leaves no dead space in the valve when it is closed, this eliminates the problem of product buildup within the valve.
- *Pinch valves* are mainly used in order to regulate the flow of slurries in certain processes and systems. Pinch valves have flexible bodies that can be shut by pinching them. They are often used when it is necessary for the slurry to

pass straight through when the valve is not pinched. They are often used for sanitary processes such as medical and pharmaceutical applications. Pinch valves can be controlled mechanically or with fluid pressure.

- *Knife valves* are used in systems that deal with slurries or powders. They are primarily used for on and off purposes; whether or not the slurry or powder flows or not. A knife gate valve can be used for fibrous material because it can cut through to close the valve.
- *Ballcock valves* are used in controlling levels in tanks. The valve is connected to a float in the tank using a lever. When the level in the tank rises, the float rises and forces the valve to be shut at the maximum level of the tank allowed. Ballcock valves are used mostly in water tanks and other low-risk areas within a certain process.
- *Solenoid valves* are used very frequently in industry. The valves have a loop of wire that is wrapped around a metal core. A current is shot through the valve creating a magnetic field, which in turn opens or closes plungers in pipelines controlling flow in that pipe. There are three types of solenoid valves:
  1. Electromechanical solenoid valves use electrical energy
  2. Pneumatic solenoid valves use compressed air
  3. Hydraulic solenoid valves use energy from pressurized fluids

Flow Characteristics

For all valves, flow rates of fluid through the valve depend on the percentage of a full valve opening. In ball and butterfly valves, the valve opening is based on rotation. In the case for butterfly valves, an open valve is a result of a 90 degree rotation. When the valve is at the open position, the flow is parallel to the valve. The flow is uninterrupted, and therefore, no pressure is on the valve. When the valve is throttling, as is common for globe valves, the flow rate is reduced or increased depending on the opening of the valve, and there is unequal pressure on the ends of the valve.

The ball valve flow stream involves at least two orifices; one for inlet and one for outlet flow. While traveling through an open ball valve, the fluid will continue in its flow straight through, with little pressure loss. When the ball valve is throttling, however, the fluid is subject to shearing and a change in flow rate in accordance with the percentage of the valve that is open. With high velocity liquids, the valve is susceptible to cavitation and erosion, and could produce noise. Also, at sonic velocities, the vena contracta will stop expanding, and create choking.

Globe valves are dependent on shape of the plug of the valve, in addition to the opening size, for flow variations. Lifting the globe valve will cause it to open. A flat plug is used for a quick opening; a conically shaped plug will create linear flow as the valve is raised; a rectangular shaped plug, where the bottom converges to one point directly in the middle of the plug, creates an equal percentage of flow through the valve.

Sizing

Once a specific valve type is chosen for a process, the next step is to figure out what size the valve will need to be. Topics to consider in valve sizing are the pressure-flowrate relationship and the flow characteristics to ensure an appropriate size for the desired valve.

Pressure-Flowrate Relationship

Control valves predominately regulate flow by varying an orifice size. As the fluid moves from the piping into the smaller diameter orifice of the valve, the velocity of the fluid increases in order to move a given mass flow through the valve. The energy needed to increase the velocity of the fluid comes at the expense of the pressure, so the point of highest velocity is also the point of lowest pressure (smallest cross section). This occurs some distance after leaving the smallest cross section of the valve itself, in a localized area known as the vena contracta.

Beyond the vena contracta, the fluid's velocity will decrease as the diameter of piping increases. This allows for some pressure recovery as the energy that was imparted as velocity is now partially converted back into pressure. There will be a net loss of pressure due to viscous losses and turbulence in the fluid.

Overview of Sizing Formulas

In order to determine the correct size of a valve for a specific system many factors must be considered. The most important factor is the capacity parameter, Cv, or the **flow coefficient**. The flow coefficient is a way of measuring how efficient a valve is at allowing fluid to flow through it and generally determined experimentally. Valve manufacturers can provide you with Cv charts for the valves they sell. To determine the valve size needed for your system, you can estimate Cv with the following equations:

Liquid Flow

$$C_v = 11.7Q\sqrt{\frac{G_f}{\Delta P}}$$

Where:
- Q = flow rate (m^3 / h)
- ΔP = pressure drop through the valve (kPa)
- G_f = specific gravity of the fluid (dimensionless)

The next equation is also for calculating the flow coefficient. It should be used when you have a design flow rate, that will be the maximum flow through the valve, and a maximum pressure drop that will occur across the valve, which will be ΔP. This equation is also for turbulent flow and the units used in the calculation are important.

$$C_v = F_{max}\sqrt{\frac{G_t}{\Delta P}}$$

Where:
- F_{max} = maximum flow through the valve (gallons/minute)
- ΔP = pressure drop across the valve (psi)
- G_t = specific gravity of the liquid (unitless)

Another important piece of information about sizing and specifications is what diameter pipe may be used with a certain flow, seen in the following equation:

$$d = \sqrt{\frac{4F_{max}}{\pi v}}$$

Where:
- d = diameter of pipe (ft)
- F_{max} = maximum flow through the valve (ft^3 / s)
- v = velocity of flow (ft/s)

Air and Gaseous Flow

(when $P_o < 0.53 P_i$)

$$C_v = Q\frac{\sqrt{G_f(T+460)}}{660 P_i}$$

(when $P_o > 0.53 P_i$)

$$C_v = Q\frac{\sqrt{G_f(T+460)}}{1360\sqrt{\Delta P(P_o)}}$$

Where:
- Q = flow rate (ft^3 / h)
- G_f = specific gravity of the gaseous fluid (dimensionless)
- ΔP = absolute pressure drop through the valve (psia)
- P_o = outlet pressure (psia)
- P_i = inlet pressure (psia)
- T = gas temperature (degrees F)

The relationship between the inlet and outlet pressure is important, as seen above, in determining which equation to use for gaseous flow. When the outlet pressure is less than 53% the inlet pressure, this is said to be a **critical** pressure drop. The gas will behave differently when the pressure drop is critical, therefore it is necessary to use the correct equation depending on the extent of pressure drop. Once you have computed the desired value of C_v, you can choose the valve. The

pH and Viscosity Control Sensors

chosen valve must have a valve coefficient greater than or equal to the computed value.

Other Sizing Information

When sizing a valve it is important not to choose a valve that is too small. Take the following tips into consideration when choosing a valve.
- Valves should not be less than half the pipe size.
- Valves are easiest to control when they are open from 10 to 80% of maximum flow.

Using the Flow Coefficient to Determine Valve Sizes

The flow coefficient, C_v, is used as a standard sizing parameter. It is used to determine valve sizes and is also used in modeling programs for large piping systems. The C_v is an essential and practical variable that is essentially required when using pipes and valves. Try searching for a valve information and you will most likely find information requiring the C_v as well as the valve type and use.

Try looking at information for specific valves at the website of American Valve.

You'll notice that every valve type and size has a listed C_v for correct sizing and function. Tables using this value are available for practically every valve imaginable and should be used to prevent over-use or under-use of valves in engineering processes.

C_v Charts

When sizing a valve, it is important to choose the correct C_v chart. Different C_v charts are used for different types of valves. Some different valve types are explained below. Note that *valve stroke* means the amount of flow through a pipe, and *valve travel* means the amount in which the valve is turned to achieve the valve stroke.

Equal Percentage Valves: Valve which produces equal valve stroke for equal increments in valve travel. This is the most common type of valve.

Linear Valves: Valves stroke is directly proportional to valve travel.

Quick Opening Valves: A small amount of valve travel produces a large valve stroke. Example: Gate Valves

Equal percentage and linear valve types offer the best throttling control. Examples are globe and butterfly valves.

Valves vary from manufacturer to manufacturer. Valve manufacturers can provide you with C_v charts to size each type of valve they manufacture. An examle C_V chart is shown below. VT stands for valve travel.

Size (NPT)	Cv @ 10% VT	Cv @ 25% VT	Cv @ 50% VT	Cv @ 75% VT	Cv @ 100% VT
1/2"	0.03	0.075	0.3	0.5	1
1/2"	0.05	0.12	0.48	0.8	1.6
1/2"	0.08	0.19	0.75	1.25	2.5
1/2"	0.13	0.3	1.2	2	4
3/4"	0.21	0.47	1.89	3.15	6.3
1"	0.33	0.75	3	5	10
1 1/4"	0.53	1.2	4.8	8	16
1 1/2"	0.83	1.88	7.5	12.5	25
2"	1.33	3	12	20	40
2 1/2"	2.10	4.73	18.9	31.5	63
3"	3.33	7.5	30	50	100
4"	5.33	12	48	80	160
5"	8.33	18.75	75	125	250
6"	13.33	30	120	200	400

Materials

It is usually not enough to simply select the type of valve suited to given process parameters. Selecting compatible materials of construction helps ensure the lifespan of the valve, as well as the safety of the workforce, the environment and the public.

Selecting the most appropriate materials of construction of control valves is guided primarily by the service of the valve, then secondarily by cost; the least expensive material that is compatible with the service will be chosen to be used. Proper material selection promotes safety by avoiding materials of construction that may react with or be corroded by the process fluid.

The principle materials that need to be selected carefully are the wetted materials, that is, the materials that come into contact with the process fluid. These generally include the ball (for ball valves), the disk (for butterfly valves), and the plug (for plug and globe valves). Also included are the seats, which is the area where the plug or disk "sits" when closed to provide the actual shut off. The seals and the valve body are usually wetted as well.

There are many resources that contain what resources are compatible with a wide variety of process fluids, such as the Cole Parmer Chemical Resistance Database and the Cat Pumps' Chemical Compatibility Guide

There are design parameters inherent in the valve designs themselves that increase safety. For high service pressures (or in case of fire) some valves are designed with initial flexible seal rings that function as the primary seals. Behind these primary seals would be a backup seal of a more durable material such as 316 stainless, inconel or hastelloy. These backup seals assist in handling the additional pressure and heat.

In the highly specialized case where the process fluid is so dangerous or unsafe that any release of process fluid is unacceptable, the valve's packing can be slightly pressurized with a barrier fluid. As long as the pressure of the barrier fluid is higher than the process fluid, any leakage between the valve and the process will leak barrier fluid into the process, and not process fluid into the environment.

Though as a side note, these applications usually require double containment piping and a whole host of other precautions beyond simply the safety of the valve. The most common barrier fluid is water or a water/antifreeze mix for freeze protection.

Some other considerations when selecting a material for valve are longevity/reliability of the valve and the temperature range of usage. If the valve is a control valve that gets constant use, it is important to select durable materials or to plan for replacement of the valve frequently. Service temperature is also important; materials need to be selected so the mechanical integrity of the valve is maintained throughout the entire service temperature.

Concerns

Noise

Noise in a system is created by the turbulence from a throttling valve. Throttling is when a valve is not used as a simple OPEN/CLOSE valve, but instead to control a flow rate.

It can be quite loud and damage people's hearing as well as the equipment. Two methods can be used to control noise. The first is to use a quiet trim that hinders system noise before it begins and the second is to treat the path (piping) where the noise already occurs. To do this, one can use thicker piping or insulation.

Cavitation

As previously mentioned, at the point where the fluid's velocity is at its highest, the pressure is at its lowest. Assuming the fluid is incompressible (liquid), if the pressure falls down to the vapour pressure of the fluid, localized areas of the fluid will vapourize forming bubbles that collapse into themselves as the pressure increases downstream. This leads to massive shock waves that are noisy and will certainly ruin process equipment. This is a process called cavitation.

For a given flow rate, a valve that has the smallest minimum flow area through the valve will have the most problems with cavitation (and choking, as mentioned in the next section).

Choking

If the fluid is compressible and the pressure drop is sufficient, the fluid could reach sonic velocity. At this point the static pressure downstream of the valve grows above design limits. This is a process known as choking, since it slows down, essentially "choking," the flow through the valve.

For a given flow rate, a valve that has the smallest minimum flow area through the valve will have the most problems with choking.

Choking is important for a gas-using process since the mass flow rate only depends on the upstream pressure and the upstream temperature. It becomes easier to control the flow rate since valves and orifice plates can be calibrated for this purpose. Choke flow occurs with gases when the ratio of the upstream pressure to the downstream pressure is greater than or equal to

$$[(k+1)/2]^{(k/(k-1))}.$$

If the upstream pressure is lower than the downstream, it is not possible to achieve sonic flow.

When choking occurs in liquids, it is usually due to the Venturi effect. If the Venturi effect decreases the liquid pressure lower than that of the liquid vapour pressure at the same temperature, it will cause the bubbles in the liquid. These bubble burst can cause enough turbulence (known as cavitations) that can lead to physical damage to pipes, valves, controllers, gauges, and/or all other equipmentslinked to that section of the flow.

As mentioned above the limiting case of the **Venturi effect** is choked flow as the smaller space in a pipe will but a cap on the total flow rate. The Venturi effect is fluid pressure resulting from an incompressible fluid travelling through a tight or constricted section of a pipe. The Venturi effect can be explained by principles such as Bernoulli's principle and the continuity equation.

Bernoulli's equation : $\dfrac{v^2}{2} + gz + \dfrac{p}{\rho} = \text{constant}$

Continuity equation: $\dfrac{\partial \rho}{\partial t} + \nabla \cdot (\rho u) = 0$

for incompressible fluids: $\nabla \cdot \rho = 0$

To satisfy the continuity equation, the velocity of the fluid must increase in the space but the pressure must decrease to agree with Bernoulli's equation, which is related to the law of conservation of energy through a pressure drop.

Venturi tubes are used to control the fluid speed. If an increase in speed is wanted, then the tube will decrease in diameter, or projected area, as the fluid moves away from the pump or energy source. If a decrease in fluid velocity is wanted, the tube will increase in diameter as it moves away from the energy source (which could be a pump). Using a venturi tube sometimes introduces air into the system which is mixed with the fluid in the pipe. This causes a pressure head at the end of the system.

Venturi tubes can be replaced by **orifice plates** in some cases. Orifice plates use the same principles as the venturi tubes but cause the system to suffer more permanent energy loss. Orifice plates, however, are less expensive to build then venturi tubes.

Upstream Pressure Increase

If a control valve is sized incorrectly and is too small, fluid velocity upstream of the valve will slow, causing an increase in pressure (much like when the end of a garden hose is partially obstructed). This increase in upstream pressure can be detrimental to certain processes, *i.e.* membrane filtration processes, where a large pressure difference across the membrane is desired. If the valve increases the pressure on the permeate side of the membrane, the driving force for separation will be reduced.

Decreasing the upstream static temperature will cause an increase in upstream static pressure therefore increasing the mass flow rate, so it is important to keep track of your temperature.

Hysteresis

Hysteresis, in general, is defined as the phenomenon where previous events can influence subsequent events in a certain process. This can create a relationship in a process going one direction, however, when carrying out the same process in the exact opposite direction, the same relationship does not occur. When processes exhibiting hysteresis are graphed there is an area between the two equally opposite curves that is used to describe the extent of hysteresis seen in the process.

The most commonly described process exhibiting hysteresis involves elastics and can be the loading and unloading of force (weight) to a stretchable rubber band. As you load more and more weight onto a rubber band, it will stretch in response to increased weight.

You can subsequently remove the weight in similar increments and observe the ability of the rubber band to compress itself. Rubber bands, however, do not obey Hooke's Law perfectly therefore as you unload, the rubber band will be longer at the corresponding weights as you loaded. Below is a graphical representation of this phenomenon.

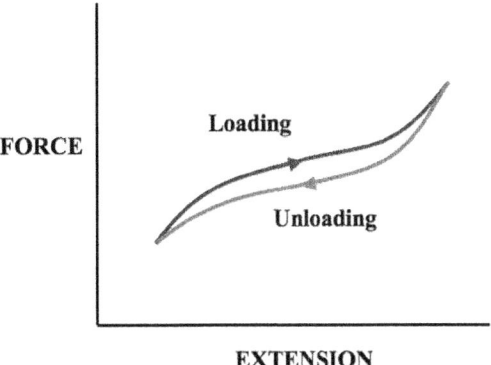

As seen in the above graph, for any given force applied to the rubber band, the extension of unloading is larger than the extension when loading. This is showing how the process of loading effected the outcome of the process of unloading. The area of the graph between the two curves is the area of hysteresis and, in this case, represents the energy dissipated as heat. The larger this area the less efficient this rubber band was at maintaining shape, and consequently the worse it obeys Hooke's Law.

Hysteresis, regarding control valves, is the principle that a control valve is dependent on the prior valve position. A common valve opening may correlate to different flow rates depending on if the valve was opened or closed to get to that position. This shows that the direction a valve is being changed may need to be accounted for in a control system to obtain a desired flow rate. If hysteresis becomes too large, it can cause the control architecture to force a system to oscillate around a desired point.

Chatter

Chatter is mainly a concern when selecting and sizing safety valves. Chatter is the rapid opening and closing of a safety valve or pressure relief device. Commonly, chatter is caused due to a large pressure drop between the vessel being relieved and the valve relieving the pressure.

If the pressure behind a closed safety valve builds up to some critical pressure, it will open. Once open, fluid will flow from the vessel being relieved out through the safety valve in order to reduce pressure within the vessel. It is possible to have enough pipe length, fittings, or other energy-reducing obstacles so that once the high pressure flow reaches the safety valve, the pressure in the fluid is again below the critical pressure of the valve.

Since there is not enough pressure in the fluid to keep the safety valve open, it closes and the vessel re-pressurizes, causing the safety valve to pop open again, locally depressurize and close again. This opening and closing is called chatter. The vibrations caused by chatter can be extremely damaging, causing unnecessary safety valve wear, possible seat-misalignment within the fittings, and even full failure in the valve or surrounding piping.

Chatter can be avoided by making sure that at critical release rates the pressure drop across the fittings to the safety valve is not large enough such that the valve will shut once fluid release is initiated.

ACTUATORS

Actuators are the mechanical equipment that supply the force necessary to open or close a valve. Actuators are, essentially, the alternative to manual operation of a valve. The method of applying the opening/closing force to a valve is what differentiates the various types of actuators. When selecting the actuator, the most important feature to specify is whether you want fail-safe open or closed. This is determined entirely by a careful analysis of the process to decide which is safer.

If all the power goes out or some other emergency occurs, the decision on the fail-safe mode of the valve is a huge factor in saving lives. For example, if a reaction is exothermic, the fail-safe mode of the cooling water in the jacket of the reactor should be fail-safe open. Pneumatic actuators have a delayed response which make them ideal for being resilient against small upsets in pressure changes of the source. Hydraulic actuators on the other hand use an incompressible fluid, so the response time is essentially instantaneous.

Pneumatic

Pneumatic actuators are the most popular type of actuators. The standard design of a pneumatic actuator consists of a pre-compressed spring that applies force against a disk on a sealed flexible chamber. The disk is usually attached to the stem of the valve it is intended to control. As the chamber is compressed with air, the chamber expands and compresses the spring further, allowing axial motion of the valve stem. Knowing the relationship between the air pressure in the chamber and the distance the stem moves allows one to accurately control flow through the valve.

The biggest advantage of the pneumatic actuators is their failsafe action. By design of the compressed spring, the engineer can determine if the valve will fail closed or open, depending on the safety of the process. Other advantages include reliability, ease of maintenance, and widespread use of such devices.

Motion Conversion

Motion conversion actuators are generally used to adapt a common translational motion from the actuator's output to a rotary valve. The rod that moves axially from the translational motion actuator is connected to a disk and the connection is pivoted. The disk itself is also pivoted about its center. This system of pivots allows the translational motion to be converted into the rotation of the disk, which would open or close the rotary valve.

The main advantage of this setup is that an inexpensive translational motion actuator can be used with rotary valves. The key drawback is that the applications in which this can be used is very limited. Specifically, this setup is useless in the common case where the rotary motion required is greater than 90°.

Hydraulic

Hydraulic actuators work using essentially the same principal as pneumatic actuators, but the design is usually altered. Instead of a flexible chamber, there is a sealed sliding piston. Also, instead of using a spring as the opposing force, hydraulic fluid is contained on both sides of the piston. The differential pressure across the area of the piston head determines the net force.

Hydraulic actuators offer the advantages of being small and yet still providing immense force. Drawbacks of hydraulic actuators are primarily the large capital cost and difficulty maintaining them.

Electric

Electric actuators typically use standard motors, powered by either AC induction, DC, or capacitor-start split-phase induction. The motor is connected to a gear or thread that creates thrust to move the valve. As a failsafe, some motors are equipped with a lock in last position on its gear. This means that the gear cannot move from forces outside of the electric motor. This helps prevent overshoot on the motor as well as helps create better positioning for the gear.

Another type of motor that can be used is called a stepper motor. It uses increments on gear reduction to alleviate problems with positioning and overshoot. The increments are in a range of 5,000 to 10,000 increments in a 90 degree rotation.

A problem with using electric actuators is that a battery operated back-up system is needed or else the system is useless during power failure. Also, the actuator needs to be in an environment that is rendered safe, meaning a non-explosive environment.

Manual

Manual actuators are usually used for overrides of power actuators described above. This is an important safety measure in case the power actuator fails. Manual actuators typically consist of either a lever or a wheel (used for larger valves) connected to a screw or thread that turns the valve.

Summary Tables

Valve Type	Description	Advantages	Disadvantages
Ball	A ball that can be spun to allow or close flow	Operated in a 1/4 turn for quick on/off application	Size limitations
Butterfly	A disk that is rotated about it's diameter to allow, throttle, or close flow	Low pressure drop	Torque issues
Globe	A plug that moves into and out of a globe to allow, throttle, or close flow	Precise throttling	High pressure drop
Plug	A plug that rotates to allow, throttle, or close flow	Quick shutoff	Limited throttling ability

Actuator	Description	Advantages	Disadvantages
Pneumatic	Compressed air fills a chamber and pushes against a spring, operating the valve	Failsafe action, reliability, low cost, and ease of maintenance	Limited force applications
Motion Conversion	Consists of a disk connected to a rod. The rod moves axially and pivots the connection with the disk, causing the disk to also pivot.	Inexpensive and can be used for rotary valves	Limited to applications of 90° or less rotations.
Hydraulic	Similar to pneumatic, but a uses an incompressible fluid	Small size with large force	High cost
Electric	Electric motor with a rotating gear that moves the valve.	Low overshoot and accurate positioning	Needs a backup battery in case
Manual	Typically either a level or wheel connected to a screw	Good for an override in case of failure of	Not efficient as primary

The table below shows a list of typical icons for different valves found in industry.

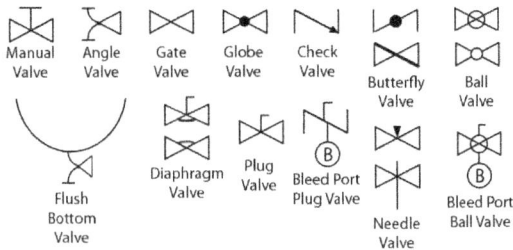

VALVE AND ACTUATOR SELECTION EXAMPLE

Note: This problem is completely fictionalized. Any relation to a real occurrence is completely coincidental.

A chemical engineer wants to use a valve at a start/stop flow of liquid water to a CSTR. The determination of the start/stop flow is a level sensor in the CSTR. Once the water reaches a certain level inside the CSTR, the flow of the water will stop. If the water level drops to a certain point, water will need to quickly flow into the CSTR. Your job is to list possible valves, and determine if each valve type can be used to start/stop the flow of water. Also, list all the potential actuators for the valves and determine which actuators *could* be used and how. Some issues to consider are choking, throttling, longevity, and reliability.

Solution

Valves:

- *Ball valve:* A ball valve can work for this example. A 2-way ball valve may not be the most ideal type of valve depending on the flow rate. If the water is moving at a high velocity, the 2-way ball valve will be susceptible to choking. If a ball valve is used, a cage valve would be the best choice, due to its durability and its on/off mechanics.
- *Globe valve:* Just like the caged ball valve, a globe valve is a conceivable solution. The plug that would best help prevent water from flowing in the CSTR quickly would be the quick opening plug. This way, there will be no throttling when closing the valve, creating a quick on/off.
- *Butterfly valve:* If a butterfly valve is selected, it would need to be a high performance valve. The benefits of this valve include a tight shutoff and improved torque over the non-high performance butterfly valves. This is important for the on/off capabilities for water flow the example calls for. Using a regular butterfly valve could lead to water leaking into the CSTR or torque issues after constant use.
- *Plug valve:* A plug valve could work, but would encounter similar problems as a 2-way ball valve with choking. Also, the friction created when constantly rotating could cause a problem in the future with the valve not closing properly.

Actuators

- *Pneumatic:* A pneumatic actuator has a good potential to be used in this example. Its failsafe action is great in case there is a failure in the system. Its low maintenance and high reliability means that the pneumatic actuator is an ideal actuator for the valve.
- *Motion Conversion:* Motion conversion actuators would be ideal for any rotating valve, like the high performance butterfly valve. Therefore, this actuator should be used with rotational valves for best reliability.
- *Hydraulic:* Hydraulic actuators have similar advantages that the pneumatic actuators have. Hydraulic actuator, however, cost more than pneumatic actuators. Therefore, if another option exists, hydraulic actuators should not be used for this example.
- *Electric:* An electric actuator will work best with rotational valves. So, if a high performance butterfly valve was chosen, then an electric actuator could possibly be used. As an effort to limit throttling issues, rotational models for ball valves would be less ideal. This is, of course, on the assumption that a back up battery would be included with the electric actuator
- *Manual:* A manual actuator could be used in conjunction with any other actuator, but would it not be a good idea to use a manual actuator alone. The manual actuator would be a good backup method if any of the other actuators fail, but tedious if used as the primary actuator.

VALVE MODELING

A valve acts as a control device in a larger system; it can be modeled to regulate the flow of material and energy within a process. There are several different kinds of valves (butterfly, ball, globe *etc.*), selection of which depends on the application and chemical process in consideration. The sizing of valves depends on the fluids processing unit (heat exchanger, pump *etc.*) which is in series with the valve. Valves need to be modeled to perform effectively with respect to the process requirements. Important components for the modeling of control valves are:

1. Flow
2. Inherent Flow Characteristics
3. Valve Coefficient, C_v
4. Pressure Drop
5. Control Valve Gain
6. Rangeability
7. Installed Characteristics

Efficient modeling of the valves can optimize the performance and stability of a process as well as reduce development time and cost for valve manufacturers.

Flow through a Valve

The following equation is a general equation used to describe flow through a valve. This is the equation to start with when you want to model a valve and it can be modified for different situations. The unfamiliar components such as valve coefficient and flow characteristics will be explained further.

$$F = C_v f(x) \sqrt{\frac{\Delta P_v}{sg}}$$

F = volumetric flow rate

Cv = valve coefficient, the flow in gpm(gallons per minute) that flows through a valve that has a pressure drop of 1psi across the valve.

ΔP_v = pressure drop across the valve

sg = specific gravity of fluid

x = fraction of valve opening or valve "lift" (x=1 for max flow)

f(x) = flow characteristic

Flow Characteristics

The inherent flow characteristic, f(x), is key to modeling the flow through a valve, and depends on the kind of valve you are using. A flow characteristic is defined as the relationship between valve capacity and fluid travel through the valve.

There are three flow characteristics to choose from:
1. f(x) = x for linear valve control
2. f(x) = $\sqrt{x}$ for quick opening valve control
3. f(x) = R^{x-1} for equal percentage valve control
   - R= valve design parameter (between 20 and 50)
   - note these are for a fixed pressure drop across the valve

Whereas a valve TYPE (gate, globe or ball) describes the geometry and mechanical characteristics of the valve, the valve CONTROL refers to how the flow relates to the "openness" of the valve or "x."

1. Linear: flow is directly proportional to the valve lift (used in steady state systems with constant pressure drops over the valve and in liquid level or flow loops)
2. Equal Percentage - equal increments of valve lift (x) produce an equal percentage in flow change (used in processes where large drops in pressure are expected and in temperature and pressure control loops)

3. Quick opening: large increase in flow with a small change in valve lift (used for valves that need to be turned either on or off frequently or where instant maximum flow is required, for example, safety systems)

For the types of valves discussed in the valve selection, the following valve characteristics are best suited:

1. Gate Valves - quick opening
2. Globe Valves - linear and equal percentage
3. Ball Valves - quick opening and linear
4. Butterfly Valves - linear and equal percentage

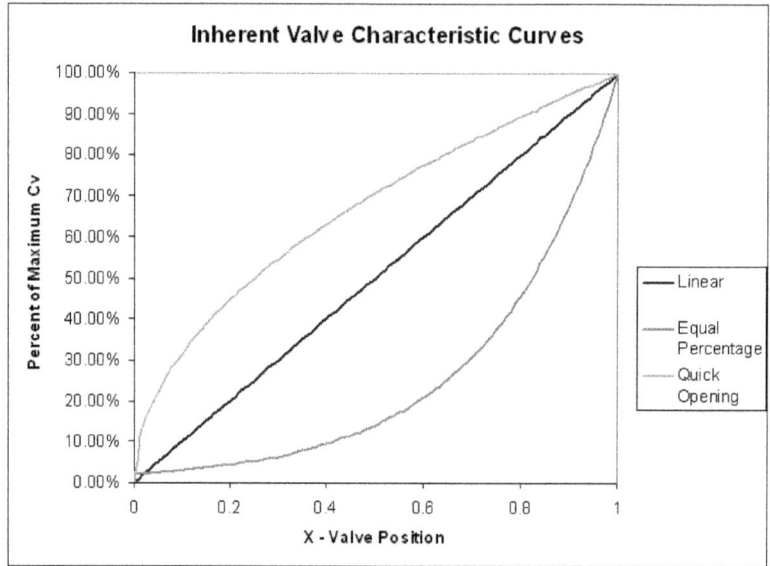

Cv_{max} depends on pipe characteristics and was chosen to be 110gpm in this example. Constant pressure throughout the pipe line is assumed and the curves are accurate when the valve position is between 5% and 95% open.

Comparing the slopes of the graphs for the quick opening and equal percentage valves, we can see that a quick opening valve would experience greater change in flow with slight change in valve position in the lower flow range. The opposite is true for the higher range of flow. The equal percentage valve experiences slighter change in flow with respect to valve position in the lower range of flow.

When selecting the appropriate control valve, it is often the goal of the engineer to choose a valve that will exhibit a linear relationship between F and x over the normal operating position(s) of the valve. This linear relationship provides the most control for the operator. The flow characteristic observed through an installed valve, and all process factors considered (*i.e.* total pressure drop, *etc.*), is termed the installed flow characteristic.

pH AND VISCOSITY CONTROL SENSORS

Therefore, it is not always the case that an inherently linear valve is desirable or even useful. An inherently linear valve is appropriate when there is a linear relationship between the valve position and the actual flow rate; however, consider the case where $\Delta P_L \neq 0$ and of significant value. In this case a valve with inherent equal percentage flow characteristic would be appropriate. The inherently non-linear valve would compensate for ΔP_L and result in an installed linear flow characteristic.

Valve Coefficient, C_v

The valve coefficient, C_v, is defined as the flow in gpm that flows through a valve with a pressure drop of 1psi across the valve ($\Delta P_v = 1\ psi$). C_v is an important parameter that comes up in other modeling equations. It is specific to the valve you are using.

$$C_v = \frac{29.9\ d^2}{\sqrt{K}}$$

d = internal diameter of the pipe in inches

K = resistance coefficient

K is specific to the pipe shape, diameter and material.

Pressure Drop

The pressure drop in the pipe line (pressure drop due to the pipe line and any other equipment in series with the valve), ΔP_L, is defined as:

$$\Delta P_L = k_L \times sg \times f^2$$

f = flow through the pipe in gallons per minute [gpm]

$k_L = [\dfrac{psi}{gpm^2}]$ = constant friction coefficient for the pipe and any equipment in series with the valve

sg = specific gravity of the liquid

The pressure drop across the valve is defined as:

$$\Delta P_v = sg \frac{f^2}{(C_v)^2}$$

So, the total pressure drop is described by the equation:

$$\Delta P_0 = \Delta P_v + \Delta P_L = \left(\frac{1}{C_v^2} + k_L\right) sgf^2$$

If the line pressure drop is negligble(constant pressure in the pipe line) then $\Delta P_L = 0$ and $\Delta P_0 = \Delta P_v$. When $\Delta P_L = 0$ a valve with a linear flow characteristic will be desirable. When $\Delta P_L \neq 0$ and of significant value, a valve with flow characteristics closer to an equal percentage or quick opening valve will be more desirable.

Control Valve Gain

The gain of a control valve (K_v) is defined as the steady-state change in output (flow through a valve, f) divided by the change in input (controller signal, m). The flow through a valve, f, can have units of gallons per minute (gpm), pounds per hour (lb/hr) or standard cubic feet per hour (scfh). The controller signal, m, usually has units of percent of controller output (%CO). The basic relationship for control valve gain is shown below.

$$K_v = \frac{df}{dm}$$

One objective when choosing a valve is to achieve "constant valve gain". The gain is a product of the dependence of valve position on controller output, the dependence of the flow on Cv, and the dependence of Cv on the valve position. The change in valve coefficient, Cv, with respect to valve position depends on the valve characteristics $f(x)$.

For linear characteristics

$$\frac{dC_v}{dx} = Cv_{max}$$

For equal percentage

$$\frac{dC_v}{dx} = (\ln R) C_v$$

Constant Pressure Drop

The dependence of flow on the C_v depends on the pressure drop, so the equation for gain is different when there is a constant pressure drop or a variable pressure drop. If the inlet and outlet pressures do not vary with flow, the gain for either liquid or gas flow in mass units is:

$$K_v = \pm \frac{\ln R}{100} W \frac{lb/hr}{\%CO}$$

%CO = percent controller output

W = mass flow rate

R = valve design parameter (usually between 20 and 50)

Note: the sign is positive if the valve fails closed (air-to-open) and negative if the valve fails open (air-to-close)

pH and Viscosity Control Sensors

Variable Pressure Drop

The valve gain for variable pressure drop is more complicated. As an example, the gain for an equal percentage is

$$K_v = \pm \frac{\ln \alpha}{100} \frac{f\, low}{(1+k_L\, C_v^2)} \frac{gpm}{\%CO}$$

k_L = constant friction coefficient for line, fittings, equipment, *etc.*

The flow term cancels some of the effect of the Cv term until the valve is fully opened, so this gain is less variable with valve opening. Therefore the installed characteristics are much more linear when compared to the inherent characteristics of an equal percentage valve.

Rangeability

Valve rangeability is defined as the ratio of the maximum to minimum controlable flow through the valve. Mathematically the maximum and minimum flows are taken to be the values when 95% (max) and 5% (min) of the valve is open.

$$\text{Rangeability} = \frac{\text{Flow at 95\% valve position}}{\text{Flow at 5\% valve position}}$$

A smaller rangeablilty correlates to a valve that has a small range of controllable flowrates. Valves that exhibit quick opening characteristics have low rangeablilty values. Larger rangeability values correlate to valves that have a wider range of controllable flows. Linear and equal percentage valves fall into this category.

Another case to consider is when the pressure drop across the valve is independent of the flow through the valve. If this is true then the flow is proportional to C_V and the rangeability can be calculated from the valve's flow characteristics equation.

Modeling Installed Valve Characteristics

When a valve is installed in series with other pieces of equipment that produce a large pressure drop in the line compared to the pressure drop across the valve, the actual valve characteristics deviate from the inherent characteristics. At large in-line pressure drops the pressure drop, and consequently the valve coefficient, varies with flow through the valve. These changes can cause changes in the rangeabilty and distorts inherent valve characteristics.

Special Considerations for the Equation describing Flow Through a Valve

Compressible Fluids: Manufacturers such as Honeywell, DeZurik, Masoneilan and Fischer Controls have modified the flow equation to model compressible flows. The equations are derived from the equation for flow through a valve but include unit conversion factors and corrections for temperature and pressure, which affect the density of the gas. It is important to remember to account for these factors if you are working with a compressible fluid such as steam.

Accuracy: This equation, and its modified forms, is most accurate for water, air or steam using conventional valves installed in straight pipes. If you are dealing with non-Newtonian, viscous or two phase systems the calculations will be less accurate.

Chapter 6

Gasification Introduction

Gasification is a technological process that can convert any carbonaceous (carbon-based) raw material such as coal into fuel gas, also known as synthesis gas (syngas for short). Gasification occurs in a gasifier, generally a high temperature/pressure vessel where oxygen (or air) and steam are directly contacted with the coal or other feed material causing a series of chemical reactions to occur that convert the feed to syngas and ash/slag (mineral residues).

Syngas is so called because of its history as an intermediate in the production of synthetic natural gas. Composed primarily of the colourless, odorless, highly flammable gases carbon monoxide (CO) and hydrogen (H_2), syngas has a variety of uses. The syngas can be further converted (or shifted) to nothing but hydrogen and carbon dioxide (CO_2) by adding steam and reacting over a catalyst in a water-gas-shift reactor. When hydrogen is burned, it creates nothing but heat and water, resulting in the ability to create electricity with no carbon dioxide in the exhaust gases.

Furthermore, hydrogen made from coal or other solid fuels can be used to refine oil, or to make products such as ammonia and fertilizer. More importantly, hydrogen enriched syngas can be used to make gasoline and diesel fuel. Polygeneration plants that produce multiple products are uniquely possible with gasification technologies. Carbon dioxide can be efficiently captured from syngas, preventing its greenhouse gas emission to the atmosphere and enabling its utilization (such as for Enhanced Oil Recovery) or safe storage.

Gasification offers an alternative to more established ways of converting feedstocks like coal, biomass, and some waste streams into electricity and other useful products. The advantages of gasification in specific applications and conditions, particularly in clean generation of electricity from coal, may make it an increasingly important part of the world's energy and industrial markets.

The stable price and abundant supply of coal throughout the world makes it the main feedstock option for gasification technologies going forward. The technology's placement markets with respect to many techno-economic and political factors, including costs, reliability, availability and maintainability (RAM), environmental considerations, efficiency, feedstock and product flexibility, national energy security, public and government perception and policy, and infrastructure will determine whether or not gasification realizes its full market potential.

The graphic below is a representation of a gasification process for coal, depicting both the feedstock flexibility inherent in gasification, as well as the wide range of products and usefulness of gasification technology.

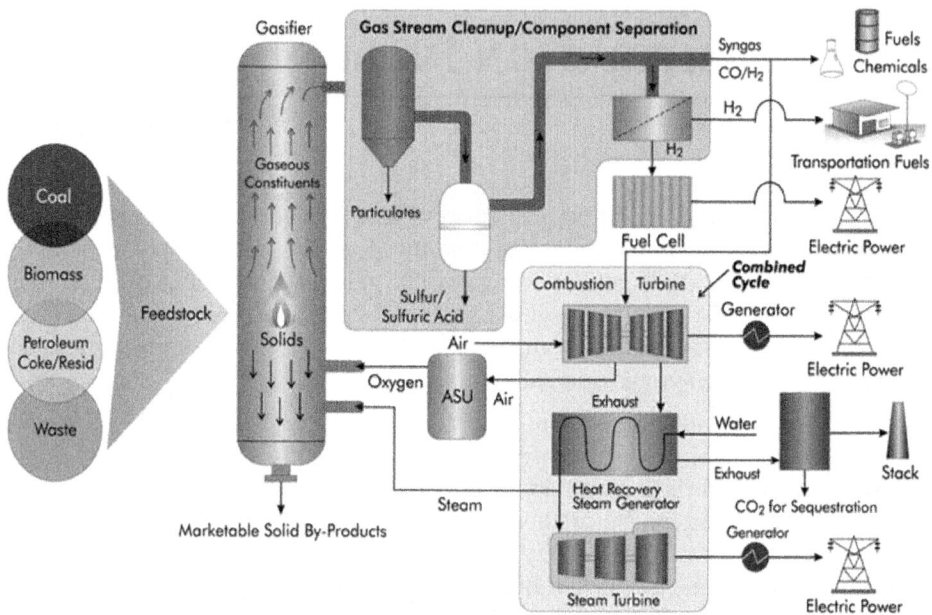

FUNDAMENTALS

Gasification is a partial oxidation process. The term partial oxidation is a relative term which simply means that less oxygen is used in gasification than would be required for combustion (*i.e.*, burning or complete oxidation) of the same amount of fuel. Gasification typically uses only 25 to 40 percent of the theoretical oxidant (either pure oxygen or air) to generate enough heat to gasify the remaining unoxidized fuel, producing syngas.

The major combustible products of gasification are carbon monoxide (CO) and hydrogen (H_2), with only a minor amount of the carbon completely oxidized to carbon dioxide (CO_2) and water. The heat released by partial oxidation provides most of the energy needed to break up the chemical bonds in the feedstock, to drive the other endothermic gasification reactions, and to increase the temperature of the final gasification products.

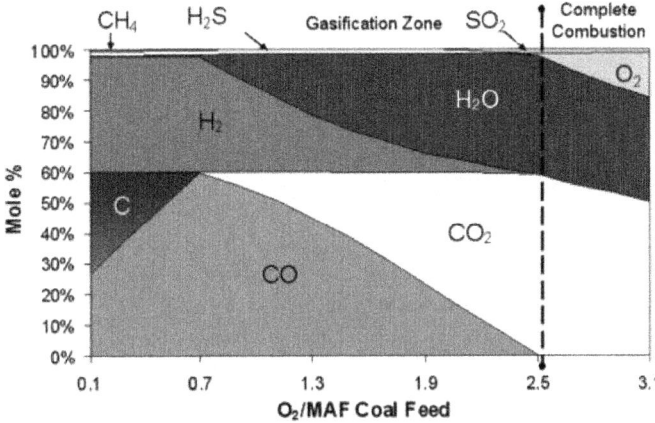

Reactions and Transformations

The chemistry of gasification is quite complex and is accomplished through a series of physical transformations and chemical reactions within the gasifier. Some of the major chemical reactions are shown in the diagram below. In a gasifier, the carbonaceous feedstock undergoes several different processes and/or reactions:

- *Dehydration* – Any free water content of the feedstock evapourates, leaving dry material and evolving water vapour which may enter into later chemical reactions.

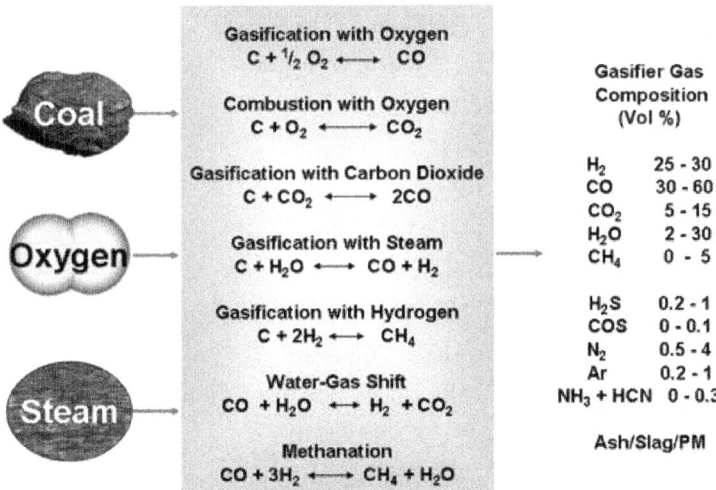

- *Pyrolysis* – This occurs as the feedstock is exposed to rising temperature in the gasifier. Devolatization and breaking of the weaker chemical bonds occurs, releasing volatile gases such as tar vapours, methane, and hydrogen, along with producing a high molecular weight char which will undergo gasification reactions.

- *Combustion* – The volatile products and some of the char react with limited oxygen to form carbon dioxide (CO_2), carbon monoxide (CO), and in doing so, provide the heat needed for subsequent gasification reactions.
- *Gasification* – The remaining char reacts with CO_2 and steam to produce CO and hydrogen (H_2).
- *Water-gas-shift and methanation* – These are separate reversible gas phase reactions taking place simultaneously based on gasifier conditions. These are minor reactions which play a small role within in the gasifier. Depending on the desired product, the syngas may undergo further water-gas shift and methanation processing downstream from the gasifiers.

Detailed Gasification Chemistry

The chemical reactions of gasification can progress to different extents depending on the gasification conditions (like temperature and pressure) and the feedstock used. Combustion reactions take place in a gasification process, but, in comparison with conventional combustion which uses a stoichiometric excess of oxidant, gasification typically uses one-fifth to one-third of the theoretical oxidant.

This only partially oxidizes the carbon feedstock. As a "partial oxidation" process, the major combustible products of gasification are carbon monoxide (CO) and hydrogen, with only a minor portion of the carbon completely oxidized to carbon dioxide (CO_2). The heat produced by the partial oxidation provides most of the energy required to drive the endothermic gasification reactions.

Within a gasification process, the major chemical reactions are those involving carbon, CO, CO_2, hydrogen (H_2), water (steam) and methane (CH_4), as follows:

The combustion reactions:
1. $C + \frac{1}{2} O_2 \rightarrow CO$ (-111 MJ/kmol)
2. $CO + \frac{1}{2} O_2 \rightarrow CO_2$ (-283 MJ/kmol)
3. $H_2 + \frac{1}{2} O_2 \rightarrow H_2O$ (-242 MJ/kmol)

Other important gasification reactions include:
4. $C + H_2O \leftrightarrow CO + H_2$ "the Water-Gas Reaction"
   (+131 MJ/kmol)
5. $C + CO_2 \leftrightarrow 2CO$ "the Boudouard Reaction"
   (+172 MJ/kmol)
6. $C + 2H_2 \leftrightarrow CH_4$ "the Methanation Reaction"
   (-75 MJ/kmol)

With the above, the combustion reactions are essentially carried out to completion under normal gasification operating conditions. And, under the condition of high carbon conversion, the three heterogeneous reactions (reactions 4 to 6) can be reduced to two homogeneous gas phase reactions of water-gas-shift and steam methane-reforming (reactions 7 and 8 below), which collectively play a key role in determining the final equilibrium synthesis gas (syngas) composition.

7. $CO + H_2O \leftrightarrow CO_2 + H_2$ "Water-Gas-Shift Reaction"
   (-41 MJ/kmol)
8. $CH_4 + H_2O \leftrightarrow CO_2 + 3 H_2$ "Steam-Methane-Reforming Reaction"
   (+206 MJ/kmol)

In the low-oxygen, reducing environment of the gasifier, most of the feedstock's sulfur coverts to hydrogen sulfide (H_2S), with a small amount forming carbonyl sulfide (COS). Nitrogen chemically bound in the feed generally converts to gaseous nitrogen (N_2), with some ammonia (NH_3), and a small amount forming hydrogen cyanide (HCN). Chlorine is primary converted to hydrogen chloride (HCl).

In general, the quantities of sulfur, nitrogen, and chloride in the fuel are sufficiently small that they have a negligible effect on the main syngas components of H_2 and CO. Trace elements associated with both organic and inorganic components in the feed, such as mercury, arsenic and other heavy metals, appear in the various ash and slag fractions, as well as in gaseous emissions, and need to be removed from the syngas prior to further use.

Gasification	versus	Combustion
CO	← C →	CO_2
H_2	← H →	H_2O
N_2	← N →	NO_x
H_2S	← S →	SO_x
	← O →	O_2

The table at right summarizes the main transformations of solid fuel constituents to gaseous species in both gasification and combustion. This shows clearly the marked differences between gasification (resulting in syngas) and combustion (resulting in exhaust gas).

Thermodynamics and Kinetics

Gasification reactions are reversible. The direction of the reaction and its conversion are subjected to the constraints of thermodynamic equilibrium and reaction kinetics. The combustion reactions

$C + \frac{1}{2} O_2 \rightarrow CO$ (-111 MJ/kmol)
$CO + \frac{1}{2} O_2 \rightarrow CO_2$ (-283 MJ/kmol)
$H_2 + \frac{1}{2} O_2 \rightarrow H_2O$ (-242 MJ/kmol)

essentially go to completion (to the right). The thermodynamic equilibrium of the gasification reactions

$CO + H_2O \leftrightarrow CO_2 + H_2$ "Water-Gas-Shift Reaction" (-41 MJ/kmol)
$CH_4 + H_2O \leftrightarrow CO_2 + 3 H_2$ "Steam-Methane-Reforming Reaction" (+206 MJ/kmol)

are relatively well defined and collectively impose a strong influence on the thermal efficiency and the produced syngas composition of a gasification process. Thermodynamic modeling has been a useful tool for estimating key design parameters for a gasification process, for example:

- Calculating of the relative amounts of oxygen and/or steam required per unit of coal feed.
- Estimating the composition of the produced syngas.
- Optimizing the process efficiency at various operating conditions.

Other deductions concerning gasification process design and operations can also be derived from the thermodynamic understanding of its reactions. Examples include:

- To produce a syngas with a low methane content, a high temperature and substantial amount of steam in excess of the stoichiometric requirement are required.
- Gasification at very high temperature, on the other hand, will increase oxygen consumption and decrease the overall process efficiency.
- To produce a syngas with a high methane content, gasification needs to be operated at low temperature (~700°C), but the methanation reaction kinetics will be poor without the presence of a catalyst.
- There is considerable advantage to carry out gasification under pressure. At a typical entrained flow gasifier operation temperature of ~2,700°F (1,500°C), the syngas composition shows very little change as a function of operating pressure, but significant savings in compression energy and cost reduction from using smaller equipment can be realized.

Relative to the thermodynamic understanding of the gasification process, its kinetic behaviour is more complex. Very little reliable kinetic information on coal gasification reactions exists, partly because it is highly depended on the process conditions and the nature of the coal feed, which can vary significantly with respect to composition, mineral impurities, and reactivity. Certain impurities, in fact, are known to have catalytic activity on some of the gasification reactions.

Syngas Composition

The figure of gasification reactions and transformations illustrated the concept of coal gasification, and noted resulting composition of syngas. This can vary significantly depending on the feedstock and the gasification process involved; however typically syngas is 30 to 60% carbon monoxide (CO), 25 to 30% hydrogen (H_2), 0 to 5% methane (CH_4), 5 to 15% carbon dioxide (CO_2), plus a lesser or greater amount of water vapour, smaller amounts of the sulfur compounds hydrogen sulfide (H_2S), carbonyl sulfide (COS), and finally some ammonia and other trace contaminants.

Wabash River Syngas Composition

The following table shows the operational syngas composition for the Wabash River Coal Gasification Repowering Project over the four year demonstration period from 1996 through 1999.

Year	1996		1997		1998		1999	
Concentration	Low	High	Low	High	Low	High	Low	High
Hydrogen, %	32.87	34.21	32.90	34.40	32.71	33.82	32.31	33.44
Carbon Dioxide, %	14.89	17.13	16.60	16.90	14.92	16.06	15.25	16.22
Carbon Monoxide, %	42.34	46.03	42.20	46.70	44.25	46.73	44.44	46.31
Methane, %	1.26	1.99	1.04	2.02	1.90	2.09	1.88	2.17
Hydrogen Sulfide, ppmv	17.28	83.36	43.08	106.50	23.48	107.2	86.32	106.0
Carbonyl Sulfide, ppmv	36.26	162.13	22.59	111.80	9.03	36.63	11.36	24.22
Heat of Combustion, Btu/sef(HHV)	256	280	254	283	268	284	267	280

The Wabash facility, using E-Gas™ gasification technology, operated on a variety of fuels over this period including two different types of coal and petroleum coke. Syngas composition remained relatively constant despite changes in coal composition. Although the gasifier is capable of handling a wide range of feedstocks, variations from the coal used as the design basis for the system can reduce syngas and steam production (differences in production will depend on the coal feed and how it differs from the design feed).

Sudden changes in feedstock can also cause disruptions in other plant processes. The syngas H_2 : CO ratio is relatively high (>0.7), typical of a slurry-feed gasification system. The high levels of COS during the earlier years of operation were a result of problems with the COS-hydrolysis unit (specifically, catalyst poisoning by trace metals and chlorides and surface area degradation by overheating). These problems were remedied over the course of the project.

Syngas Composition Variability

The linked table summarizes data from several demonstrations, analyses, and reports on syngas composition across a variety of gasifiers, and types of coal feedstocks, showing the wide variations that occur. Caution must be used in examining this table, as many factors may be missing that have significant impact on the syngas produced. For example, to accurately detail effects on syngas composition a detailed account of the coal used (ultimate and proximate analysis) would be required.

Gasifier operating mode, operating conditions, *etc.*, would also need to be specified (for example, the GE gasifier can be operated in three different modes: total quenched; radiant & convective cooling; radiant followed by partial quenched—the syngas could be quite different for each). This table, therefore, serves as a general overview and is not detailed enough for design work.

The data shows that a wide range of syngas compositions can be obtained by varying the gasifier type, feedstock, and operation parameters. The typical design process of a gasification facility entails selecting a readily available feedstock for the potential facility site and a suitable gasification technology, and then using variations on downstream processes to optimize the syngas composition for creating the desired end product.

Syngas Optimized for Intended Products

The discussion of syngas composition is of considerable importance considering the varying requirements on composition and impurities demanded according to final uses of the syngas. The following table shows the widely varying characteristics desirable for the principal uses of syngas, including use as fuel gas to fire boilers or turbines in power cycles, use of syngas as feedstock for production of synthetic fuels such as gasoline, use as feedstock for methanol synthesis, and use as feedstock for production of hydrogen.

Table. Desirable Syngas Characteristics for Different Applications.

Product	Synthetic Fuels	Methanol	Hydrogen	Fuel Gas	
	FT Gasoline			Boiler	Turbine
Hydrogen to carbon monoxide ratio (H_2/CO)	0.6[a]	~2.0	High	Unimportant	Unimportant
Carbon Dioxide	Low	Low[c]	Not Important[b]	Not Critical	Not Critical
Hydrocarbons	Low[d]	Low[d]	Low[d]	High	High
Nitrogen	Low	Low	Low	Note[e]	Note[e]
Water Vapour	Low	Low	High[f]	Low	Note[g]
Contaminants	<1 ppm Sulfur Low Particulates	<1 ppm Sulfur Low Particulates	<1 ppm Sulfur Low Particulates	Note[k]	Low Part. Low Metals
Heating Value	Unimportant[h]	Unimportant[h]	Unimportant[h]	High[i]	High[i]
Pressure, bar	~20-30	~50(liquid phase) ~140(vapour phase)	~28	Low	~400
Temperature, °C	200-300[j] 300-400	100-200	100-200	250	500-600

(a) Depends on catalyst type. For iron catalyst, value shown is satisfactory; for cobalt catalyst, near 2.0 should be used.

(b) Water gas shift will have to be used to convert CO to H_2; carbon dioxide (CO_2) in syngas can be removed at same time as CO_2 generated by the water gas shift reaction.

(c) Some CO_2 can be tolerated if the H_2/CO ratio is above 2.0(as can occur with steam reforming of natural gas); if excess H_2 is available, the CO_2 will be converted to methanol.

(d) Methane and heavier hydrocarbons need to be recycled for conversion to syngas and represent system inefficiency.

(e) Nitrogen (N_2) lowers the heating value, but level is unimportant as long as turbine or boiler system efficiencies are satisfactory. Presence of excess N_2 may be unacceptable in carbon capture scenarios, however.

(f) Water is required for the water gas shift reaction.

(g) Can tolerate relatively high water levels; steam sometimes added to moderate combustion temperature to control nitrogen oxides (NO_x) formation.

(h) As long as H_2/CO and impurities levels are met, heating value is not critical.

(i) Efficiency improves as heating value increases.

(j) Depends on catalyst type; iron catalysts typically operate at higher temperatures than cobalt catalysts.

(k) Small amounts of contaminants can be tolerated.

Choice of coal and the gasification technology utilized will of course have significant impact on syngas composition. Additionally, downstream processes of various types for conditioning the syngas to meet its final application characteristics are available:

- Water-Gas Shift – Water-gas shift is used to adjust the ratio of H_2 to CO.
- Acid Gas Removal – Used to remove sulfur (typically in the form of hydrogen sulfide) and CO_2.
- Hydrogen Separation – Separates high purity H_2 from syngas streams.
- Methanation – Converts carbon oxides and hydrogen in the syngas into methane and water.
- Gas Cooling
- Syngas Cleanup and Trace Component Removal

Staged Gasification

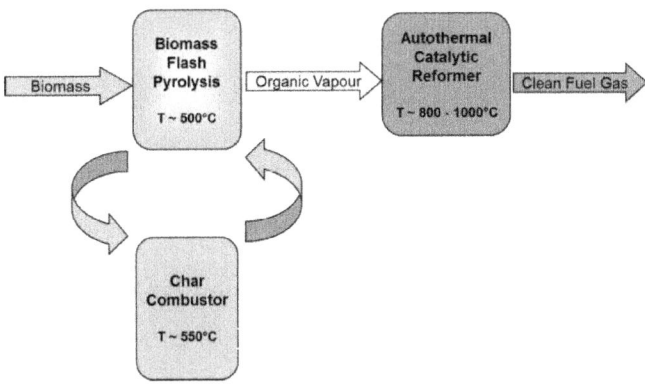

In BTG's two-stage gasification process, biomass and residues are converted at low temperature in an organic vapour, and subsequently the vapour is catalytically reformed into a clean fuel gas. The low temperature stage is to a large extent based on BTG's pyrolysis process. In the gasification stage a catalyst can be applied.

Background

Biomass gasification systems are investigated and developed for a long period with some emphasis on tar and tar reduction. The presence of tar in the fuel gas hampers troublefree operation of prime movers. Additionally, it would be a significant advantage if systems can convert a wide range of possible feedstocks, *i.e.*, multi-fuel systems. Worldwide, huge amounts of biomass residues and waste streams are available like *e.g.* agricultural residues.

Typically, these streams have a low bulk density and contain significant amounts of minerals. When used in conventional systems these minerals may cause ash melting problems or result in high emissions. For example, the gas phase concentrations of K and Cl significantly increase at temperatures above 700 °C.

In BTG's two-stage gasification process biomass is "vapourized" at low temperatures, and in the second stage the vapours are reformed. In the second stage a catalyst may be applied. Typical features of the system are:

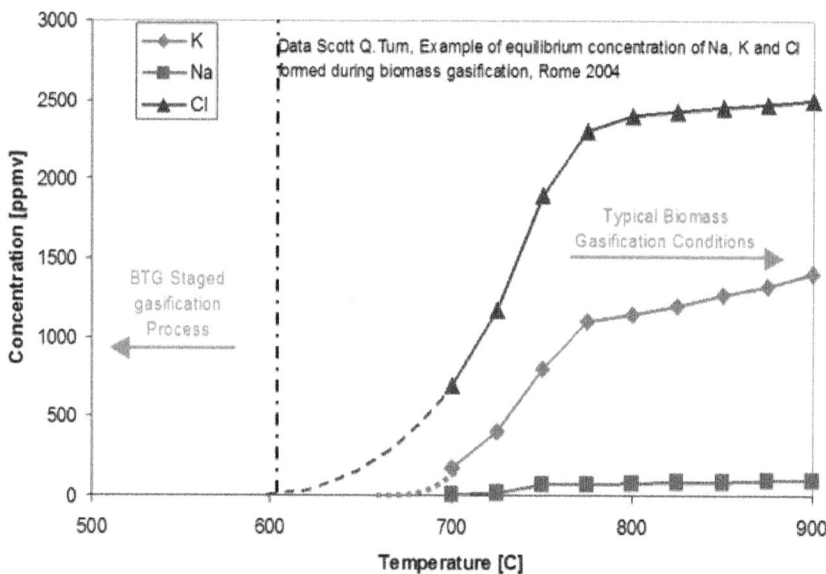

- Organic vapours do not contain minerals (*i.e.* catalyst poisons) and the use of catalysts becomes feasible;
- Reforming of a vapour is much easier to control than gasification of solids;
- Ammonia originating from fuel nitrogen will be converted to nitrogen and hydrogen in case a Ni catalyst is applied;

- Due to the low temperature in the pyrolysis stage a high fuel flexibility is achieved
- To some extend the system is self controlling with respect to the water content of the feedstock without affecting the gas quality of the fuel gas.

The charcoal produced in the pyrolysis stage is used to preheat the biomass and to evapourate the water. Charcoal is not used in the high temperature reforming stage. Minerals will not reach a high temperature in the process. Due to the application of a fast pyrolysis stage, the amount of charcoal is limited.

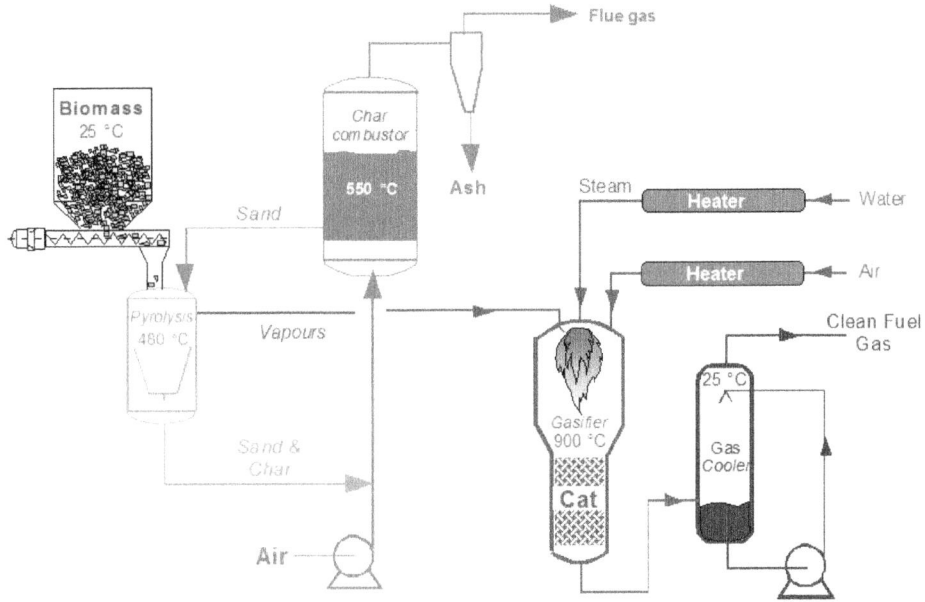

Process Description

Biomass is fed to the fast pyrolysis reactor, where organic vapours are produced. Whereas in the pyrolysis process the vapours are condensed, in the two-stage gasifier the vapours are reformed into a clean fuel gas. In the top section of the 'gasifier' the vapours are mixed with (preheated) air to increase the temperature to 800 – 950 °C.

The bottom part can be filled with a reforming catalyst to convert remaining tar and ammonia. In the last stage the gas is cooled to ambient. An overall, cold-gas efficiency in the range of 65-80% is expected.

Process Description Biomass is fed to the fast pyrolysis reactor, where organic vapours are produced. Whereas in the pyrolysis process the vapours are condensed, in the two-stage gasifier the vapours are reformed into a clean fuel gas. In the top section of the 'gasifier' the vapours are mixed with (preheated) air to increase the temperature to 800 – 950 °C.

The bottom part can be filled with a reforming catalyst to convert remaining tar and ammonia. In the last stage the gas is cooled to ambient. An overall, cold-gas efficiency in the range of 65-80% is expected.

Gasification of Wood Chips

Gasification is an alternative to traditional combustion plants as it is possible to generate more efficient electricity in small plants and thereby reduce the fuel input.

Gasification is a reliable and clean energy technology that can turn biomass or any material containing carbon into synthetic gas. This gas can then be used in a gas engine for the production of electricity and heat. A significant benefit is that the equipment is compact, which enables a plant to be built in small communities where electricity, steam or heat is needed.

Small Gasification Plants Can Obtain a High Electrical Efficiency

Gasification is an alternative to traditional combustion plants as it is possible to generate more electricity than with the existing solutions based on steam in small district heating and industrial plants.

In small combustion systems (up to10MWe), energy output is limited to production of hot water (*e.g.* for district heating) or steam for relatively low-efficiency processes. In gasification systems efficiency may exceed 30%.

The technology plays a key role in the development of CO_2-neutral energy supplies. Because it emits less CO_2 and NO_x pr.-produced KWhe, a gasification plant significantly reduces the environmental impact compared to traditional technologies. And because less fuel is needed, the whole production chain is smaller. The plant can be placed decentralised to minimise losses in the grid network and secure local workers' jobs for sustainable energy production.

Profitable Feed-in Tariffs for Smaller Plants

Feed-in tariffs, or renewable energy payments, vary from country to country. But generally governments support smaller plants, which are local and frequently use locally produced fuel. This by itself is more environmentally sound since transport is closer.

The Rapid Turn-down Ratio is Suitable for the Electrical Grid of the Future

Biomass plants are designed to deliver a certain amount of output. Turning that down, or feeding the plant less fuel, can be difficult as it gets harder to stay within emission limits. With gasification it's easier to stay within the limits even when you turn down the amount of fuel. This is especially important during times when you need less electricity. The rapid turn-down ratio is very suitable for the electrical grid of the future. Anticipating that much of our electricity will come from wind energy, a gasification plant can secure electrical supply to the grid, when the wind is not blowing.

No Specialised Pre-treatment of Fuel is Needed

The great advantage to this technology is that you do not need to pre-treat your wood chips. Our fuel is primarily wood chips and comes directly from the woods where it is chopped. This saves time and money because specialized treatment such as drying or sorting of fractions is unnecessary.

Tailored Gasification Technology to Meet Your Needs

We can tailor our gasification technology to suit several different concepts, *e.g.*: power plants, combined heat and power plants (CHP), combined cycle gasification, heat, and syngas. The latter can be used as fuel for an external superheater at a waste-to-energy plant (WasteBoost™). This concept is designed to increase steam parameters and thereby electrical efficiency.

Today we market a CHP design of a 2MWe plant with a power efficiency of 28 %. This concept is based on virgin wood chips coming directly from the woods without any pretreatment or drying.

Capital investment cost figures for gasification plants are relatively high compared to traditional steam cycle concepts. However, where high feed-in tariffs are provided, an evaluation over a period of Total Cost of Ownership (TCO) gives the optimal feasibility for the investor. This is mainly due to the much higher electrical efficiency.

Capital investment cost figures for the "electro-mechanical" part of the plants are approx. 13,500,000 EUR for the 2MWe design. Add approx. 20% extra cost for the turnkey prices.

This cost level is only feasible in a few countries, mainly in Europe, where you have a high tariff for power and have the possibility of selling the heat. Depending upon your required payback time and fuel costs, *etc.*, this will form the basis for the evaluation for the TCO.

How Gasification Works

In the gasification process, the unhomogeneous biomass such as wood (bark, branches, leaves, woodchips, *etc.*) goes through a thermal chemical process using moisturised air.

The moist biomass is fed at the top and descends though gases rising in the reactor.

In the upper zone a drying process occurs, below which pyrolysis is taking place. Following this, the material passes through a reduction zone (gasification) and in the zone above the grate an oxidation process is carried out (combustion). To supply air for the combustion process and steam for the gasification process, moist, hot air is supplied at the bottom of the reactor.

Combustible gas at a low temperature is discharged at the top of the reactor, and inert ash from the heat-generating combustion process is extracted from the reactor bottom through a water lock.

The process breaks down the unhomogeneous biomass to the molecular level and converts it into a homogeneous fuel: synthetic gas (syngas).

The syngas can be used to create a variety of valuable products or you can burn it in a gas engine, which allows you to produce more electricity than any other available technology. The flue gas created in the process can be used to produce steam or heat water that can be provided to a district heating grid.

Our Fully Automatic Gasification Plants Can Save Labour Cost

Gasification plants from us are fully automatic and require minimal monitoring. This means that the operating and maintenance costs can be kept at a very reasonable level. Our demo plant at Harboøre, Denmark, is in unmanned operation evenings, nights, and weekends and to date the gas engines have been in operation for more than 100,000 hours, which is considered outstanding. Because it's an unmanned operation, you save labour costs.

Residues from Gasification: Extremely Low Unburned Carbon

The ash produced in the gasification process makes up just 1% of the biomass that is fed into the plant. In this ash there is very low TOC (Total Organic Carbon) which means that there is no energy loss. The ash is so pure that it can be used as fertilizer on farming fields.

As with All our Technology, our Gasification Process is Highly Reliable

The water produced in the process is cleaned of tar which is converted into bio-oil. The bio-oil can be stored in a tank so that when the facility is closed down for revision you can still produce energy and provide the community with heat and electricity from biomass instead of being forced to supplement with fossil fuels.

The bio-oil is also useful in times of special need, during a very cold winter, for instance, when you need more hot water than usual for the district heating grid.

This makes our gasification plants extremely practical as well as reliable.

For more information please contact our gasification specialist Robert Heeb, (+45)76143596.

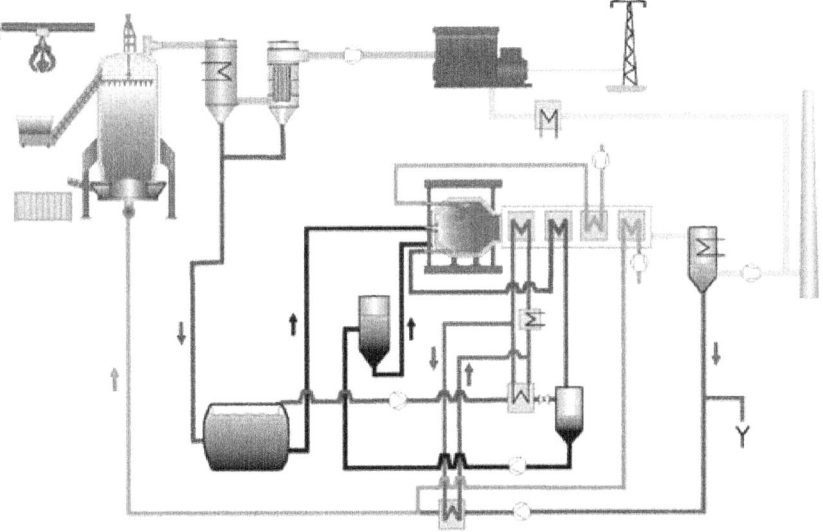

Fig. : Process illustration of a gasification plant.

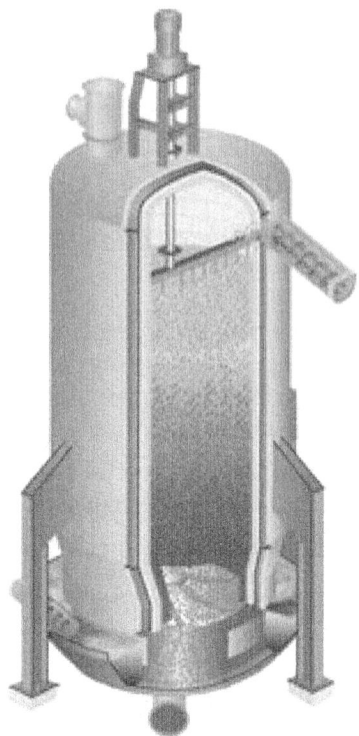

Fig. : The gasifier.

GASIFICATION TECHNOLOGY

Clean coal research and development is focused on developing and demonstrating advanced power generation and carbon capture, utilization and storage technologies for existing facilities and new fossil-fueled power plants by increasing overall system efficiencies and reducing capital costs. In the near-term, advanced technologies that increase the power generation efficiency for new plants and technologies to capture carbon dioxide (CO_2) from new and existing industrial and power-producing plants are being developed. In the longer term, the goal is to increase energy plant efficiencies and reduce both the energy and capital costs of CO_2 capture and storage from new, advanced coal plants and existing plants. These activities will help allow coal to remain a strategic fuel for the nation while enhancing environmental protection.

Carbon Capture, Utilization and Storage Research

The Carbon Capture, Utilization and Storage research and development program advances safe, cost effective, capture and permanent geologic storage and/or use of CO_2. The technologies developed and large-volume injection tests conducted through this program will be used to benefit the existing and future fleet of fossil fuel power generating facilities by creating tools to increase our un-

derstanding of geologic reservoirs appropriate for CO_2 storage and the behaviour of CO_2 in the subsurface.
- Carbon Capture
- Industrial Carbon Capture and Storage
- Carbon Storage/Sequestration

Advanced Energy Systems

Advanced Energy Systems program focuses on improving the efficiency of coal-based power systems, enabling affordable CO_2 capture, increasing plant availability, and maintaining the highest environmental standards. The program supports gasification research and development to convert coal into synthesis gas that can be converted into electricity, chemicals, hydrogen, and liquid fuels.

The program also advances hydrogen turbine designs to improve the performance of pre-combustion CO_2 capture systems and supports advanced combustion systems through research and development focused on new high-temperature materials and the continued development of oxy-combustion technologies.
- Hydrogen Turbines
- Gasification Technologies
- Advanced Combustion
- Solid Oxide Fuel Cells
- Hydrogen from Coal
- Coal to Liquids

The Turbines of Tomorrow

Combustion (gas) turbines are key components of advanced systems designed for new electric power plants in the United States. With gas turbines, power plants will supply clean, increasingly fuel-efficient, and relatively low-cost energy.

Typically, a natural gas-fired combustion turbine-generator operating in a "simple cycle" converts between 25 and 35 percent of the natural gas heating value to useable electricity. Today, most new smaller power plants also install a recuperator to capture waste heat from the turbine's exhaust to preheat combustion air and boost efficiencies. In most of the new larger plants, a "heat recovery steam generator" is installed to recover waste heat in the exhaust to generate steam for a steam turbine-generator. This configuration is called a "combined cycle."

How Gas Turbine Power Plants Work

In 1992, the U.S. Department of Energy's Fossil Energy program began an intensive effort to break through technical barriers that had essentially capped gas turbine efficiencies. Within eight years, this program produced turbine systems that could operate at temperatures in excess of 2600 degrees F (300 degrees hotter than conventional turbines) and achieve efficiencies above 60 percent, a mark once thought unachievable.

At the same time, new combustion techniques were developed to limit the formation of nitrogen oxide (NO_x) emissions (the principal air pollutant released by gas turbines). As a result, high-efficiency natural gas turbines continue to be among the cleanest ways to generate electricity from fossil fuels. Gas turbines being developed under the Department of Energy's Fossil Energy Program will be able to achieve high efficiencies with NO_x emissions at less than 2 parts per million.

The use of gases produced from coal as gas turbine fuel offers an attractive means for efficiently generating electric power from our Nation's most abundant fossil fuel resource. The adaptation of gas turbine technologies to use with fuels produced from coal gasification has been demonstrated under the Clean Coal Technology Program.

The Department of Energy's Fossil Energy Program is developing key technologies that will enable advanced turbines to operate cleanly and efficiently when fueled with coal derived synthesis gas and hydrogen fuels. Developing this turbine technology is critical to the creation of near-zero emission power generation technologies.

The Federal turbine research and development program is an investment in secure U.S. electric power production that is clean, efficient, affordable and fuel flexible, and will make possible the continued use of coal--our Nation's largest domestic fossil energy resource.

Gasification Technology R&D

Coal gasification offers one of the most versatile and clean ways to convert coal into electricity, hydrogen, and other valuable energy products.

Coal gasification electric power plants are now operating commercially in the United States and in other nations, and many experts predict that coal gasification will be at the heart of future generations of clean coal technology plants.

Rather than burning coal directly, gasification (a thermo-chemical process) breaks down coal - or virtually any carbon-based feedstock - into its basic chemical constituents. In a modern gasifier, coal is typically exposed to steam and carefully controlled amounts of air or oxygen under high temperatures and pressures. Under these conditions, molecules in coal break apart, initiating chemical reactions that typically produce a mixture of carbon monoxide, hydrogen and other gaseous compounds.

Gasification, in fact, may be one of the most flexible technologies to produce clean-burning hydrogen for tomorrow's automobiles and power-generating fuel cells. Hydrogen and other coal gases can also be used to fuel power-generating turbines, or as the chemical "building blocks" for a wide range of commercial products.

The Energy Department's Office of Fossil Energy is working on coal gasifier advances that enhance efficiency, environmental performance, and reliability as well as expand the gasifier's flexibility to process a variety of coals and other feedstocks(including biomass and municipal/industrial wastes).

Environmental Benefits

The environmental benefits of gasification stem from the capability to achieve extremely low SO_x, NO_x and particulate emissions from burning coal-derived gases. Sulfur in coal, for example, is converted to hydrogen sulfide and can be captured by processes presently used in the chemical industry. In some methods, the sulfur can be extracted in either a liquid or solid form that can be sold commercially. In an Integrated Gasification Combined-Cycle (IGCC) plant, the syngas produced is virtually free of fuel-bound nitrogen.

NO_x from the gas turbine is limited to thermal NO_x. Diluting the syngas allows for NO_x emissions as low as 15 parts per million. Selective Catalytic Reduction (SCR) can be used to reach levels comparable to firing with natural gas if required to meet more stringent emission levels. Other advanced emission control processes are being developed that could reduce NO_x from hydrogen fired turbines to as low as 2 parts per million.

The Office of Fossil Energy is also exploring advanced syngas cleaning and conditioning processes that are even more effective in eliminating emissions from coal gasifiers. Multi-contaminant control processes are being developed that reduce pollutants to parts-per-billion levels and will be effective in cleaning mercury and other trace metals in addition to other impurities.

Coal gasification may offer a further environmental advantage in addressing concerns over the atmospheric buildup of greenhouse gases, such as carbon dioxide. If oxygen is used in a coal gasifier instead of air, carbon dioxide is emitted as a concentrated gas stream in syngas at high pressure. In this form, it can be captured and sequestered more easily and at lower costs. By contrast, when coal burns or is reacted in air, 79 percent of which is nitrogen, the resulting carbon dioxide is diluted and more costly to separate.

Efficiency Benefits

Efficiency gains are another benefit of coal gasification. In a typical coal combustion-based power plant, heat from burning coal is used to boil water, making steam that drives a steam turbine-generator. In some coal combustion-based power plants, only a third of the energy value of coal is actually converted into electricity.

A coal gasification power plant, however, typically gets dual duty from the gases it produces. First, the coal gases, cleaned of impurities, are fired in a gas turbine - much like natural gas - to generate one source of electricity. The hot exhaust of the gas turbine, and some of the heat generated in the gasification process, are then used to generate steam for use in a steam turbine-generator.

This dual source of electric power, called a "combined cycle," is much more efficient in converting coal's energy into usable electricity. The fuel efficiency of a coal gasification power plant in this type of combined cycle can potentially be boosted to 50 percent or more.

Future concepts that incorporate a fuel cell or a fuel cell-gas turbine hybrid could achieve efficiencies nearly twice today's typical coal combustion plants. If any of the remaining heat can be channeled into process steam or heat, perhaps for nearby factories or district heating plants, the overall fuel use efficiency of future gasification plants could reach 70 to 80 percent.

Higher efficiencies translate into more economical electric power and potential savings for ratepayers. A more efficient plant also uses less fuel to generate power, meaning that less carbon dioxide is produced. In fact, coal gasification power processes under development by the Energy Department could cut the formation of carbon dioxide by 40 percent or more, per unit of output, compared to today's conventional coal-burning plant.

The capability to produce electricity, hydrogen, chemicals, or various combinations while eliminating nearly all air pollutants and potentially greenhouse gas emissions makes coal gasification one of the most promising technologies for energy plants of the future.

Solid Oxide Fuel Cells

Fuel cells are an energy user's dream: an efficient, combustion-less, virtually pollution-free power source, capable of being sited in downtown urban areas or in remote regions that runs almost silently and has few moving parts.

A fuel cell is a galvanic cell that has active materials (*e.g.*, fuel and oxidizer), which are continuously supplied from a source external to the cell and the reaction products continuously removed converting chemical energy to electrical energy. Over a dozen types of fuel cells exist. Developments continue as motivated by the desirability of bigger sizes, more endurance, more power density, less emissions, or lower cost to list a few. The Office of Fossil Energy concentrates its fuel cell research, development, and deployment on Solid Oxide Fuel Cells (SOFC) to be fueled with gasified solid hydrocarbons.

SECA

The Solid State Energy Conversion Alliance (SECA), founded in the fall of 1999, is collaboration between the Federal Government, private industry, academic institutions and national laboratories devoted to the development of low-cost, modular, and fuel-flexible solid oxide fuel cell technology suitable for a variety of power generation applications.

HYDROGEN FROM COAL

Hydrogen from coal research supports goals of increasing energy security, reducing environmental impact of energy use, promoting economic development, and encouraging scientific discovery and innovation by researching and developing novel technologies that convert the nation's abundant coal resources into hydrogen. The use of coal — America's largest domestic fossil energy resource — offers the potential to economically produce hydrogen and capture carbon dioxide emissions for the generation of low-carbon electricity.

Hydrogen can be produced from coal by gasification (*i.e.*, partial oxidation). Coal gasification works by first reacting coal with oxygen and steam under high pressures and temperatures to form synthesis gas, a mixture consisting primarily of carbon monoxide and hydrogen. The synthesis gas is cleaned of impurities and the carbon monoxide in the gas mixture is reacted with steam *via* the water-gas shift reaction to produce additional hydrogen and carbon dioxide.

Hydrogen is removed by a separation system and the highly concentrated CO_2 stream can subsequently be captured and sequestered. The hydrogen can be used in a combustion turbine or solid oxide fuel cell to produce power, or utilized as a fuel or chemical feedstock.

Gasification of coal is a promising technology for the co-production of electric power and hydrogen from integrated gasification combined-cycle (IGCC) technology. However, there currently are no commercial demonstrations of these joint power and hydrogen plants. Conceptual plants have been simulated using computer models to estimate technical and economic performance of co-production facilities.

To reduce costs, novel and advanced technology must be developed throughout the entire system that produces hydrogen from coal. For example, carbon dioxide produced in the hydrogen production process could be sequestered by technologies now being developed in DOE's Carbon Sequestration Program and eventually demonstrated in other activities by the Office of Clean Coal.

Research and Development Needs

The hydrogen from coal production research and development activities include: advanced water-gas shift technologies, hydrogen separation, process intensification, and demonstrations.

- Advanced water-gas shift technologies will focus on the development of more active and impurity-tolerant shift catalysts and technologies that integrate water-gas shift and hydrogen separation into a single step
- Advanced hydrogen separations will explore technology for advanced pressure swing adsorption (PSA), membranes, solvents, reverse selective systems, and other technology alternatives. Areas of focus will be the identification of low-cost materials, stabilization of membranes, membrane

seal and fabrication technologies methods for module preparation and scale-up, and analysis of current status and preferred separation options.
- Process intensification is the concept of integrating several processes into one step, such as synthesis gas clean-up, water-gas shift, and hydrogen separation will be investigated. Novel, "out-of-the-box" concepts will also be studied that produce hydrogen from coal.
- Demonstrations will be performed to test advanced technologies to confirm laboratory, bench-scale, and pre-engineering module results.

The research and development activities performed under the Hydrogen from Coal Program will develop advanced technology for use in future electric power and co-production plants.

Benefits of Producing Hydrogen from Coal

The United States has an abundant, domestic resource in coal — nearly a 250-year supply based on current estimates. The production of hydrogen from coal offers efficiency and environmental benefits when integrated with advanced technologies in coal gasification, power production, and carbon sequestration. The integration of these technologies facilitates the capture of multiple pollutants such as sulfur oxides, nitrogen oxides, mercury, and particulates, as well as greenhouse gases such as carbon dioxide.

Coal to Liquids

The Hydrogen and Clean Coal Fuels Program supports DOE's strategic goals of increasing energy security, reducing environmental impact of energy use, promoting economic development, and encouraging scientific discovery and innovation by researching and developing novel technologies that convert the nation's abundant coal resources into hydrogen and other clean fuels. The use of coal — America's largest domestic fossil energy resource — offers the potential to economically produce hydrogen and capture carbon dioxide emissions for the generation of low-carbon electricity.

In a co-production concept, coal can also be converted into power and other fuels such as gasoline, diesel and aviation fuel; chemicals; and substitute natural gas (SNG) which can help reduce reliance on imported oil and natural gas. Carbon capture and storage can be integrated with these processes and, because these processes typically utilize gasification, biomass can be co-utilized to further reduce the carbon footprint while allowing biomass to take advantage of the economies of scale associated with coal.

As part of DOE's overall effort to research, develop, and demonstrate novel technologies to produce hydrogen for clean power and low-carbon fuels, the Hydrogen and Clean Coal Fuels Program is composed of three key elements:

Hydrogen Production from Coal. America's abundant coal resources offer an attractive option for producing hydrogen to meet the goal of affordable, low-carbon power. The Hydrogen and Clean Coal Fuels Program is investigating innovative technologies and methods to produce, deliver, and utilize hydrogen from coal.

Coal and Biomass to Liquids

Over the last several decades, the Office of Fossil Energy performed RD&D activities that made significant advancements in the areas of coal conversion to liquid fuels and chemicals. Technology improvements and cost reductions that were achieved led to the construction of demonstration-scale facilities. The program is now supporting work to reduce the carbon footprint of coal derived liquids by incorporating the co-feeding of biomass and carbon capture.

In the area of direct coal liquefaction, which is the process of breaking down coal to maximize the correct size of molecules for liquid products, the U.S. DOE made significant investments and advancements in technology in the 1970s and 1980s. Research enabled direct coal liquefaction to produce greater yield, quality, and safety of the liquids and enabled the successful demonstration of continuous operation of plants such as the 200 tons of coal per day Ashland Synthetic Fuels plant in Catlettsburg, KY.

DOE had also successfully partnered with industry to develop liquid phase methanol technology. The project was successfully demonstrated at a capacity of 250 tons of methanol per day at a facility located in Tennessee.

During its demonstration period, the facility produced nearly 104 million gallons of methanol from coal-derived synthesis gas with a demonstrated plant capacity in excess of 300 tons of methanol per day, more than 15 percent greater than the plant's design rate. As a result, the technology is still being utilized to provide a portion of the facility's methanol feedstock requirements.

Research and Development Needs

Based on past research, many of the technical challenges associated with conversion of coal to low-carbon fuel alternatives have been addressed. However, achieving greater reductions in carbon dioxide emissions is one of the key challenges that remain. One option being explored is the co-feeding of biomass with coal.

Co-feeding allows biomass to take advantage of the economies of scale associated with coal, while helping to reduce the carbon dioxide emissions of coal by integrating it with a renewable feedstock. Studies have shown that the life cycle carbon emissions of coal derived liquids are below the petroleum baseline when implementing biomass co-feeding and/or carbon capture, utilization, and storage.

While there has been some commercial testing of co-gasification of coal and biomass, these tests have not been conducted on the range of coal and biomass types available and at varying concentrations. Specifically, the impacts of coal-biomass mixtures on gasification kinetics, operating conditions, gasifier effluent components and concentration, downstream processes, and capability to feed the mixture across a pressure gradient and into a gasifier, must be addressed.

Benefits

Production of low-carbon fuels from coal and coal-biomass mixtures offers several benefits. These fuels will reduce the amount of petroleum that the United States imports because they are produced from domestic resources. These fuels are also considered "drop-in" fuel substitutes, which can be used in existing engines with little or no modification required. Coal and biomass to liquids offers an effective way to provide quality, reliable fuels while reducing the amount of CO_2 produced in the transportation sector.

Other Low-Carbon Fuel Alternatives from Coal

Liquid fuels, chemicals, and SNG can be produced from coal and coal-biomass to reduce reliance on imports while also addressing CO_2 emissions.

Systems Analysis. Systems analysis provides guidance and support for the Hydrogen and Clean Coal Fuels Program by analyzing, measuring, and validating the progress and benefits of the Program's research and development portfolio.

CO_2 Utilization Focus Area

Carbon dioxide (CO_2) utilization efforts focus on pathways and novel approaches for reducing CO_2 emissions by developing beneficial uses for the CO_2 that will mitigate CO_2 emissions in areas where geologic storage may not be an optimal solution. CO_2 can be used in applications that could generate significant benefits. It is possible to develop alternatives that can use captured CO_2 or convert it to useful products such chemicals, cements, or plastics.

Revenue generated from the utilized CO_2 could also offset a portion of the CO_2 Processes or concepts must take into account the life cycle of the process to ensure that additional CO_2 is not produced beyond what is already being removed from or going into the atmosphere. Furthermore, while the utilization of CO_2 has some potential to reduce greenhouse gas emissions to the atmosphere, CO_2 has certain disadvantages as a chemical reactant.

Carbon dioxide is rather inert and non-reactive. This inertness is the reason why CO_2 has broad industrial and technical applications. Each potential use of CO_2 has an energy requirement that needs to be determined; and the CO_2 produced to create the energy for the specific utilization process must not exceed the CO_2 utilized.

The figure below illustrates most of the current and potential uses of CO_2. However, many of these uses are small scale and typically emit the CO_2 to the atmosphere after use, resulting in no reduction in overall CO_2 emissions. Some of the more significant current and potential uses of CO_2 are highlighted in the research underway in this focus area.

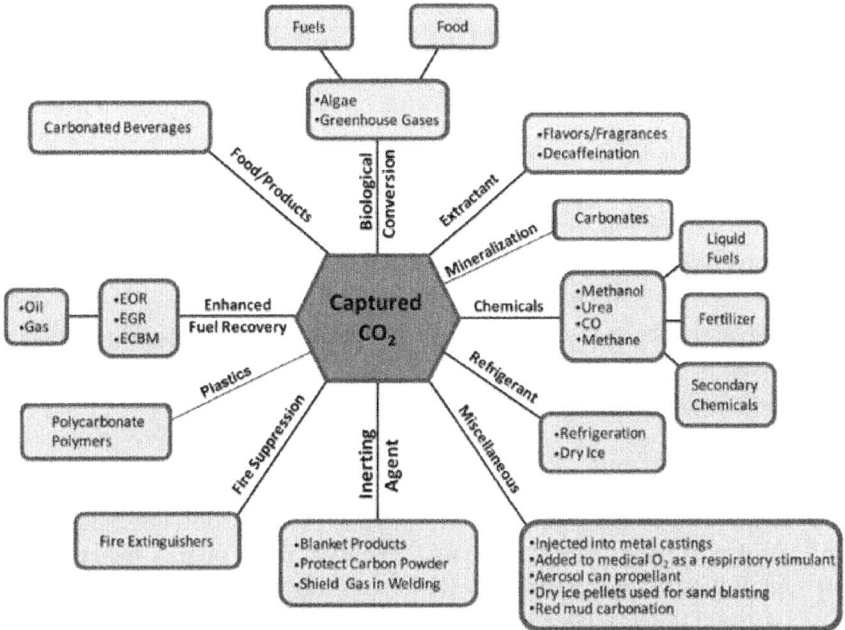

Fig. : Schematic Illustrating the Uses of CO_2.

CO_2 Utilization Goals

CO_2 Utilization covers a broad area of research with different technical challenges. The goals of the Carbon Storage Program are set to achieve successful implementation of various applications at different time horizons.

In general, the area of CO_2 utilization for carbon storage is relatively new and less well-known compared to other storage approaches, such as geologic storage. Thus, more exploratory technological investigations are needed to discover new applications and new reactions. Many challenges exist for achieving successful CO_2 utilization, including the development of technologies capable of economically fixing CO_2 in stable products for indirect storage.

CO_2 Utilization Technologies

Four main CO_2 utilization research focus areas the Carbon Storage Program supports include:

- Cement

    Description

    o Instead of the traditional energy intensive steam curing technology, develop a concrete curing process that consumes substantial amounts of waste CO_2 from onsite flue gases and local combustion sources. The produced concrete products should exhibit material performance equal to that of the traditional curing process, while using less energy. This use of CO_2 should sequester the carbon for many years. The transition between demonstration and commercial scale should be rapid, since the new process and technology is anticipated to require limited modification to the existing curing process.

Research Focus
- o Improve curing rates and CO_2 yield to increase efficiency of CO_2 use.
- o Develop curing processes based on carbonation chemistry rather than hydration chemistry to reduce energy requirements and CO_2 emissions.
- o Develop cement to meet American Standard Test Method (ASTM) standards.

- Polycarbonate Plastics

  Description
  - o Traditional monomers, such as ethylene and propylene, can be combined with CO_2 to produce polycarbonates, such as polyethylene carbonate and polypropylene carbonate. The advantage of this process is that it copolymerizes CO_2 directly with other monomers without having to first convert the CO_2 to carbon monoxide (CO) or some other reactive species. This significantly reduces energy requirements. There are many potential uses for polycarbonate plastics, including coatings, plastic bags, and laminates. Depending on the final fate of the plastic, such as a landfill, this use could represent semi-permanent storage of carbon. This promising CO_2 utilization technology needs to be proven at pilot scale.

  Research Focus
  - o Utilize waste energy or alternative energy sources to convert CO_2.
  - o Develop catalysts to reduce energy requirements.
  - o Develop stabilizers to inhibit degradation of plastics.

- Mineralization

  Description
  - o Carbonate mineralization refers to the conversion of CO_2 to solid inorganic carbonates. Naturally occurring alkaline and alkaline-earth oxides react chemically with CO_2 to produce minerals, such as calcium carbonate ($CaCO_3$) and magnesium carbonate ($MgCO_3$). These minerals are highly stable and can be used in construction or disposed of without concern that the CO_2 they contain will release into the atmosphere. One problem is that these reactions tend to be slow, and unless the reactions are carried out in situ, there is a large volume of rock to move. Carbonates can also be used as filler materials in paper and plastic products.

  Research Focus
  - o Reduce energy requirements for grinding feedstock materials needed for the mineralization process.

- o   Utilize waste streams to obtain oxides from existing mining operations.
- o   Develop chemicals or catalysts to speed reaction rates and reduce thermal and pressure requirements.
- o   Meet industrial standards for building materials.
- Enhanced Hydrocarbon Recovery
  Description
  - o   Enhanced Recovery (ER) involves the injection of CO_2 into a depleted oil or gas bearing field to increase production. This could involve injecting CO_2 into clastic, carbonate, coal, or organic shale formations.

  Research Focus
  - o   Maximize the amount of CO_2 that could be stored as well as maximize hydrocarbon production as part of these ER operations.

Advanced Combustion Technologies

The workhorse of America's electric power sector is the coal-fired power plant. Today, coal combustion plants account for more than half of the Nation's electric power generation. Largely because of these plants, U.S. consumers benefit from some of the most affordable power rates in the world.

The technology of burning coal has made remarkable advances in the last quarter century, and much of this progress is due to federal research and development partnerships with private sector developers.

In the 1990s, fluidized bed combustion - a process that removes pollutants inside the coal boiler - was termed "the commercial success story of the last decade" by a major power industry publication. The first new coal-fired power plant to be built in Illinois in more than 15 years will employ a new type of "low-emission boiler" technology developed in the federal government's energy program. Innovations in burner designs, refractory materials, and high-temperature heat exchanges are all products of the Department of Energy's research program into cleaner, more efficient ways to burn coal.

Oxycombustion

Oxy-fuel combustion is a promising near-term technology for CO_2 capture and storage (CCS) from pulverized coal (PC) fired boilers for power or industrial applications with ultra-low emissions. The basic concept of oxycombustion is to replace the combustion air with a mixture of oxygen and recycled flue gas and/or water for temperature control.

The remainder of the flue gas, that is not recirculated, is rich in carbon dioxide and water vapour, and is easily separated, producing a stream of CO_2 ready for utilization or sequestration. Prior research on PC oxy-fuel has shown that, in order to maintain the oxygen/ recirculated flue gas flame so that the oxy-combustion system has heat transfer characteristics similar to that of an air-fired system, an oxygen level of about 30-35 percent is required in the gas entering the boiler.

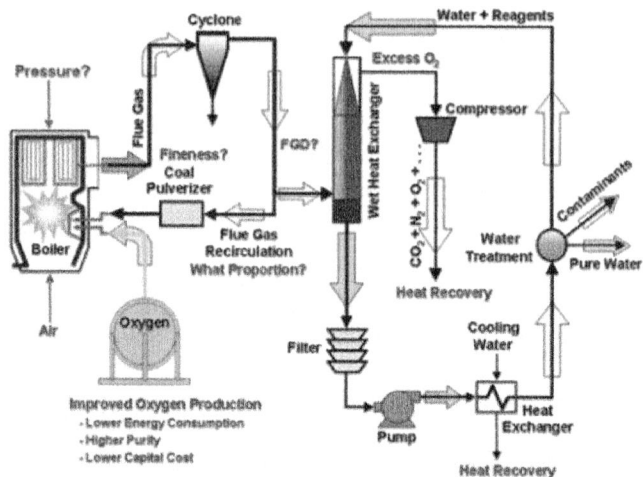

Fig. : **Schematic of oxycombustion process.**

Multiple oxycombustion facilities at various scales are being constructed or are in operation around the world. DOE is continuing research to design an optimized oxycombustion power plant by:

- developing a better understanding of the oxycombustion flame, and of heat and mass transfer in oxycombustion systems;
- developing an understanding of the character and distribution of ash and slag in PC oxycombustion systems;
- developing solutions for the potential low-pressure steam turbine imbalance in PC retrofit applications;
- supporting development of improved systems and CFD models and modeling tools.

Chemical Looping

Another breakthrough oxycombustion concept under development is the chemical looping combustion process. Chemical looping splits combustion into separate oxidation and reduction reactions. A metal (*e.g.*, iron, nickel, copper, or manganese) oxide is used as an oxygen carrier, which then releases the oxygen in a reducing atmosphere and the oxygen reacts with the fuel.

The metal is then recycled back to the oxidation chamber where the metal oxide is regenerated by contact with air. The advantage of using two chambers for the combustion process is that the CO_2 is concentrated, once the water is removed, and not diluted with nitrogen gas. The benefit of the process is that no air separation plant or external CO_2 separation equipment is required.

Air Separation

Oxycombustion cannot be simply substituted for air combustion in existing fossil-fueled power plants due to differences in combustion characteristics. For oxycombustion to be a cost-effective power generation option, a low-cost supply of pure oxygen is required In order for oxycombustion to be utilized in existing plants, a thermal diluent is required to replace the nitrogen in air.

In the most frequently proposed version of this concept, a cryogenic air separation unit is used to supply high purity oxygen to the boiler. This commercially available technology is both capital and energy-intensive and could raise the cost of electricity from coal-fired plants considerably, in addition to degrading the overall plant efficiency. However, novel technologies currently under development, such as oxygen and ion transport membranes, have the potential to reduce the cost of oxygen production.

Chapter 7

MULTIPLE INPUT MULTIPLE OUTPUT CONTROL

DECOUPLE

A system of inputs and outputs can be described as one of four types: SISO (single input, single output), SIMO (single input, multiple output), MISO (multiple input, single output), or MIMO (multiple input, multiple output).

Multiple input, multiple output (MIMO) systems describe processes with more than one input and more than one output which require multiple control loops. Examples of MIMO systems include heat exchangers, chemical reactors, and distillation columns. These systems can be complicated through loop interactions that result in variables with unexpected effects. Decoupling the variables of that system will improve the control of that process.

An example of a MIMO system is a jacketed CSTR in which the formation of the product is dependent upon the reactor temperature and feed flow rate. The process is controlled by two loops, a composition control loop and a temperature control loop. Changes to the feed rate are used to control the product composition and changes to the reactor temperature are made by increasing or decreasing the temperature of the jacket.

However, changes made to the feed would change the reaction mass, and hence the temperature, and changes made to temperature would change the reaction rate, and hence influence the composition. This is an example of loop interactions. Loop interactions need to be avoided because changes in one loop might cause destabilizing changes in another loop. To avoid loop interactions, MIMO systems can be decoupled into separate loops known as single input, single output (SISO) systems.

Decoupling may be done using several different techniques, including restructuring the pairing of variables, minimizing interactions by detuning conflicting control loops, opening loops and putting them in manual control, and using linear combinations of manipulated and/or controlled variables. If the system can't be decoupled, then other methods such as neural networks or model predictive control should be used to characterize the system.

There are two ways to see if a system can be decoupled. One way is with mathematical models and the other way is a more intuitive educated guessing method. Mathematical methods for simplifying MIMO control schemes include the relative gain array (RGA) method, the Niederlinski index (NI) and singular value decomposition (SVD). This chapter will discuss the determination of whether a MIMO control scheme can be decoupled to SISO using the SVD method. It will also discuss a more intuitive way of decoupling a system using a variation of the RGA method.

Definitions of Input and Output System Types

1. **SISO-** *Single Input, Single Output*

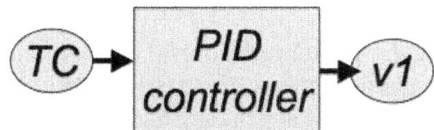

These systems use data/input from one sensor to control one output. These are the simplest to design since they correspond one sensor to one actuator. For example, temperature (TC) is used to control the valve state of v1 through a PID controller.

2. **SIMO-** *Single Input, Multiple Output*

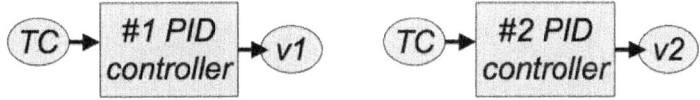

These systems use data/input from one sensor to control multiple outputs. For example, temperature (TC) is used to control the valve state of v1 and v2 through PID controllers.

3. **MISO-** *Multiple Input, Single Output*

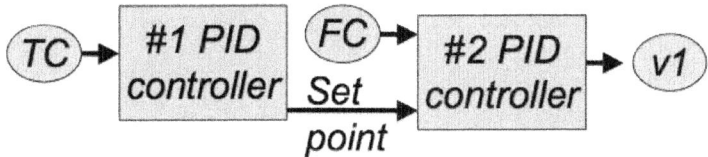

These systems use data/input from multiple sensors to control one ouput. For example, a cascade controller can be considered MISO. Temperature (TC) is used in a PID controller (#1) to determine a flow rate set point *i.e.* FCset. With the FCset and FC controller, they are used to control the valve state of v1 through a PID controller (#2).

4. MIMO- *Multiple Input, Multiple Output*

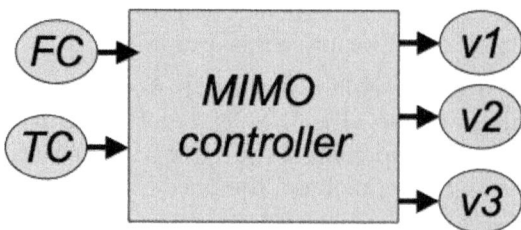

These systems use data/input from multiple sensors to control multiple outputs. These are usually the hardest to design since multiple sensor data is integrated to coordinate multiple actuators. For example, flow rate (FC) and temperature (TC) are used to control multiple valves (v1, v2, and v3). Often, MIMO systems are not PID controllers but rather designed for a specific situation.

Singular Value Decomposition

Singular value decomposition is a useful mathematical way to determine if a system will be prone to interactions due to sensitivity problems that result from small errors in process gains. It is a matrix technique that determines if a system is able to be decoupled.

Two Input Two Output System

The SVD method starts with the steady state gain matrix, which is obtained from the relative gain array (RGA) method for a given 2 input-2 output control loop system:

$$[G] = \begin{bmatrix} g_{1,1} & g_{1,2} \\ g_{2,1} & g_{2,2,} \end{bmatrix},$$

Using [G], we obtain the eigenvalues for the system. This can be done through two methods; either the eigenvalues can be obtained through numerical software as described in the Eigenvalues and Eigenvectors section or the eigenvalues can be calculated by hand. The hand calculations are shown here. [G]gives the following values for the system, which have been derived from theoretical work and are used to find the eigenvalues:

$$b = g_{11}^2 + g_{12}^2 \tag{1}$$
$$c = g_{11}g_{21} + g_{12}g_{22} \tag{2}$$
$$d = g_{21}^2 + g_{22}^2 \tag{3}$$

b, c, and d are parameters created for this method that do not have a direct physical meaning. Using the formulas below with the values of b, c, and d gives the eigenvalues for the system:

$$\lambda_1 = s_1^2 = \frac{b+d+\sqrt{(b-d)^2 + 4c^2}}{2} \tag{4}$$

MULTIPLE INPUT MULTIPLE OUTPUT CONTROL

$$\lambda_2 = s_2^2 = \frac{bd - c^2}{s_1^2} \tag{5}$$

s_1 and s_2 are the positive square roots of the respective eigenvalues. The condition number CN is defined as the ratio of the larger of the two values to the smaller value:

$$CN = \frac{s_1}{s_2} \text{ if } s_1 \geq s_2, \text{ or } CN = \frac{s_2}{s_1} \text{ if } s_2 \geq s_1$$

The greater the CN value, the harder it is for the system in question to be decoupled. As a rule of thumb, a system with a CN number of more than 50 is impossible to decouple. Such a system has manipulated variables which all have significant and similar relative impact on the controlled variables they affect and hence the control equations for the system cannot be restructured to obtain a simpler control system.

In the worst case scenario for a two input two output system, using the exact same control variable twice would give a CN number of infinity, because both control variables would have the same effect on the controlled variables. An ideal system would have a CN number of one, where each control variable controls a single distinct output variable.

MIMO Systems with Two or More Inputs and Outputs

To use this method, it is necessary to know how to create a transpose matrix. A transpose matrix is created by writing the columns of the original matrix as the rows of the transpose matrix. Mathematically, the transpose matrix is denoted by $[A]^T$, where $[A]$ is the original matrix. An example is shown here:

$$\begin{bmatrix} i & j \\ k & l \end{bmatrix}^T = \begin{bmatrix} i & k \\ j & l \end{bmatrix}$$

To use singular value decomposition on systems with two or more inputs and outputs, the m x m steady state gain matrix [G] of the form

$$[G] = \begin{bmatrix} g_{1,1} & g_{1,2} & \cdots & g_{1,m} \\ g_{2,1} & g_{2,2} & \cdots & g_{2,m} \\ \vdots & & & \\ g_{n,1} & g_{n,2} & \cdots & g_{n,m} \end{bmatrix}$$

is broken down into the form:

$G = U\Sigma V^T$

Where:

*U is an m x m matrix, the column vectors of which are the unit eigenvectors of the m x m matrix GG^T, G^T being the transpose matrix of G.

*V is an m x m matrix, the column vectors of which are the unit eigenvectors of the m x m matrix G^TG.

*Σ is an m x m diagonal matrix containing singular values, where the values are arranged in descending order from the top left most corner to the bottom right corner.

$$\Sigma = \begin{bmatrix} s_1 & 0 \\ 0 & s_2 \end{bmatrix}$$ for a 2 input 2 output system if $s_1 \geq s_2$, or $\Sigma = \begin{bmatrix} s_2 & 0 \\ 0 & s_1 \end{bmatrix}$ for a 2 input 2 output system if $s_2 \geq s_1$.

The CN number can then be determined by the ratio of the largest value in Σ to the smallest value. Notice in the previous two input two output example, there was no need to calculate the eigenvalue matrix Σ for a two by two matrix, since the eigenvalues can be calculated directly with the parameters b,c, and d.

Mathematica can also be used to find U, Σ and V for a given m x m Matrix G. This is accomplished by typing in the command

Map[MatrixForm, {u, w, v}=SingularValueDecomposition[G]],

giving U, Σ and V in that order.

Intuitive Decoupling using the RGA

With this technique, the first step is to construct the RGA by placing all of the controllers in manual, changing the value of the output manually, and then recording the final value of the controller after making the change. This will give the gain matrix, which is then multiplied by its inverse to give the relative gain matrix.

The goal here is to pair measurements and control valves so that the relative gain elements are close to 1, and all other combinations are close to zero. A values between 0 and 1 tells that the net effect of the other loops on this pairing is to change the measurement signal in the same direction as the control valve signal (cooperation). A value greater than one means that the measurement signal will go in the opposite direction of the control valve signal (conflict).

Finally a relative gain element less than zero means that the net effect is conflict, but here the gain will change sign, not the measurement signal. Therefore, if a loop is stable when the other loops are in manual, it will probably not be stable when the other loops are in automatic, and vice versa.

To intuitively decouple this system then would require a guess and check of different pairings, with the goal being to get the pairings to a relative gain element close to one. If this is not possible and no pairings give a value close to one, then the system will need to be decoupled using decoupling control.

Decoupling a System Using Decoupling Control

The goal of decoupling control is to eliminate complicated loop interactions so that a change in one process variable will not cause corresponding changes in other process variables. To do this a non-interacting or decoupling control scheme is used. In this scheme, a compensation network called a decoupler is used right before the process. This decoupler is the inverse of the gain array and allows for all measurements to be passed through it in order to give full decoupling of all of the loops. This is shown pictorially below.

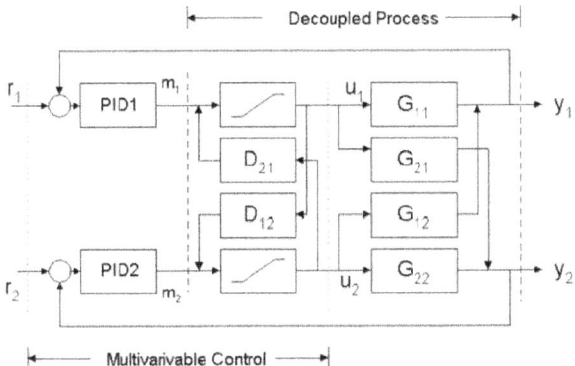

For a simple jacketed reactor example, if changing the flow rate of reactant to obtain a desired composition also increases the reactor temperature by five degrees, the decoupler would decrease the temperature of the jacket to decrease the temperature inside the reactor by five degrees, offsetting the original change. This jacket temperature decrease also changes the composition inside the reactor, so a properly designed controller accounts for both effects in the control scheme to obtain the desired composition without changing the reactor temperature.

RELATIVE GAIN ARRAY

Single variable Input or Single variable Output (SISO) control schemes are just one type of control scheme that engineers in industry use to control their process. They may also use MIMO, which is a **Multi-Input-Multi-Output** control scheme. In MIMO, one or more manipulated variables can affect the interactions of controlled variables in a specific loop or all other control loops.

A MIMO control scheme is important in systems that have multiple dependencies and multiple interactions between different variables- for example, in a distillation column, where a manipulated variable such as the reflux ratio could directly or indirectly affect the feed flow rate, the product composition, and the reboiler energy. Thus, understanding the dependence of different manipulated and controlled variables in a MIMO control scheme could be extremely helpful in designing and implementing a control scheme for a process.

One method for designing and analyzing a MIMO control scheme for a process in steady state is with a **Relative Gain Array (RGA)**. RGA is useful for MIMO systems that can be decoupled. For systems that cannot be decoupled, model predictive control or neural networks are better choices of analysis tool than RGA. A good MIMO control scheme for a system that can be decoupled is one that can control a process variable without greatly affecting the other process variables.

It must also be stable with respect to dynamic situations, load changes, and random disturbances. The RGA provides a quantitative approach to the analysis of the interactions between the controls and the output, and thus provides a method of pairing manipulated and controlled variables to generate a control scheme.

What is RGA?

Relative Gain Array is an analytical tool used to determine the optimal input-output variable pairings for a multi-input-multi-output (MIMO) system. In other words, the RGA is a normalized form of the gain matrix that describes the impact of each control variable on the output, relative to each control variable's impact on other variables. The process interaction of open-loop and closed-loop control systems are measured for all possible input-output variable pairings. A ratio of this open-loop 'gain' to this closed-loop 'gain' is determined and the results are displayed in a matrix.

$$RGA = \Lambda = \begin{bmatrix} \lambda_{11} & \lambda_{12} & \cdots & \lambda_{1n} \\ \lambda_{21} & \lambda_{22} & \cdots & \lambda_{2n} \\ \vdots & & & \\ \lambda_{n1} & \lambda_{n2} & \cdots & \lambda_{nn} \end{bmatrix}$$

The array will be a matrix with one column for each input variable and one row for each output variable in the MIMO system. This format allows a process engineer to easily compare the relative gains associated with each input-output variable pair, and ultimately to match the input and output variables that have the biggest effect on each other while also minimizing undesired side effects.

Understanding the Results of the RGA

- The closer the values in the RGA are to 1 the more decoupled the system is
- The maximum value in each row of the RGA determines which variables should be coupled or linked
- Also each row and each column should sum to 1
- Example

The table below includes the RGA results. The values highlighted in red are the maximum values in the row. These values indicated that the valve for that row should be used to control the variable that is listed in the corresponding column.

RGA

P	T	F	Ca	Cb	Valves
0.903911	0.015328	0.021940	0.015397	0.043423	V1
0.084549	0.228626	-0.0313343	0.230685	0.487474	V2
0.001492	0.139518	-0.002314	0.763468	0.097835	V3
-0.027496	0.584721	0.029974	-0.025906	0.438707	V4
0.037542	0.031807	0.981733	0.016356	-0.067439	V5

V1->P
V2->Cb
V3->Ca
V4->T
V5->F

Calculating RGA

There are two main ways to calculate RGA:

(1) Experimentally determine the effect of input variables on the output variables, then compile the results into an RGA matrix.

(2) Use a steady-state gain matrix to calculate the RGA matrix.

Method (1) should be used when it is possible to carry out the experiments as detailed in the **Calculating RGA with Experiments** section. This method will generally yield the most accurate RGA matrix of the system because it is based on actual data taken from the operating control system. If performing these extensive experiments is not possible, method (2) can be used. If a process model is available, method (2) can be used with no experimental data. If there is no process model available, some experimental data must be taken (though less extensively than in method (1)) and used in conjunction with method (2).

Method 1: Calculating RGA with Experiments

This method of calculating the RGA can be used when it is possible to run experiments on each of the input-output pairings. Below is a step-by-step explanation of how to experimentally compile the RGA for a simple MIMO system.

The simplest MIMO system is one that has two inputs and two outputs. Remember that by definition, a change in one of the inputs in a MIMO system will change *both* of the outputs. This system can be expressed mathematically as written below.

$$y_1 = a_{11}m_1 + a_{12}m_2$$
$$y_2 = a_{21}m_1 + a_{22}m_2$$

Where y_i is the output for loop i, the m variables are the inputs for each loop, and the a variables are the transfer functions. It also helps to see this system as a control diagram, as shown here.

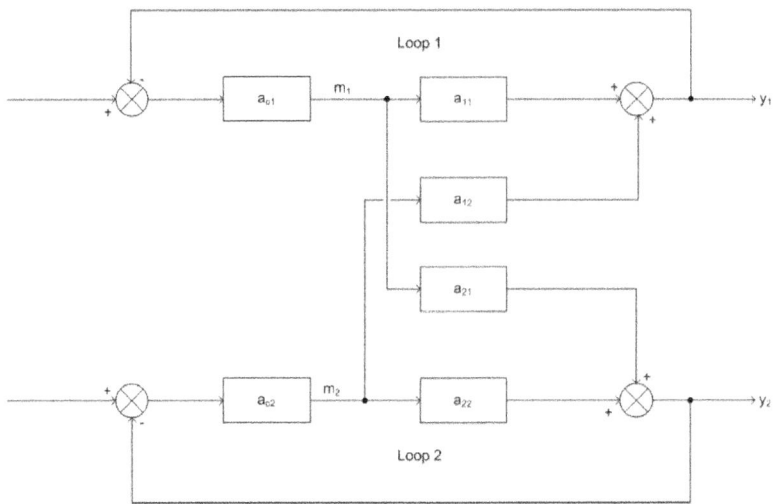

Clearly, both y_1 and y_2 are affected by both inputs (m_1 and m_2). Because of this, there are two choices for control. y_1 could be controlled by m_1, or y_1 could be controlled by m_2. y_2 would then be controlled by the input that is not controlling y_1. For a MIMO system with n input variables, there are $n!$ possible control configurations.

The question is: Which configuration will give you the best control? The answer can be determined by finding the relative gain array (RGA). There are a few ways to determine the RGA, one of which is by doing two experiments, repeated for every possible configuration. The RGA can be assembled from the data collected. These experiments are explained below using the above drawing and the configuration where m_1 controls y_1.

Experiment 1

For the first experiment, the objective is to observe the effect of m_1 on y_1. To do this, *all* of the loops must be open. In other words, the feedback loop is removed and the system is run manually with no control. This configuration is shown below.

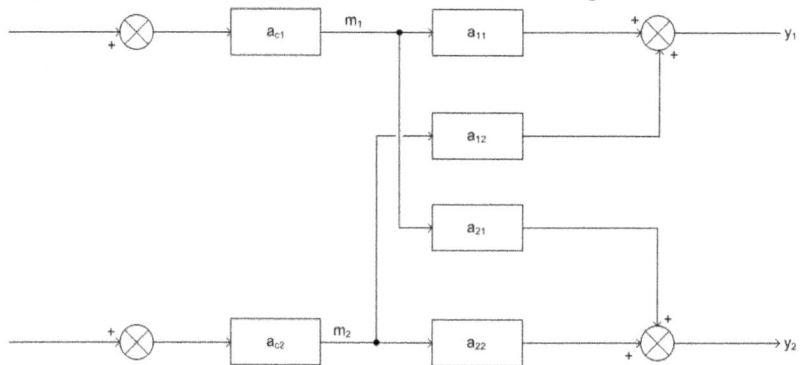

Now, since the system is under manual control, it is possible to introduce a step change Δm_1 while keeping m_2 constant. Because m_2 is held constant, the change in y_1, Δy_1, can be attributed entirely to m_1. Although a change in m_1 will also affect y_2, it is the relationship between m_1 and y_1 that must be observed at this point. Then, the gain of y_1 with respect to m_1 with all loops open is defined as g_{11}. The calculation of g_{11} is shown below.

$$g_{11} = \frac{\Delta y_1(all-loops-open)}{\Delta m_1(all-loops-open)}$$

The final objective of experiment one is obtaining the value g_{ij} where the controlled variable i is controlled by manipulated variable j.

Experiment 2

The goal of experiment two is to determine the effect of m_2 on y_1. To do this, loop 1 must remain open, but all other loops remain closed. This configuration is shown below.

Multiple Input Multiple Output Control

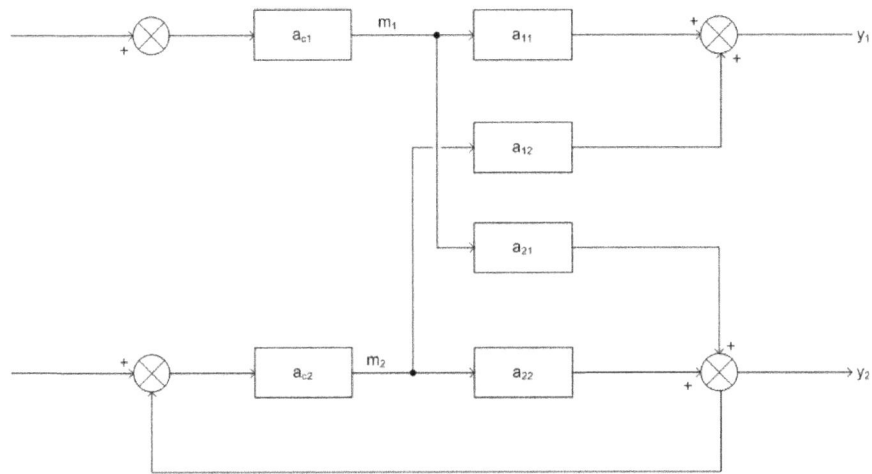

Now, the *same* step change that was introduced to m_1 in experiment one must be introduced again. The difference is, now loop 2 is in *perfect control* which means that when m_1 changes, it affects y_2, so the control loop on loop two will adjust m_2 in order to compensate and keep y_2 constant. When m_2 changes it, in turn, affects y_1. It is this change that is being observed. The amount y_1 changes is Δy_1 and the amount m_1 was stepped up is Δm_1. Now:

$$g_{11}^* = \frac{\Delta y_1(only-loop-one-open)}{\Delta m_1(only-loop-one-open)}$$

The objective of experiment two is to obtain this value of g^*_{ij} where controlled variable i is being controlled by manipulated j.

Compiling the Array

Once the experiments are run for every possible configuration, the results can be used to compile the relative gain array (RGA). To do this, we first have to find the *relative gain*. The relative gain is the ratio of g_{ij} to g^*_{ij} and is represented by λ_{ij}.

$$\lambda_{ij} = \frac{g_{ij}}{g_{ij}^*} = \frac{\left(\frac{\delta y_i}{\delta m_j}\right)_{all-loops-open}}{\left(\frac{\delta y_i}{\delta m_j}\right)_{only-loop-i-open}}$$

The value for λ must be computed for every possible combination of i and j. After every value is found, the RGA matrix can be formed:

$$RGA = \Lambda = \begin{bmatrix} \lambda_{11} & \lambda_{12} & \cdots & \lambda_{1n} \\ \lambda_{21} & \lambda_{22} & \cdots & \lambda_{2n} \\ \vdots & & & \\ \lambda_{n1} & \lambda_{n2} & \cdots & \lambda_{nn} \end{bmatrix}$$

Method 2: Calculating RGA with Steady-State Gain Matrix

Sometimes it is not convenient or possible to perform the experiments described above on every control pairing possibility in the system. If that is the case, a *steady-state gain matrix* can be used to determine the RGA. If a process model is available, the steady-state gain matrix can be calculated from the model equations.

If a process model is not available, the steady-state gain matrix can be calculated from experimental data (fewer experiments are required than when the RGA is calculated directly from experimental data as above). Once the steady-state gain matrix is calculated, it can be used to compute the RGA. Beware that a RGA has to have the same number of controlled variables and manipulated variables (same number of inputs and outputs) to be in an $m \times m$ matrix form.

Process Model Available

If a process model is available, the steady-state gain matrix relates the manipulated variables to the controlled variables according to the following equation:

$$y = Gm$$

where y is a vector of the controlled variables:

$$\mathbf{y} = \begin{bmatrix} y_1 \\ y_2 \\ \vdots \\ y_n \end{bmatrix}$$

m is a vector of the manipulated variables:

$$\mathbf{m} = \begin{bmatrix} m_1 \\ m_2 \\ \vdots \\ m_n \end{bmatrix}$$

and G is the steady-state gain matrix:

$$G = \begin{bmatrix} g_{11} & g_{12} & \cdots & g_{1n} \\ g_{21} & g_{22} & \cdots & g_{2n} \\ \vdots & & & \\ g_{n1} & g_{n2} & \cdots & g_{nn} \end{bmatrix}$$

The values of g_{ij} are calculated by taking partial derivatives of the equations governing the controlled variables. Specifically:

$$g_{ij} = \frac{\partial y_i}{\partial m_j}$$

During differentiation, hold constant all manipulated variables other than m_j. To fill in matrix G with numerical values, simply substitute the appropriate steady state values for the process into the expressions derived from differentiation.

Now that the steady-state gain matrix has been found, the RGA can be calculated with just a little bit more math. However, now that you have the steady state gain matrix, this might be a good time to determine if your system can even be decoupled! Assuming SVD gives a condition number of less than about 50, we can proceed to calculation of the RGA. First, define a matrix R to be the transpose of the inverse of matrix G:

$$R = (G^{-1})^T$$

The calculation of the **inverse** of a matrix is quite complicated for anything larger than a 2x2 matrix. Fortunately Mathematica will do this with the following command for a square matrix M:

$$\text{Inverse}[M]$$

The **transpose** of a matrix is when the rows become columns and the columns become rows. For a square matrix, this just means reflecting across the diagonal running from the top left to the bottom right corner of the matrix. The **transpose** can be found in Mathematica with the following command:

$$\text{Transpose}[M]$$

The RGA can now be obtained one element at a time according to this equation:

$$\lambda_{ij} = g_{ij} r_{ij}$$

Note that this is not your usual matrix multiplication! Here you multiply corresponding elements of the G and R matrices to get the corresponding element of the RGA. This is the type of multiplication Mathematica does with the standard multiplication operator.

Process Model Not Available

In case there is no process model available and it is not feasible to determine the RGA by carrying out both experiments necessary for full experimental determination, it still may be possible to develop a steady-state gain matrix experimentally which can then be used to derive the RGA. In this case, just carry out experiment 1 as described above to determine the elements of the steady-state gain matrix:

$$g_{ij} = \frac{\Delta y_i}{\Delta m_j} \text{ with all loops open}$$

Each element of the steady-state gain matrix can be determined this way, and then the RGA can be calculated from the steady-state gain matrix as shown in the previous section.

Interpreting the RGA

There are some important properties and guidelines in understanding and analyzing the RGA, and what the different values of the RGA mean:

1) **All elements of the RGA across any row, or down any column will sum up to one:**

$$\sum_{i=1}^{n} \lambda_{ij} = \sum_{j=1}^{n} \lambda_{ij} = 1$$

This makes calculating the RGA easier because:
- in 2×2 case, only 1 element must be calculated to determine all elements
- in 3×3 case, only 4 elements must be calculated to determine all elements and so on.

2) The λ_{ij} calculated from steady-state matrix is dimensionless and unaffected by scaling.

3) Each of the rows in the RGA represent one of the outputs. Each of the columns represent a manipulated variable.
   - If $\lambda_{ij} = 0$: The manipulated variable (m_j) will have no effect on the output or the controlled variable (y_i).
   - If $\lambda_{ij} = 1$: The manipulated variable m_j affects the output y_i without any interaction from the other control loops in the system. From the definition of λ_{ij}, this implies that the gain loop with all loops open is equal to the gain loop with all other loops closed, ie: $g_{11} = g^*_{11}$.
   - If $\lambda_{ij} < 0$: The system will be unstable whenever m_j is paired with y_i, and the opposite response in the actual system may occur if other loops are opened in the system.
   - If $0 < \lambda_{ij} < 1$: This implies that other control loops (m_j- y_i) are interacting with the manipulated and controlled variable control loop.

Three different relationships based on $\lambda=0.5$ imply different interpretations of pairing and the RGA:
- If $\lambda_{ij} = 0.5$: The control pairing effect is equal to the retaliatory effect of other loops.
- If $\lambda_{ij} < 0.5$: The other control loops are influencing the control pair, and the influence from the other control loops are greater than the influence from the control pair.
- If $\lambda_{ij} > 0.5$: This means that the control pair has a greater influence on the system than the other control loops.
- If $\lambda_{ij} > 1$: The open-loop gain of the control pair is greater than the gain with all other loops closed, ie: $g_{11} > g^*_{11}$. The positive value of RGA indicates that the control pair is dominant in the system, but the other loops are still affecting the control pair in the opposite direction. The higher the

value of λ_{ij}, the more correctional effects the other control loops have on the pair.

λ_{ij}	Possible Pairing
$\lambda_{ij} = 0$	Avoid pairing m_j with y_i
$\lambda_{ij} = 1$	Pair m_j with y_i
$\lambda_{ij} < 0$	Avoid pairing m_j with y_i
$\lambda_{ij} =$ or <0.5	Avoid pairing m_j with y_i
$\lambda_{ij} > 1$	Pair m_j with y_i

NI Analysis with RGA

NI, the Niederlinski Index, is a calculation used to analyze the stability of the control loop pairings using the result of the RGA, evaluated at Steady State:

$$NI = \frac{|G|}{\prod_{i=1}^{n} g_{ij}}$$

A negative NI value indicates instability in the control loop. For a 2 × 2 matrix, a positive NI value indicates stability in the pairings, but this is not necessarily true for larger matrices! For matrices larger than 2 × 2, a conclusion can only be drawn from a negative NI, which indicates instability. NI should not be used for systems that have time delays (dead time), as the NI stability predictions assume immediate feedback to the controller.

Here's an example NI calculation: given the steady state gain matrix

$$G = \begin{bmatrix} -0.002 & 0.001 \\ 0.002 & 0.003 \end{bmatrix}$$

the NI can be calculated from the following expression:

$$NI = \frac{(-0.002 \times 0.003) - (0.001 \times 0.002)}{-0.002 \times 0.003} = 1.333$$

Since this is a 2 × 2 matrix, the positive value of the NI indicates stability in the control loop pairings.

Optimizing a MIMO Control Scheme: Pairing Rules

The goal of the RGA and NI analysis is to quantitatively determine the optimal variable pairing for a given process. Some basic rules to remember when attempting to obtain an optimal pairing of control loops in a system are:

- **Rule 1:** The **positive** RGA elements that are closest to 1.0 should have the corresponding manipulated and controlled variables paired. When the

CN number is large, implying a less decoupled system, one should look for the max RGA elements.
- **Rule 2:** If the NI value is negative, the loop pairing for that control system configuration is unacceptable.

MODEL PREDICTIVE CONTROL

This chapter will describe how to control a system with multiple inputs and outputs using model predictive control (MPC). MPC is a linear algebra method for predicting the result of a sequence of control variable manipulations. Once the results of specific manipulations are predicted, the controller can then proceed with the sequence that produces the desired result. One can compare this controller method to "look ahead" in chess or other board games.

In look ahead, you foresee what an action might yield some time in the future using specific knowledge of the process (or game in the case of chess), and are thereby able to optimize your actions to select for the best long term outcome. MPC methods can prevent an occurrence with conventional PID controllers in which actions taken achieve short term goals, but end up very costly in the end. This phenomenon can be described as "winning the battle but losing the war."

The open ended nature of MPC allows the process control engineer use MPC to control any system for which models can be generated.

MPC is a widely used means to deal with large multivariable constrained control issues in industry. THe main aim of MPC is to minimoze a performance criterion in the future that would possibly be subject to constraints on the manipulated inputs and outputs, where the future behaviour is computed according to a model of the plant.

The model predictive controller uses the models and current plant measurements to calculate future moves in the independent variables that will result in operation that honors all independent and dependent variable constraints. The MPC then sends this set of independent variable moves to the corresponding regulatory controller set-points to be implemented in the process.

MPC uses the mathematical expressions of a process model to predict system behaviour. These predictions are used to optimize the process over a defined time period. An MPC controller can operate according to the following algorithm.

1. Development of a process model by the control engineers.
2. At time t, previous process inputs and outputs are used, along with the process model, to predict future process outputs $u(f)$ over a "prediction horizon."
3. The control signals that produce the most desired behaviour are selected.
4. The control signal is implemented over a pre-defined time interval.
5. Time advances to the next interval, and the procedure is repeated from step 2.

This is one of the many algorithm possibilities, which can be applied to systems with any number of inputs or outputs. The process model can be variable as well. Examples include physical models, input-output models, and state models which are all derived from the specific system being controlled.

When comparing predicted behaviour to desired behaviour, there are multiple techniques. A common procedure is to generate a second mathematical model that describes your desired behaviour. When process behaviour is predicted in step 2 of the MPC algorithm, the control signals which produce the predicted behaviour that minimize deviations from your desired behaviour over the [t,t+h] interval, are selected. The generation of models and optimization process is repeated continuously as the algorithm is repeated.

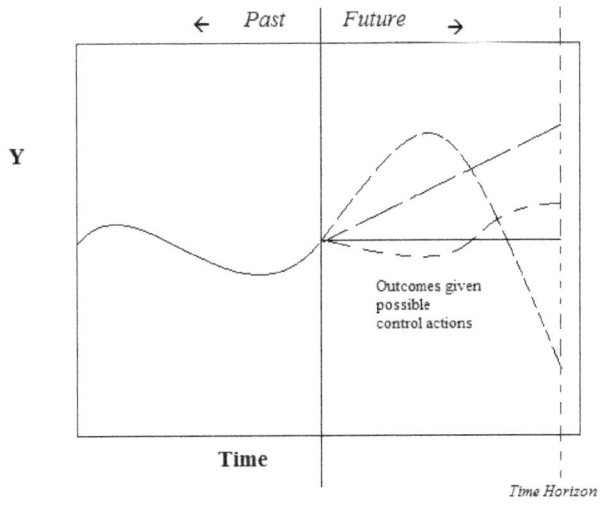

As seen in the figure above, depending on the algorithm, MPC may generate possible outcomes given possible controller action. These generations are either based on past process outputs, or the process model. After many possible outcomes are generated, the controller can pick one based on the optimization goals. This generation and optimization process is repeated at every time step.

The flow diagram below depicts the flow of information used by the controller.

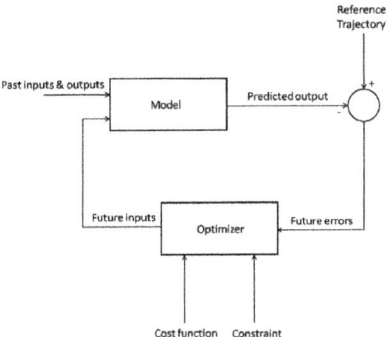

The figure above shows the basic structure of a Model Predictive Controller. The model takes data from past inputs and outputs, and combines it with the predicted future inputs, and gives a predicted output for the time step. This predicted output is combined with the reference trajectory, giving the predicted future errors of the system.

These errors are fed into an optimizer, which enforces the constraints of the system (for instance, ensuring that a flow rate calculation for the model is not greater than the maximum flow) on the predicted outputs and minimizes the operating cost function. This gives the predicted future inputs, which are fed back into the main model, restarting the cycle.

Motivation

The motivation for the development and implementation of MPC controllers is compelling. MPC is very simple for sampled systems in which the control signal is constant over the interval [t,t+h]. The value of h can then be taken as the sampling interval, and the prediction horizon can become a small number of sampling intervals. This can reduce the computational power needed to implement a model predictive controller.

A highly desired feature of MPC is that constraints can be implemented in the controller. These constraints include heaters and valves that have a finite operating range, actuators with finite states (on/off or low/high), and cost or energy limits for the process. MPC can incorporate these constraints and eliminate the possibility of variables exceeding their limits. This helps the process operate efficiently, prevents damage to equipment, and prevents the system from running away by continuouly increasing a variable's setting when the limit for the equipment has already been reached.

Another benefit of MPC controllers incorporating system constraints is the ability of the model to dynamically react to system changes. For example, if a valve is stuck open, it can be added as a constraint, and the model will compensate accordingly. This allows the controller to continue effectively controlling the system after an equipment malfunction. In a simpler control, such as PID control, this would not be possible.

MPC is a highly specific method for controlling a process. Each controller is specific to the system it was designed for and the model equations, constraints, and set points will change for different systems. This means that a controller developed for a tank reactor will not be able to control an evapouration unit as the process model will be very different. This weakness in MPC brings out its biggest strength. The specificity and customizable nature of the controller will empower you with the freedom to design for exactly what is desired.

MPC is a flexible control technique that uses discrete time segments and is the most commonly applied advanced control technique in the chemical process industry. MPC helps to simplify or completely eliminate controller design and instead works as a system modeling controller. With MPC the designer does not have to worry about optimizing control parameters such as with PID control. The ability of MPC controllers to handle constraints in an optimal fashion is also a contributor to its success.

Model Predictive Control Example

To demonstrate the concepts of MPC, a general example for the development of a sampled process and a first-order system will be shown.

General Model

Take equation 1 as our process model:
$$y(t) + a_1 y(t-h) + \ldots + a_n y(t-nh) = b_1 u(t-h) + b_2 u(t-2h) + \ldots + b_n(t-nh)$$

Where u is controller input, y is process output, and h is the time interval. This is a general equation that relates previous process output $y(t - h)$ and previous controller input $u(t - h)$. In some situations this equation may be created using fitted experimental data, but is most often a derivation using knowledge of your specific system and fundamentals of chemical engineering. At time t, total previous behaviour y_p is shown as equation 2.
$$y_p = f(y(t), y(t-h), \ldots, u(t-h), u(t-2h), \ldots)$$

Future process output y_f can be predicted using current and future control signals, $u(t)$ and $u(t + h)$ respectively:
$$y_f = f(u(t), u(t-h), \ldots, u(t+Nh))$$

Both y_p and y_f could possibly be created by fits of experimental data, but are more likely to be derived from specific equations related to your system. Deviations from the desired behaviour y_d, either specified by another mathematical model or reference trajectory, produce an error function

$e(t) = y(t) - y_d(t)$

for increments of control actions

$\Delta u(t) = u(t) - u(t - h)$.

The loss function J to be minimized is shown as equation.
$$J(u(t), u(t-h), \ldots, u(t+Nh)) = \sum_{k=1}^{t+N} e(t+kh)^2 + \rho(\Delta u(t+(k-1)h))^2$$

The control inputs that minimize equation are then applied to the system by the controller over the time interval, and the process is repeated. The control input function F in equation is determined implicitly by the optimization.
$$u(t) = F(y(t), y(t-h), \ldots y(t-nh), u(t-h), y(t-2h), \ldots, u(t-nh))$$

This general model is meant to be a guideline, and the equations listed representative, for the thought process required to create a model predictive controller.

First-Order System Example

Take the process model to be equation:
$$\Delta y(t + h) = -a\Delta y(t) + b\Delta u(t)$$

Let us define
$$\Delta y(t) = y(t) - y(t - h)$$

and
$$\Delta u(t) = u(t) - u(t - h).$$

Let us also define our desired system behaviour y_d as a function which starts at y(t) and exponentially approaches a set point y_{sp} with time constant T. Our desired behaviour y_d then becomes equation:

$$y_d(t + h) = y(t) + (1 - e^{\frac{-h}{T}})(y_{sp} - y(t))$$

Assuming that our controller can take as much action as needed to produce the desired behaviour, the desired behaviour can be realized in the next sampling period. This is done by setting $y(t + h)$ equal to $y_d(t + h)$, and can be seen in equation:

$$y(t+h) = y(t) + \Delta y(t+h) = y(t) - a\Delta y(t) + b\Delta u(t) = y(t) + (1 - e^{\frac{-h}{T}})(y_{sp} - y(t))$$

Solving equation for $u(t)$ gives equation:

$$\Delta u(t) = \frac{a}{b}\Delta y(t) + \frac{1 - e^{\frac{-h}{T}}}{b}(y_{sp} - y(t))$$

Upon examination of this result, you can see that we have produced a PI controller with gains

$$k = \frac{a}{b}$$

And

$$k_i = \frac{1 - e^{\frac{-h}{T}}}{b}.$$

It should be noted that the proportional gain k will only depend upon the developed process model, and the integral gain k_i depends on both the process model and the desired response rate T.

This process can be modified to include multiple inputs and outputs *via* the process model and desired behaviour.

Question: is the above expression a PI control system or more of a PD control system?

The first term is k*(y(present0)-y(past)) which seems more like a derivative term.

The second term is more like k*(y(set)-y(present)) which seems more like a P term.

Differences from Other Controllers Types

Model predictive control uses a mathematical model to simulate a process. This model then fits the inputs to predict the system behaviour. In this way, MPC is a type of feed forward control. It uses system inputs as a basis of control. MPC is more complex than most other feed forward control types because of the way these predictions are used to optimize a process over a defined amount of time.

Most feed forward control types do not take into account the process outputs much past a residence time. The MPC algorithm will compare predicted outputs to desired outputs and select signals that will minimize this difference over the time selected. This control type can see ahead into multiple time steps in the future in order to optimize the process. Normal PID type controllers use mathematical expressions based on error from a set point. The governing equation for MPC controllers are based on set points, system properties, and desired outcomes and optimization.

MPC is very specific to the process it is modeling. Unlike ratio or cascade control set ups, where it is simple to implement and change set points in various situations, MPC will model one specific process and optimize it. As previously mentioned this can either be an advantage or disadvantage. MPC is great for selecting one type of operation on one system and perfecting it to the desired conditions. This also has a downside in that the model equation will work for one and only one situation.

Limitations of MPC

Advantages of MPC
1. MPC can be used to handle multivariable control programs.
2. MPC can consider actuator limitations.
3. MPC can increase profits by allowing for operation close to the system constraints.
4. MPC can perform online computations quickly.
5. MPC can be used for non-minimal phase and unstable processes.
6. MPC is easy to tune.
7. MPC is able to handle structural changes.

Disadvantages of MPC
1. Several MPC models are limited to only stable, open-loop processes.
2. MPC often requires a large number of model coefficients to describe a response.
3. Some MPC models are formulated for output disturbances, and they may not handle input disturbances well.
4. Some forms of MPC use a constant output disturbance assumption. This corrects for the fact that the output predicted by the model is not exactly

equal to the actual measured output. This method assumes the correction term is constant in the future, which may not yield a good performance if there is a real disturbance at the plant input.

5. If the prediction horizon is not formulated correctly, control performance will be poor even if the model is correct.
6. Some systems have a wide range of operating conditions that change frequently. Some examples of this include exothermic reactors, batch processes, and any systems where different consumers have different product specifications. An MPC linear model will not be able to handle the dynamic behaviour of these processes. A nonlinear model must be used for better control performance.

Industrial MPC Applications

There are many industrial applications that incorporate model predictive control in order to effectively control a multivariable system. In order to effectively do this, one needs to set up a working model by testing many different parameters in a plant. This is usually done by starting up a plant, varying many different parameters, and having the MPC program analyze the data.

In order to test the plant, one may vary parameters such as:

1. feed flow rate/composition
2. steam pressure
3. heat duty
4. recycle ratio
5. reactor temperatures

This is only a small sample of parameters that can be changed. Once these parameters have been changed, the data is analyzed, and downstream effects of these parameters are characterized as a function of these variables. This relationship can be a combination of many different relationships (linear, nonlinear, logarithmic, exponential, power, *etc.*).

By testing the plant thoroughly and coming up with a robust model, the engineers ensure that an MPC controller will be able to much more effectively run the plant. Barring any major process changes, this model should be accurate for normal use.

Some industrial MPC applications are:

1. Model Predictive Heuristic Control by Richard et al. 1976(Adersa)
2. Dynamic Matrix Control (DMC) by Cutler and Ramaker1979(Shell Oil)
3. Quadratic-Program Dynamic Matrix (QDMC) Control by Cutler et al. 1983(Shell Oil)
4. IDCOM-M by Setpoint, Inc(part of ASPEN Technology)
5. Generalized Predictive Control (GPC)

Implementing MPC using Excel

In MPC, values of the control variables will be optimized for a given time interval in order to best tell the system how it should act. The control variables will be optimized by optimizing some characteristic. Usually, this characteristic is simply the least squared error between an actual state and a "set" or desired state. This can easily be done using the Solver tool in Excel.

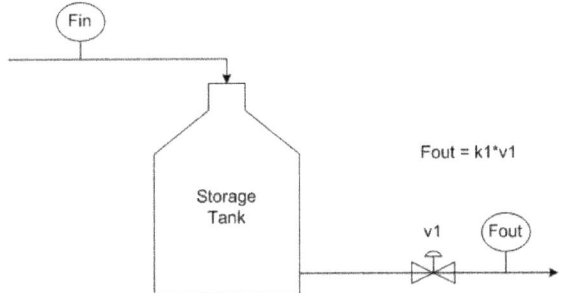

To describe this system in words, there is a variable feed, Fin, with time that pours into a storage tank. The flow out of the tank, Fout, is controlled by the valve, v1, multiplied by a constant of proportionality, k1 = 50. The tank volume is to be maintained at 150.

There are three scenarios contained in the example Excel file. In the first scenario, the valve is continuous and in the second, it is binary (can be open or closed only). In the third scenario, the valve is still binary; however, Solver cannot recognize this setup even though this setup may seem intuitive (this is discussed further below). As it turns out, it doesn't really matter if a binary or continuous valve is used. Both types of valves yield very similar results; however, the continuous valve yields slightly tighter control over the tank volume (see screenshot below).

As you can see from this screen shot, the sum of the least squares between the actual volume and the set volume (Row 22) is lowest for the 1st scenario, when the valve is continuous. This measure of deviation from the set value is only slightly lower than in the 2nd scenario, when the valve is binary. In the 3rd scenario, Solver didn't work properly, resulting in all of the v1 variables (Column P) to be set to 0.

At the start of the simulation, the volume of the tank is 0, so v1 stays closed until time = 5.4 when the volume is very close to the set volume. If the deviations that occurred during the time v1 was closed (which is the same time length for all 3 scenarios) is removed from the sum of the least squares (Row 23) the 1st scenario suddenly looks much better at controlling the volume than the 2nd scenario.

There are a couple things to note while using Solver in Excel to perform MPC:

1. This note only applies if your system has a binary control variable. If you intend to use a conditional statement (such as an IF() function) that depends on the optimization of the control variable, you must leave open the possibility that this control variable can be any real number, whether it's binary or not. This may seem unnecessary because Solver will output only optimized discreet numbers. To put this in the context of the example Excel file, you might think you could have simply done the following to control Fout:
   - = IF(v1 = 1, k1*1, 0)

(The variables "v_1" and "k_1" are not recognized by Excel. Instead of typing the variables into Excel, cells containing the variable values must be referenced.)

This statement says that if the valve is open, allow a flow of k1(or 50), otherwise, the flow is 0. This would not allow Solver to work properly. If Solver chose a value of v1 = 0.1, measured the target cell, and then it chose v1 = 0.2, it wouldn't see any difference because in both cases the flow would be 0. Solver needs to be able to see a change in the system, whether it's practical or not, to determine the actual gradient of your system. This gradient cannot be determined if an IF() statement turns the continuous output to discreet output.

The proper way to handle this is to program a constraint into solver that allows only binary outputs of v1 values (select bin from the pull down menu when adding the constraint). Fout can then be programmed as:
   - = v1 * k1

Using this method, Solver can calculate the gradient of the system to define how it would react if v_1 changed to a non-binary value. By using this method, Solver will output only binary values for the valve state as is desired.

2. Solver can only handle manipulating 200 variables at a time, so if your model requires looking at more than 200 time steps, you will have to optimize your system in sections.

NEURAL NETWORKS

Multiple Input Multiple Output (MIMOs)are systems that require multiple inputs and generate multiple outputs. MIMOs are controlled by controllers that combine multiple input readings in an algorithm to generate multiple output signals. MIMOs can be used with a variety of algorithms. The most versatile algorithm used to date is the neural network. Neural networks, which were initially designed to imitate human neurons, work to store, analyze, and identify patterns in input readings to generate output signals. In chemical engineering, neural networks are used to predict the ouputs of systems such as distillation columns and CSTRs.

MIMOs

As mentioned, Multiple Inputs Multiple Outputs (MIMOs) are systems that require multiple inputs and generate multiple outputs.

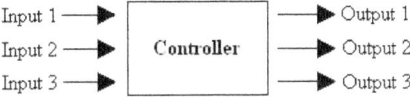

Fig. : Schematic diagram of MIMOs.

As shown in the figure, MIMOs are controlled by controllers that combine multiple input readings in an algorithm to generate multiple output signals. Typically, MIMOs do not require that the the number of inputs and outputs be the same. Instead, any number of input readings could be used to generate any number of output signals.

Various types of controllers can be used to control MIMOs. One of the most accurate and versatile controllers used for MIMOs is the neural network. Neural networks are controllers that crudely imitate the human neuron. Initially, these networks were designed to model neural brain activity.

However, as people began to recognize the advantages of neural networks, these networks were applied to controller algorithm design. Like a human neuron, neural networks work to store, analyze, and identify patterns in data by performing learning tasks. The ability of these networks to "learn" parallels the human neuron's ability to learn, making these automatic controllers the closest analog to a human controller.

Neurons

Like neurons in the body, network neurons receive inputs, store this data, and transmit outputs to either another neuron or directly to the MIMO. In order to transmit this data, the neuron must relate the multiple inputs to the multiple outputs. A simple mathematical representation of this relationship is shown below.

$$y = f(w_1 a_1 + w_2 a_2 + w_3 a_3 + \ldots + w_n a_n) = f(\sum_{i=1}^{n} w_i a_i)$$

w_i = weight
a_i = input
y = output
f = sigmoid function (any nonlinear function)

According to this relationship, the multiple input parameters are each multiplied by the corresponding weight factor, w_i. These weight factors "weigh" the significance of each input, scaling each input proportionally to the effect it will have on the output. These weighted inputs are then added and the sum is input into the sigmoid function to generate an output. This output can then be sent to multiple neurons that, in turn, each generate its own output.

The sigmoid function in this relationship is a nonlinear, empirical function that relates the input readings to the output signals. This empirical function can take on many forms depending on the data set. The equation that is best able to predict the outputs for the given system will be used (polynomial, sine, logarithmic, *etc.*). For example, one form this function may take is the hyperbolic sine function, where

$f(x) = \sinh(\alpha x)$

$x = \text{sum of weighted inputs} = \sum_{i=1}^{n} w_i a_i$

α = empirical parameter

In this sigmoid function, α is an empirical parameter that adjusts the function outputs. The effect of α on this sigmoid function is shown in figure.

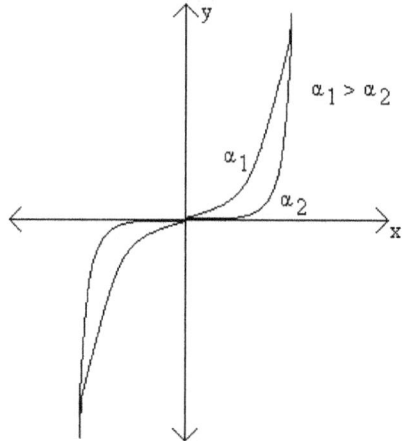

Fig. : Plot of hyperbolic sine sigmoid function with varying αx

As shown in Figure, increasing α increases the output for this particular sigmoid function. Like most empirical functions, the hyperbolic sine sigmoid function can only be used within a specified range of values (x values between the vertical asymptotes). In this case, the range would depend on the value of α.

Combining Neurons into Neural Networks

Once neurons have been programmed to correlate input and output data, they can be connected in a feedforward series to produce a neural network, or neural net (NN). A schematic diagram of a neural network is shown in figure.

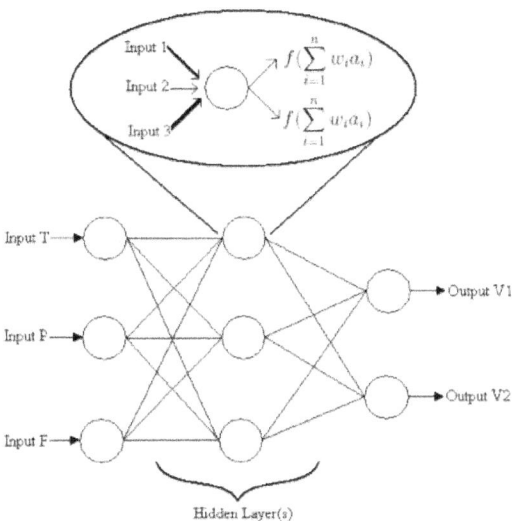

Fig. : Schematic diagram of a neural network.

Figure shows MIMO parameters, such as temperature, pressure, and flow readings, are first processed in the first layer of neurons. The outputs of the first layer of neurons then serve as the inputs to the second layer. The outputs of the second layer then become the inputs to the third layer, and so on, until the final output of the network is used to directly affect MIMO controls such as valves. The layers of neurons between the initial and final layers are known as hidden layers.

The way neurons within hidden layers correlate the inputs and outputs is analogous to the way individual neurons correlate these variables. As shown in the flowchart inset of the diagram, a neuron within the network receives multiple inputs. Each of these input parameters are then weighted, having the most effective parameters weighted more heavily. These weighted input values are added and input into the particular sigmoid function the neuron is programmed to follow.

The output of this function is then sent to other neurons as input. These input values are then reweighed for that particular neuron. This process is continued until the final output of the neuron network is used to adjust the desired controls. Although the diagram shows only one hidden layer, these neuron networks can consist of multiple layers.

Although almost all continuous functions can be approximated by a single hidden layer, incorporating multiple hidden layers decreases the numbers of weights used. Since more layers will result in more parameters, each of these individual parameters will be weighted less. With more parameters and less weights, the system becomes more sensitive to parameter change (a greater "rippling" effect within the network).

The "rippling" effect of a neuron network makes the system difficult to model analytically. A small change in a single input variable would result in multiple changes throughout the the entire network. Although modelling these complex networks is beyond the scope of the class, only a basic, qualitative understanding of how neural networks function is necessary to analyze neural network controllers and their effects on input and output parameters.

Learning Process

The ability of neural networks to learn distinguishes them from most automatic controllers. Like humans, neural networks learn by example, and thus need to be trained. Neural networks are usually configured to specific applications and have the ability to process large amounts of data. Complex trends and patterns can be detected by neural networks that would otherwise be imperceptible to humans or other computing programs.

Within neural networks, there are learning procedures that allow the device to recognize a certain pattern and carry out a specific task. These learning procedures consist of an algorithm that enables the network to determine the weighting parameters in order to match the given data (inputs and outputs) with a function. In this iterative procedure, the initial input values are used to generate initial output values.

Based on these input and output values, the weights within the network are adjusted to match the data. These adjusted weights are then used to correlate the next pair of input and output values. Again, these values are used to adjust weights. This process continues until the network obtains a good fit for the data. A flow chart summarizing this iterative process is shown in figure.

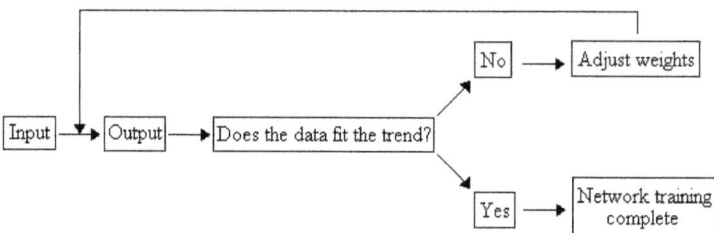

Fig. : Flowchart for training a network.

The trajectories shown in these videos demonstrate an important characteristic of neural networks. In these videos, the neural network can only predict outputs for objects that behave similar to the output it encountered during training. For instance, if a person was to start running forward, then stop, spin in a circle, and start running backwards, the neural network output would be unpredictable.

Once the neural network becomes capable of predicting system outputs, it is run like any other controller. In real-time, this often entails the chemical process and a computer system of some sort. Often, such as in LabVIEW, the chemical engineer is presented with a user-friendly data acquisition program that allows the user to set desired temperatures, flowrates, *etc.* and displays the system's input and outputs.

Although the neural network continually sends signals to the system controllers (such as valves), the network algorithm is embedded in the acquisition program.

Advantages and Disadvantages

Given the interesting, human-like behaviour of neural networks, one would expect all process applications to be controlled by neural networks. However, the advantages and disadvantages of neural networks limit their use in applications. The following lists summarize these advantages and disadvantages.

Advantages

- Neural networks are very general and can capture a variety of patterns very accurately
- The static, non-linear function used by neural networks provide a method to fit the parameters of a particular function to a given set of data.
- A wide variety of functions can be used to fit a given set of data
- Neural networks do not require excessive statistical training
- There is no need to assume an underlying input data distribution when programming a neural network
- Neural networks can detect all possible, complex nonlinear relationships between input and outputs
- Neural networks are the closest thing to having an actual human operate a system (*i.e.* they can "learn")

Disadvantages

- Neural networks are difficult to design. One must determine the optimal number of nodes, hidden layers, sigmoid function, *etc.*
- Neural networks are diffcult to model analytically because a small change in a single input will affect the entire network
- The operation of neural networks is limited to the training process. If the network is trained poorly, then it will operate poorly and the outputs cannot be guaranteed.
- There is a great computational burden associated with neural networks
- Neural networks require a large sample size in order to empirically fit data
- Neural networks have a "black box" nature. Therefore, errors within the complex network are difficult to target.
- Outside of their data training range, neural networks are unpredictable. This occurs because neural networks may "overfit" data. For instance, during training, a neural network may fit a 10th order polynomial to only 5 data points. When using this model to make predictions of values outside this 5-point range, the neural network behaves unpredictably.

Applications of Neural Networks

Each neural network will generate a different algorithm based on the inputs and outputs to the system. Because neural networks fit a specific function to the given data, they can be used in a variety of applications. Within chemical engineering, neural networks are frequently used to predict how changing one input (such as pressure, temperature, *etc.*) on a distillation column will influence the compositions and flow rates of the streams exiting the column.

The network training is performed on various inputs to the column, and thus can predict how changing one input will affect the product streams. Within a CSTR, neural networks can be used to determine the effect of one input parameter such as temperature or pressure on the products.

In addition to its applications to chemical equipment, neural networks can also be applied to model material responsess as a function of various loads under various conditions. These models can then be used in product development to create a device for a particular application, or to improve an existing device. For example, by modeling the corrosion of steel under different temperature and pH conditions, implanted biomedical devices can be manufactured or improved.

Neural networks are also often used in biology and biological applications to predict the outcome of a certain event. For example, neural networks can be used to predict the growth of cells and bacteria in cell culture labs, given a set of varying conditions, such as temperature and pH. In addition, neural network models have been used to predict the mortality rate in intensive care units in hospitals.

Data was collected from different patients, and a neural network model was created to predict the mortality of future patients given a set of specified conditions. Neural networks have also been used to diagnose breast cancer in patients by predicting the effects of a tumor provided specified input conditions of the patient.

Neural networks are also used in applications beyond the chemical aspect of controllers. For instance, neural networks are used to predict travel time based on different travel conditions. Signs on the highways that give estimated travel times are examples of neural networks. They predict the amount of time required to reach a certain destination given the varying traffic volume, road conditions, and weather conditions.

Neural networks are not limited to the applications listed above. They can be used to model most predictable events, and the complexity of the network will increase depending on the situation.

Chapter 8

FACILE SYNTHESIS OF MONO-DISPERSED POLYSTYRENE (PS)/AG COMPOSITE MICROSPHERES VIA MODIFIED CHEMICAL REDUCTION

Wen Zhu[1,2], Yuanyuan Wu[1], Changhao Yan[1], Chengyin Wang [1,*], Ming Zhang[1,*] and Zhonglian Wu[2]

[1] College of Chemistry and Chemical Engineering, Yangzhou University, Yangzhou 225002, China; E-Mails: czzwen@163.com (W.Z.); 1747168992@qq.com (Y.W.); 250178276@qq.com (C.Y.)

[2] College of Materials Science and Engineering, Jiangsu University of Technology, Changzhou 213001, China; E-Mail: 871765383@qq.com (Z.W.)

* Author to whom correspondence should be addressed; E-Mails: wangcy@yzu.edu.cn (C.W.); lxyzhangm@yzu.edu.cn (M.Z.); Tel.: +86-514-8799-0926 (C.W); +86-514-8215-8781 (M.Z.); Fax: +86-514-8797-9244; (C.W); +86-514-8797-5244 (M.Z.).

ABSTRACT

A modified method based on *in situ* chemical reduction was developed to prepare mono-dispersed polystyrene/silver (PS/Ag) composite microspheres. In this approach; mono-dispersed PS microspheres were synthesized through dispersion polymerization using poly-vinylpyrrolidone (PVP) as a dispersant at first. Then, poly-dopamine (PDA) was fabricated to functionally modify the surfaces of PS microspheres. With the addition of $[Ag(NH_3)_2]^+$ to the PS dispersion, $[Ag(NH_3)_2]^+$ complex ions were absorbed and reduced to silver nanoparticles on the surfaces of PS-PDA microspheres to form PS/Ag composite microspheres. PVP acted both as a solvent of the metallic precursor and as a reducing agent. PDA also acted both as a chemical protocol to immobilize the silver nanoparticles at the PS surface and as a reducing agent. Therefore, no additional reducing agents were needed. The resulting composite microspheres were characterized by TEM, field emission scanning electron microscopy (FESEM), energy-dispersive X-ray spectroscopy (EDS), XRD, UV-Vis and surface-enhanced Raman spectroscopy (SERS). The results showed that Ag nanoparticles (NPs) were homogeneously immobilized onto the PS microspheres' surface in the presence of PDA and PVP. PS/Ag composite microspheres were well formed with a uniform and compact shell layer and were adjustable in terms of their optical property.

Keywords

Silver nanoparticle; PS/Ag composite; dopamine; PVP; hollow microsphere.

1. INTRODUCTION

As shown in recent studies, various methods have been developed to expand the application area of noble metal nanoparticles and to control the morphology and the behavior. Composite microspheres with noble metal nanoshells have great advantages in catalysis, optics, conductivity, chemical sensors, and so on [1-5]. Moreover, these hybrid materials can prevent noble metal nanoparticles from agglomerating without the use of a stabilizer and can be easily retrieved, owing to the relatively large size of the supports. In particular, silver, with its outstanding combination of properties, has continued to be of great interest in terms of its potential applications in composite materials.

To date, considerable efforts have been put in to integrate noble metals into support particles. One approach is to coat silica or polystyrene with silver nanoshells. The prepared methods can be divided into two main categories [6,7]. The first one is described as follows. Template particles are treated by some chemicals or a chemical process to make the surfaces of the template particles bear functional groups. Then, the precursors of noble metals are added to the template particles. By adding the reducing agent, the precursors are reduced to zero-valent metals. Thus, the composite microspheres are obtained. The second process, called the layer-by-layer (LBL) self-assembly technique, is also widely used to prepare noble metal composite microspheres [8,9]. Mayer *et al.* prepared polystyrene/silver composite microspheres through an *in situ* reduction method [10], but this approach had shortcomings, such as incomplete coverage, rough surfaces, nonuniformity in shell thickness and aggregation of polystyrene (PS)/Ag microspheres. Another promising approach is to fabricate composite microspheres containing homogeneously dispersed metal nanoparticles. Wang and Asher had prepared silica spheres containing dispersed silver particles in a micro-emulsion [11]. However, it is difficult to control the diameter in a wide range, and the size distribution is often too wide. Therefore, improving the traditional technology of preparing mono-dispersed composite microspheres with uniform and complete Ag nanoshells is urgently needed.

In this work, a facile method for preparing PS/Ag composite microspheres was presented. The emphasis of this work was on the control of the monodispersity, shell thickness, average density and regularity in the morphology of PS/Ag composite microspheres. The poly-vinylpyrrolidone (PVP) and poly-dopamine (PDA) with hydroxyl and amino groups played an extremely important role, anchoring silver ions into the PS matrix. On the basis of the experimental results, the effects of PVP and PDA in synthesizing PS/Ag composite microspheres were discussed in detail. The silver nanoparticles (Ag NPs) were homogeneously doped into the composite microspheres. The PS/Ag composite microspheres showed the typical surface plasma resonance (SPR) peak of nanosized silver. The effects of silver nanoparticles on the morphology and optical properties of the composite microspheres were studied by TEM and UV-Vis spectra. The experimental approach of this work was simple and easy to operate.

2. EXPERIMENTAL SECTION

2.1. Materials

Styrene (St) was purchased from Shanghai Chemical Reagent Co. (China) and purified by treating with 5% NaOH aqueous solution to remove the inhibitor. Silver nitrate ($AgNO_3$, 99%), ammonia hydroxide (25%), poly-vinylpyrrolidone (PVP K30), absolute ethanol (EtOH), 2, 2'-azobisisobutyronitrile (AIBN, 98%), dopamine (DA) and trihydroxymethyl aminomethane (Tris) were also purchased from Shanghai Chemical Reagent Co. (China) and used as received. Deionized water was used for all the experimental processes.

2.2. Synthesis of Mono-Dispersed PS Microspheres

Mono-dispersed PS microspheres were synthesized by dispersion polymerization using PVP as the dispersant in the mixture of EtOH and water [12]. In a typical process, St (10.0 g), PVP (5.0 g), AIBN (1.0 g), deionized water (10.0 g) and EtOH (70.0 g) were charged into four-neck round flask equipped with a mechanical stirrer, an N_2 inlet, a thermometer with a temperature controller, a condenser and a thermostatic water bath. This mixture was deoxygenated by bubbling with nitrogen gas at room temperature for about 30 min, followed by heating to 70 °C. Under a stirring rate of 300 rpm, the polymerization was continued for 6 h.

2.3. Preparation of PS-PDA Microspheres

The synthesized PS microspheres were modified with PDA solution (0.02 g dopamine, 0.12 g Tris and 10.0 g deionized water). A 1 mL quantity of as-prepared PS dispersion was immersed in 10 mL PDA solution and reacted for about 24 h under magnetic stirring. The PDA-functionalized PS microspheres were obtained.

2.4. Preparation of PS/Ag Composite Microspheres

An appropriate amount of silver nitrate was dissolved into double distilled water to obtain $AgNO_3$ (0.12 M) aqueous solution. Subsequently, ammonia (2%) was gradually added into the $AgNO_3$ solution until the generated precipitates vanished. A 10 mL quantity of PDA-decorated PS dispersion was added to the 20 mL freshly as-prepared $[Ag(NH_3)_2]^+$ solution. The mixtures were stirred for 30 min at room temperature to ensure that the $[Ag(NH_3)_2]^+$ ions were absorbed to the PDA-decorated PS microspheres. Subsequently, the mixtures were heated to 80 °C with stirring for about 60 min. Then, plenty of brown PS/Ag composites were obtained in the solutions. After filtering and fully washing, the as-obtained products were collected and stored in ethanol for further examination.

2.5. Characterization

Transmission electron microscopy (TEM) characterization was performed on a Tecnai 12 electron microscope with an operating voltage of 120 kV. Field emission scanning electron microscopy (FESEM) was carried out with a Hitachi S-4800 scanning electron microscope operating at an acceleration voltage of 15 kV. Elemental mapping images were acquired by energy-dispersive

X-ray spectroscopy (EDS) using a Tecnai G2 F30 S-TWIN electron microscope equipped with a scanning transmission electron microscopy (STEM) unit and an Inca Energy 250 detector. Powder X-ray diffraction (XRD) patterns were recorded on a Bruker D and Advance diffractometer with Cu K á radiation. UV-Vis absorption spectra were measured by a Shimadzu UV-2501 spectrophotometer. Surface-enhanced Raman spectroscopy (SERS) was performed with a Renishaw Raman spectrometer.

3. RESULTS AND DISCUSSION

3.1. Synthesis and Morphology of Spheres

The surface morphologies of the samples before and after being coated with Ag NPs are studied by TEM and FESEM, which are shown in Figure 1. Figure 1a and b present the typical TEM and FESEM images of the bare PS microspheres. It can be seen clearly that the formed PS microspheres are uniform, with smooth surfaces and a diameter of *ca.* 850 nm. Figure 1c and d demonstrate the typical TEM and FESEM images of Ag-coated composites fabricated using 0.12 M $[Ag(NH_3)_2]^+$ solution. After being coated, the surfaces of the PS beads become rough. The shape of Ag NPs is near-spherical, and all the Ag NPs evenly coat the PS beads. The sizes of the Ag NPs are estimated to be about 40 nm. As shown in Figure 1d, although the surfaces roughened, the monodispersity and spherical shape of PS/

Ag composites are mostly preserved. Therefore, mono-dispersed PS microspheres coated with uniform Ag NPs can be successfully synthesized.

Figure 2 shows X-ray diffraction patterns of pristine PS, PS-PDA and PS/Ag composites. The strong reflection at $2\theta = 20°$ is assigned to amorphous PS. The typical XRD pattern of PS/Ag exhibits peaks at 2θ angles of 38.1°, 44.3°, 64.4°, 77.4° and 81.5° corresponding to the reflections of the (111), (200), (220), (311) and (222) crystal plane of the face-centered cubic (*fcc*) structure of Ag (Joint Committee on Powder Diffraction Standarda Card 04-0783). It further confirms that silver nanoparticles with crystallinity could be obtained successfully by reducing $[Ag(NH_3)_2]^+$ ions.

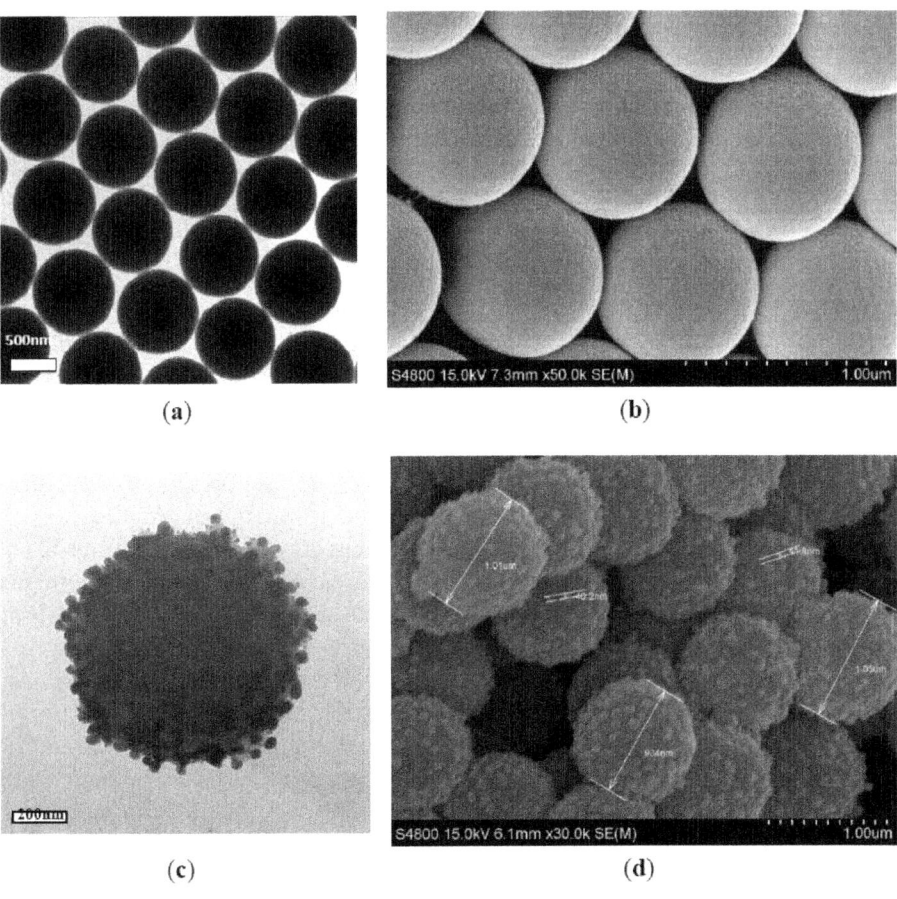

Figure 1. (a,b) TEM and field emission scanning electron microscopy (FESEM) images of the polystyrene (PS) microspheres with a diameter of 850 nm synthesized by dispersion polymerization; (c,d) TEM and FESEM images of the PS/Ag composites.

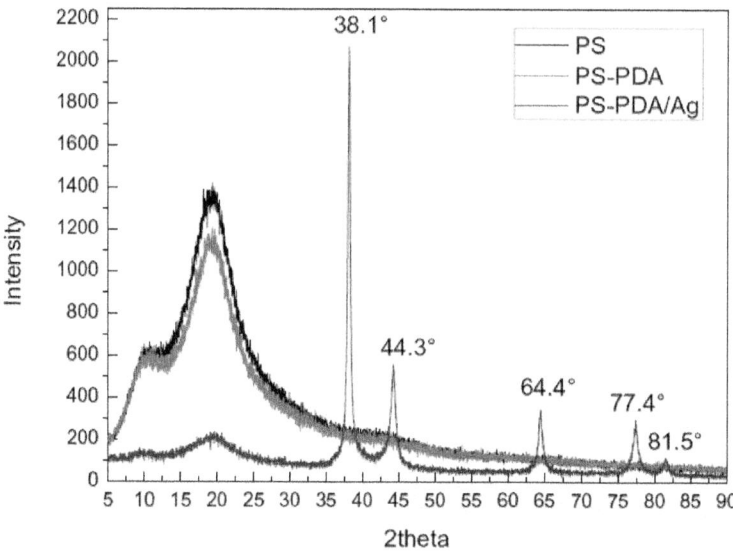

Figure 2. XRD patterns of pristine PS, PS- poly-dopamine (PDA) and PS/Ag composite microspheres.

3.2. Effects of PVP and PDA on the Formation of PS/Ag Composite Structures

In our experiments, PVP first acted as a dispersant to stabilize the formed PS microspheres in the synthesis of PS dispersion. Figure 3 shows the FTIR spectrum of unwashed PS microspheres (A) and washed PS microspheres (B) prepared by dispersion polymerization. As shown in curve A, the absorption band at 1670 cm^{-1} is the typical band of PVP, which decreases significantly, but not completely disappearing in curve B. This is good evidence that the effective stabilizer is PVP containing PS. Some ungrafted PVP adsorbs to the particles and aids in the steric stabilization of the particles during the reaction [13]. A PVP macromolecule in solution, which most likely adopts a pseudorandom coil conformation, may take part in some form of association with the metal atoms, thus increasing the probability of nucleus formation [14]. It has been widely proven by experiments that PVP could be effective in protecting composite spheres from aggregation and in modifying the morphology of composites [15–17]. This role of the PS surface in the formation step of silver nanoparticles is similar with that of seed in the heterogeneous nucleation technique [18]. PVP also importantly acted as the protection agent in the reduction of $[Ag(NH_3)_2]^+$ ions to Ag NPs during our experiments. The inserted higher magnification image (Figure 3) of part of a PS/Ag composite further shows that the narrow size distribution of individual Ag NPs is clear evidence of the separation of nucleation and growth steps caused by the nucleation site role of the PS surface. In addition, immobilization of Ag NPs onto the PS surface acts as a stabilization mechanism for Ag NPs.

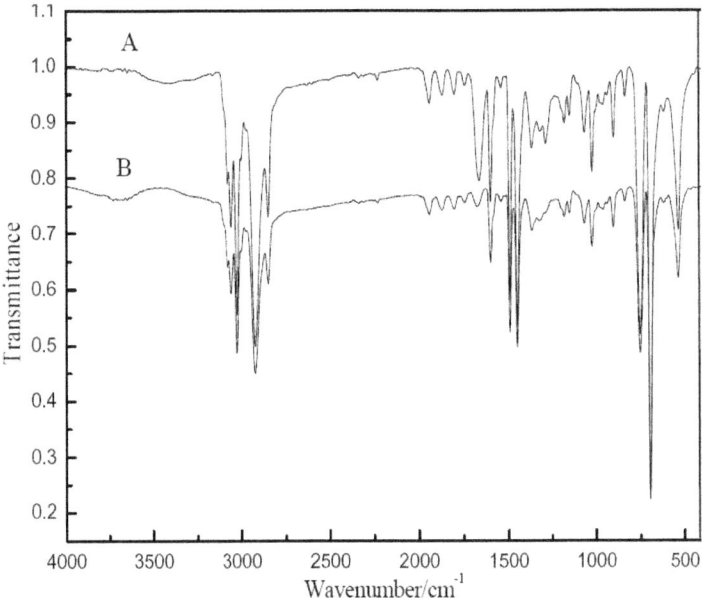

Figure 3. FTIR spectra of (A) unwashed PS microspheres and (B) washed PS microspheres prepared by dispersion polymerization.

To our knowledge, few works have discussed in detail the mechanism of mono-dispersed metal-doped composite microsphere synthesis. PDA is formed by *in situ* spontaneous oxidative polymerization of dopamine and is introduced to the PS microsphere surface. PDA improves the dispersion of hydrophobic PS microspheres in aqueous solution, because of its hydrophilicity [19,20]. More importantly, the metal-binding ability of phenolic hydroxyl groups present in the PDA structure is exploited to absorb $[Ag(NH_3)_2]^+$ ions onto the PDA-coated PS microsphere surface. The absorbed $[Ag(NH_3)_2]^+$ ions are reduced to zero-valent silver by the reducibility of PVP and PDA, and silver nuclei are formed on the PS microsphere surfaces. Nuclei are created at the silver ions bound to PDA-coated PS microspheres by the nucleation site role of Ag^+ ions bound to PDA, similar to the role of seed materials. Finally, Ag NPs are formed on PS microspheres surfaces by the growth of nuclei as the thermal energy supplied to the system increases to a given temperature [21]. Reduced silver species in solution are deposited to silver nuclei immobilized onto the surfaces of PS microspheres, which are attributed to the slower reduction rate of the polyol process than that of the general chemical reduction method using a reducing agent [22]. Elemental mapping analyses on a single PS/Ag composite microsphere are given in Figure 4. The images of nitrogen (Figure 4a) and oxygen (Figure 4b) confirm the presence of PDA on the surface of the PS microsphere. Consequently, this fact means that PDA successfully acts as a chemical protocol between the Ag NPs and the PS surfaces. The image of silver (Figure 4c) indicates the formation of Ag particles loaded onto the PS surfaces.

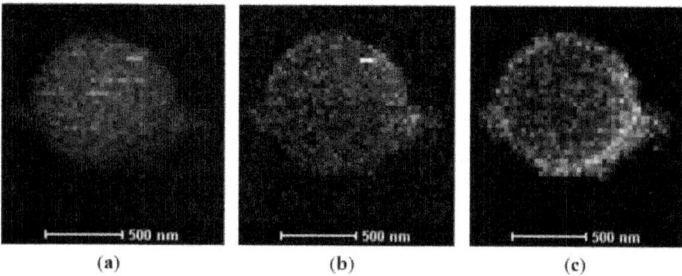

Figure 4. Elemental mapping images of (a) N; (b) O and (c) Ag of a PS/Ag composite microsphere with PDA fabricated using 0.12 M [Ag(NH$_3$)$_2$]$^+$ solution.

To confirm the important role of PVP and PDA that could attract [Ag(NH$_3$)$_2$]$^+$ ions onto the surfaces of PS microspheres, control experiments were performed. The other components were experimented on using reducing agent, but without PVP and PDA; a 10 mL washed PS dispersion was added to the 20 mL 0.12 M [Ag(NH$_3$)$_2$]$^+$ solution. The other components were experimented on without PDA; just a 10 mL quantity of unwashed PS dispersion was added to the 20 mL 0.12 M [Ag(NH$_3$)$_2$]$^+$ solution. As shown in Figure 5a and b, it can be found that there are Ag NPs dispersed in the system, but they had not been attracted onto PS microspheres to form composite microspheres. Obviously, without the presence of PDA, few [Ag(NH$_3$)$_2$]$^+$ ions are attracted onto the surfaces of PS microspheres; after the formation of zero-valent silver, Ag NPs dispersed in the system separate from PS microspheres. With the appropriate amount of PDA addition, Ag NPs are homogeneously distributed in PS microspheres (Figure 5c). In this system, PDA acts as a nucleation-prompting agent for Ag NPs. It is noticeable that the surfaces of the resultant composite microspheres become rougher.

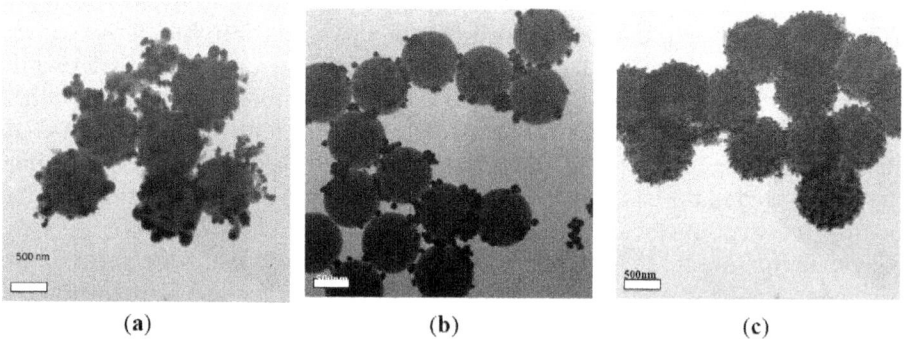

Figure 5. TEM images of the PS/Ag composites (a) prepared without PVP and PDA; (b) prepared without PDA; and (c) prepared with PDA modification.

3.3. Concentration Changing of the [Ag(NH$_3$)$_2$]$^+$ Ions

Based on the *in situ* self-catalytic synthesis method, changing the concentration of the [Ag(NH$_3$)$_2$]$^+$ ions can control the size and coating status of Ag NPs [23].

Figure 6 shows the FESEM images of a series of PS/Ag composites fabricated using different concentrations of [Ag(NH$_3$)$_2$]$^+$ ions (0.06 M, 0.12 M, 0.18 M and 0.24 M). As shown in Figure 6a, the shape of the Ag NPs is near spherical, and an incomplete nanoshell formed because of an insufficient amount of Ag precursor. Upon increasing the concentration of the [Ag(NH$_3$)$_2$]$^+$ ions, the coverage rate of the Ag NPs are elevated, because more [Ag(NH$_3$)$_2$]$^+$ ions were reacted and enhanced the yield of Ag, and the average size of Ag NPs on the PS surface increase from ~20 nm to ~200 nm. However, when the concentration was increased to 0.24 M, the Ag NPs over the PS become nonuniform, and there are many large Ag NPs. With the excess of silver ions in the solution, it is favorable for them to aggregate.

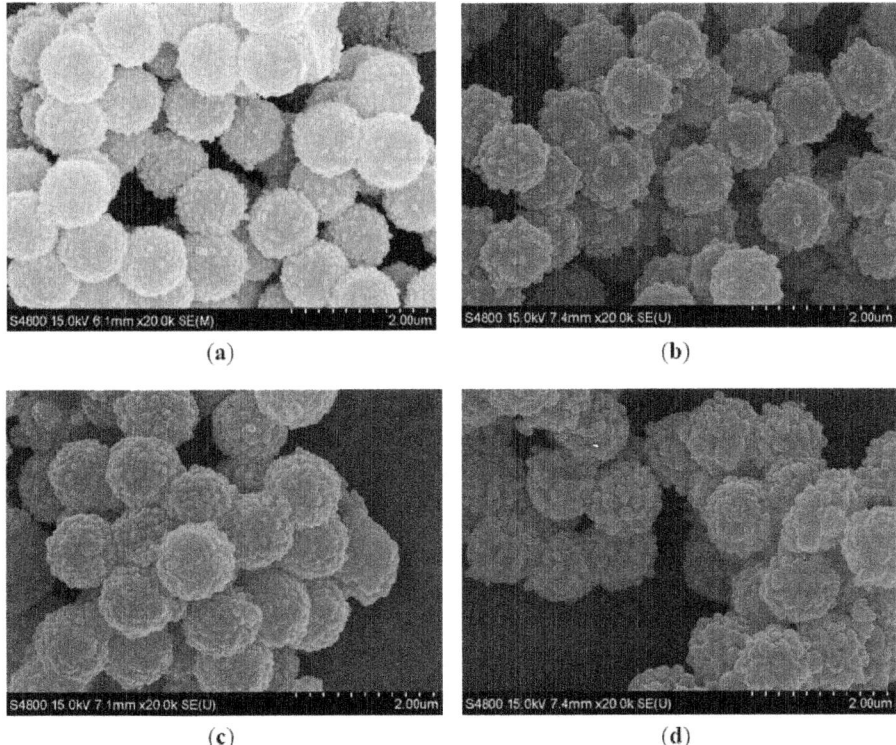

Figure 6. FESEM images of the PS/Ag composites prepared with different concentrations of [Ag(NH$_3$)$_2$]$^+$ ions: (**a**) 0.06 M; (**b**) 0.12 M; (**c**) 0.18 M; and (**d**) 0.24 M.

Thermogravimetric Analysis (TGA) is used to measure the weight percentage of silver in the composites. The pure PS completely decomposes to H$_2$, CH$_4$ and other gases from 350 to 450 °C, so the residual weight should be that of silver [24]. Figure 7 shows the TGA curves of the samples. It can be seen that the weight loss of the PS/Ag composites took place in the temperature ranges from 400 to 500 °C. According to the TGA curves, the silver contents prepared with various concentrations of the [Ag(NH$_3$)$_2$]$^+$ ions (0.06 M, 0.12 M, 0.18 M and 0.24 M) are found to be 38.92%, 67.65%, 76.16% and 77.33%, respectively. An increase in the silver shell thickness of PS/Ag composites is observed when the concentration of the $[Ag(NH_3)_2]^+$ ions was increased under the same reaction conditions.

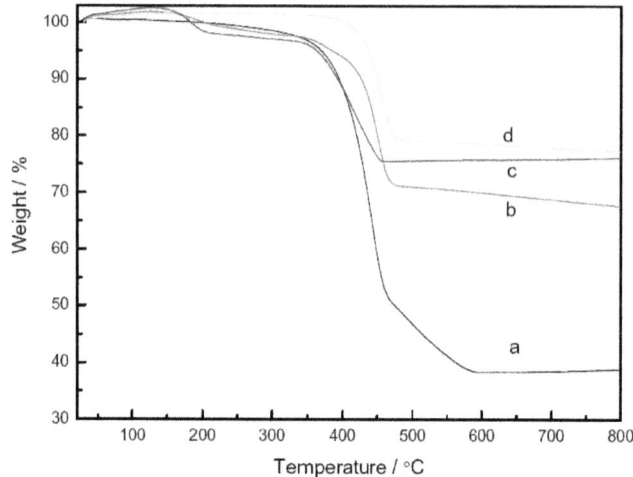

Figure 7. Thermogravimetric Analysis (TGA) curves of the PS/Ag composites prepared with various concentrations of the $[Ag(NH_3)_2]^+$ ions: (a) 0.06 M; (b) 0.12 M; (c) 0.18 M; and (d) 0.24 M.

It is clear that homogenous and complete silver shells form by dissolving the PS cores in tetrahydrofuran (THF). Figure 8 shows the TEM images of hollow silver microspheres. Complete and compact hollow structures formed upon increasing the concentration of the $[Ag(NH_3)_2]^+$ ions to 0.18 M; the close-packed Ag NPs form a shell. A complete hollow silver microsphere in Figure 8b is the result of the high surface coverage.

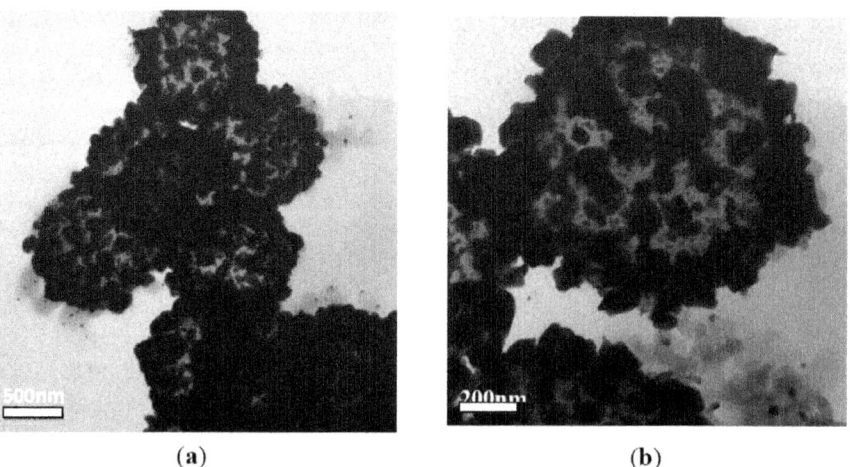

Figure 8. TEM images of the hollow Ag microspheres prepared with 0.12 M $[Ag(NH_3)_2]^+$ ions. (a) magnification 26500; (b) magnification 59000.

3.4. Optical Properties of PS/Ag Composites

Silver nanoparticles are well known for their surface plasma resonance (SPR) properties, which originate from the collective oscillation of conduction electrons in response to optical excitation [25]. The UV-Vis absorption spectra of the PS/Ag composites are sensitive to size, shape, aggregation state and local environment of the Ag NPs. To investigate the optical properties of the Ag NPs doped in the PS matrix, the samples of PS/Ag composites were diluted with deionized water for measurement of the absorption spectrum. Figure 9 shows the UV-Vis absorption spectra of the PS/Ag composites fabricated using different concentration of the $[Ag(NH_3)_2]^+$ ions. At relatively low silver concentrations, a broad peak is observed at ~530 nm (curve a), which was assigned to the localized surface plasma resonance (LSPR) of Ag NPs bound to the surface of PS. In addition, there is no SPR peak of the isolated Ag NPs at approximately 420 nm, which confirms that few free Ag NPs appeared in dispersions [17,26]. The position and width of the SPR peak are linked with the size and shape of the metal particles, also connected with its own dielectric constant and that around it. With changing the anisotropy of particles, the peak would vary in the visible and near-infrared spectrum regions [27-29]. However, as the silver coverage increased from a low to a high level, the Ag NPs density and the contacting area between Ag NPs and PS supports would increase; so, the absorption spectra of PS/Ag composites are mainly red-shifted and broadening (curves b-d). Curves b-d are not pronounced peaks; the results may be explained as follows. As the silver coverage increases to a high level, the PS/Ag composites mainly show the collective absorption behavior of the Ag NPs. This finding is consistent with the FESEM images shown in Figure 6a-d. The same phenomena had been observed with other works [30,31].

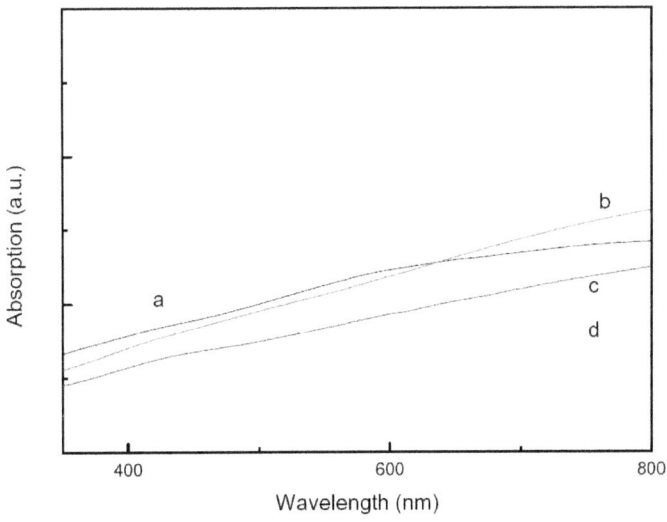

Figure 9. UV-Vis absorption spectra of PS/Ag composites prepared with different concentration of $[Ag(NH_3)_2]^+$ ions: (**a**) 0.06 M; (**b**) 0.12 M; (**c**) 0.18 M; and (**d**) 0.24 M.

3.5. SERS Properties of PS/Ag Composites

Surface-enhanced Raman scattering (SERS) is a type of abnormal optical enhanced effect based on the rough surface at the nano-scale, the particle morphology and the granular lumps [32]. Silver is one of the most SERS-active metals. The PS/Ag composites with high Ag NPs coverage were chosen for SERS measurements. The SERS results are shown in Figure 10. As for the 1×10^{-3} mol/L Rhodamine 6G (R6G) without PS/Ag composites, the Raman lines for the R6G molecules can be seen from curve a; no peaks appear. However, when the 1×10^{-3} mol/L R6G was deposited on the selected PS/Ag composites, the peaks can be seen clearly; almost all the distinctive peaks corresponded to the Raman lines for the R6G molecules. The observed peaks included the v(C–H) out-of plane bend mode at 774 cm^{-1} and the v(C–C) stretching mode at 1360 cm^{-1}, 1508 cm^{-1}, which agree well with the literature [33]. With an increase of Ag coverage, the Raman signals are more intensified; this can be seen in curve b–d. As known from the literature, the SERS effects depend strongly on the roughness of the metal nanostructure used as the substrate [34,35].

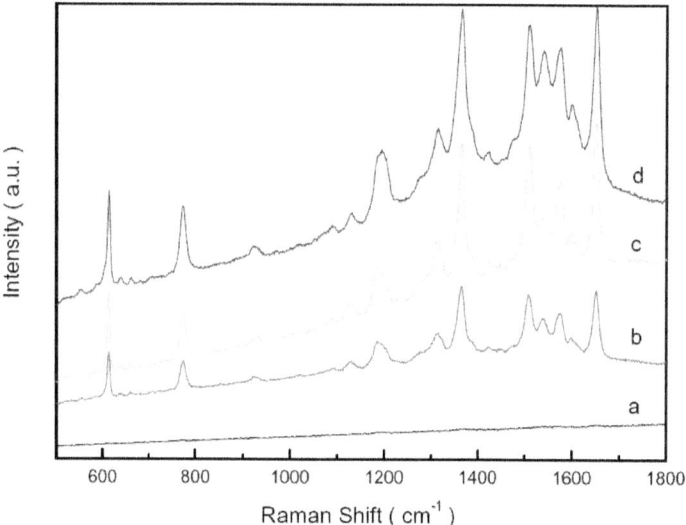

Figure 10. SERS spectra of 1×10^{-3} mol/L R6G without PS/Ag composites (**a**) and with PS/Ag composites prepared with different concentration of [Ag(NH$_3$)$_2$]$^+$ ions: (**a**) 0.06 M; (**b**) 0.12 M; (**c**) 0.18 M; and (**d**) 0.24 M.

4. CONCLUSIONS

In summary, we have demonstrated that mono-dispersed PS/Ag composites could be prepared by a convenient *in situ* reduction of [Ag(NH$_3$)$_2$]$^+$ complex ions. In this method, PVP took both the dispersant and protection agent at the same time. On the basis of experimental results, a polystyrene surface was modified with PDA as a nucleation site during the Ag NP formation period, which is similar to the role of seed materials in the heterogeneous nucleation technique. Therefore, PVP and PDA played an important role in bridging the polystyrene surface and the silver nanoparticles. TEM, FESEM and XRD confirm the formation of PS/Ag composites, which had both the bulk properties of the polystyrene and the surface properties of the Ag NPs. A UV-Vis spectrometer shows that the PS/Ag composites possess good optical properties. Raman spectra indicate that the PS/Ag composites as SERS substrates have excellent SERS performance.

This method provides an alternative approach for the preparation of metal doped polystyrene microspheres. It is also believed that the as-prepared PS/Ag composites have other potential applications, such as catalysis, antibacterial action and the optical detection of macromolecules. Antibody-antigen interaction and DNA hybridization can be dynamically detected from the shift of the SPR peaks [36].

Acknowledgments

This work was supported by the National Natural Science Foundation of China (Grant No. 50873085 and No. 21375116), a project funded by the Priority Academic Program Development of Jiangsu Higher Education Institutions (PAPD).

Conflicts of Interest

The authors declare no conflict of interest.

REFERENCES

1. Jackson, J.B.; Halas, N.J. Silver nanoshells: Variations in morphologies and optical properties. *J. Phys. Chem.* B **2001**, *105*, 2743–2746.
2. Pol, V.G.; Srivastava, D.N.; Palchik, O.; Palchik, V.; Slifkin, M.A.; Weiss, A.M.; Gedanken, A. Sonochemical deposition of silver nanoparticles on silica spheres. *Langmuir* **2002**, *18*, 3352–3357.
3. Sau, T.K.; Rogach, A.L. Nonspherical noble metal nanoparticles: Colloid-chemical synthesis and morphology control. *Adv Mater.* **2010**, *22*, 1781–1804.
4. Zhu, Y.; Huang, X.B.; Li, W.Z.; Fu, J.W.; Tang, X.Z. Preparation of novel hybrid inorganic-organic microspheres with active hydroxyl groups using ultrasonic irradiation via one-step precipitation polymerization. *Mater. Lett.* **2008**, *62*, 1389–1392.
5. Huang, X.H.; El-Sayed, I.H.; Qian, W.; El-Sayed, M.A. Cancer cell imaging and photothermal therapy in the near-infrared region by using gold nanorods. *J. Am. Chem. Soc.* **2006**, *128*, 2115–2120.
6. Dong, A.G.; Wang, Y.J.; Tang, Y.; Ren, N.; Yang, W.L.; Gao, Z. Fabrication of compact silver nanoshells on polystyrene spheres through electrostatic attraction. *Chem. Commun.* **2002**, *4*, 350–351.
7. Lu, Y.; Mei, Y.; Schrinner, M.; Ballauf, M.; Möller, M.W.; Breu, J. In situ formation of Ag nanoparticles in spherical polyacrylic acid brushes by UV irradiation. *J. Phys. Chem.* C **2007**, *111*, 7676–7681.
8. Cassagneau, T.; Caruso, F. Contiguous silver nanoparticle coatings on dielectric spheres. *Adv. Mater.* **2002**, *14*, 732–736.
9. Kim, K.; Lee, H.B.; Park, H.K.; Shin, K.S. Easy deposition of Ag onto polystyrene beads for developing surface-enhanced-Raman-scattering-based molecular sensors. *J. Colloid Interface Sci.* **2008**, *318*, 195–201.

10. Mayer, A.B.R.; Grebner, W.; Wannemacher, R. Preparation of silver-latex composites. *J. Phys. Chem.* B **2000**, *104*, 7278–7285.

11. Wang, W.; Asher, S.A. Photochemical incorporation of silver quantum dots in monodisperse silica colloids for photonic crystal applications. *J Am. Chem. Soc.* **2001**, *123*, 12528–12535.

12. Zhang, J.H.; Chen, Z.; Wang, Z.L.; Zhang, W.Y.; Ming, M.B. Preparation of monodisperse polystyrene spheres in aqueous alcohol system. *Mater. Lett.* **2003**, *57*, 4466–4470.

13. Song, J.S.; Chagal, L.; Winnik, M.A. Monodisperse micrometer-size carboxyl-functionalized polystyrene particles obtained by two-stage dispersion polymerization. *Macromolecules* **2006**, *39*, 5729–5737.

14. Lee, J.M.; Jun, Y.D.; Kim, D.W.; Lee, Y.H.; Oh, S.G. Effects of PVP on the formation of silver-polystyrene heterogeneous nanocomposite particles in novel preparation route involving polyol process: Molecular weight and concentration of PVP. *Mater. Chem. Phys.* **2009**, *114*, 549– 555.

15. Silvert, P.Y.; Herrera-Urbina, R.; Tekaia-Elhsissen, K. Preparation of colloidal silver dispersions by the polyol process. Part 2—Mechanism of particle formation. *J. Mater. Chem.* **1997**, *7*, 293– 299.

16. Lee, J.M.; Kim, D.W.; Lee, Y.H.; Oh, S.G. New approach for preparation of silver-polystyrene heterogeneous nanocomposite by polyol process. *Chem. Lett.* **2005**, *34*, 928–929.

17. Deng, Z.W.; Chen, M.; Wu, L.M. Novel method to fabricate SiO_2/Ag composite spheres and their catalytic, Surface-Enhanced Raman Scattering properties. *J. Phys. Chem.* C **2007**, *111*, 11692– 11698.

18. Silvert, P.Y.; Herrera-Urbina, R.; Duvauchelle, N.; Vijayakrishnan, V.; Tekaia-Elhsissen, K. Preparation of colloidal silver dispersions by the polyol process. Part 1—Synthesis and characterization. *J.Mater. Chem.* **1996**, *6*, 573–577.

19. Zhang, J.; Liu, J.; Wang, S.; Zhan, P.; Wang, Z.; Ming, N. Facile methods to coat polystyrene and silica colloids with metal. *Adv. Funct. Mater.* **2004**, *14*, 1089–1096.

20. Kobayashi, Y.; Tadaki, Y.; Nagao, D.; Konno, M. Deposition of gold nanoparticles on silica spheres by electroless metal plating technique. *J.Colloid Interface Sci.* **2005**, *283*, 601–604.

21. Caruso, F. Nanoengineering of particle surfaces. *Adv. Mater.* **2001**, *13*, 11–22.

22. Lee, H.; Dellatore, S.M.; Miller, W.M.; Messersmith, P.B. Mussel-inspired surface chemistry for multifunctional coatings. *Science* **2007**, *318*, 426–430.

23. Hu, Y.G.; Zhao, T.; Zhu, P.L.; Sun, R. Preparation of monodisperse polystyrene/silver composite microspheres and their catalytic properties. *Colloid Polym Sci.* **2012**, *290*, 401–409.

24. Kawahashi, N.; Shiho, H. Copper and copper compounds as coatings on polystyrene particles and as hollow spheres. *J. Mater. Chem.* **2000**, *10*, 2294–2297.

25. Yong, K.T.; Sahoo, Y.; Swihart, M.T.; Prasad, P.N. Synthesis and plasmonic properties of silver and gold nanoshells on polystyrene cores of different size and of gold–silver core–shell nanostructures. *Coll. Surf. A Physicochem. Eng. Asp.* **2006**, *290*, 89–105.

26. Sai, V.V.R.; Gangadean, D.; Niraula, I.; Jabal, J.M.F.; Corti, G.; McIlroy, D.N.; Aston, D.E.; Branen, J.R.; Hrdlicka, P.J. Silica nanosprings coated with noble metal anoparticles: Highly active SERS substrates. *J. Phys. Chem.* C **2011**, *115*, 453–459.

27. Oldenburg, S.J.; Averitt, R.D.; Westeott, S.L.; Halas, N.J. Nanoengineering of optical resonances. *Chem. Phys. Lett.* **1998**, *288*, 243–247.

28. Tian, C.; Mao, B.; Wang, E.; Kang, Z.; Song, Y.; Wang, C.; Li, S. Simple strategy for preparation of core colloids modified with metal nanoparticles. *J. Phys. Chem.* C **2007**, *111*, 3651–3657.

29. Liang, Z.J.; Susha, A.S.; Caruso, F. Gold nanoparticle-based core-shell and hollow spheres and ordered assemblies thereof. *Chem. Mater.* **2003**, *15*, 3176–3183.

30. Ma, Y.H.; Zhang, Q.H. Preparation and characterization of monodispersed PS/Ag composite microspheres through modified electroless plating. *Appl. Surf. Sci.* **2012**, *258*, 7774–7780.
31. Zhang, S.F.; Ren, F.; Wu, W.; Zhou, J.; Sun, L.L.; Xiao, X.H.; Jiang, C.Z. Modified *in situ* and self-catalytic growth method for fabrication of Ag-coated nanocomposites with tailorable optical properties. *J. Nanopart Res.* **2012**, *14*, 1105–1118.
32. Li, J.F.; Huang, Y.F.; Ding, Y. Shell-isolated nanoparticle-Enhanced Raman spectroscopy. *Nature* **2010**, *464*, 392–395.
33. Hildebrandt, P.; Stockburger, M. Surface-enhanced resonance Raman spectroscopy of Rhodamine 6G adsorbed on colloidal silver. *J. Phys. Chem.* **1984**, *88*, 5935–5944.
34. Qian, L.H.; Yan, X.Q.; Fujita, T.; Inoue, A.; Chen, M.W. Surface enhanced Raman scattering of nanoporous gold: Smaller pore sizes stronger enhancements. *Appl. Phys. Lett.* **2007**, *90*, 153120–153123.
35. Lang, X.Y.; Guan, P.F.; Zhang, L.; Fujita, T.; Chen, M.W. Size dependence of molecular fluorescence enhancement of nanoporous gold. *Appl. Phys. Lett.* **2010**, *96*, 073701:1–073701:3.
36. Liu, S.H.; Zhang, Z.H.; Han, M.Y. Gram-scale synthesis and biofunctionalization of silica-coated silver nanoparticles for fast colorimetric DNA detection. *Anal. Chem.* **2005**, *77*, 2595–2600.

This page left intentionally blank.

INDEX

A

Absorption, 162
Accuracy Issues, 63
Actuator Type: Thermal Systems, 110
Advanced Energy Systems, 241
Advantages and Disadvantages of Cascade Control, 55
Advantages and Disadvantages, 281
Advantages with the Doppler Effect Ultrasonic Flowmeter, 152
Air and Gaseous Flow, 208
Air Separation, 253
Alkylating Agents, 23
Alkylation, 23
Altitude Valves, 129
Applications of Neural Networks, 282

B

Ball Valves, 197
Bellows Spring Loaded Safety Relief Valve, 201
Benefits of Producing Hydrogen from Coal, 246
Benefits of Visual Level Sensors, 125
Bimetal Thermometer, 99
Blend Stations, 62
Bourdon Tube Gauges, 113
Bromination, 16

Buoyancy Types, 127
Butterfly Valves, 198

C

Calculating RGA, 261
Calorimetric Flow Meter, 155
Capacitance Level Sensing, 132
Carbene Alkylating Agents, 24
Carbon Capture, Utilization and Storage Research, 240
Cavitation, 211
CCL Liquid, 69
CCL Reactors, 85
CCL Temperature, 73
Chemical Dehalogenation, 17
Chemical Looping, 252
Chlorination, 15
Chromatography, 171
Classes of the Reaction, 7
Closed Loop Control versus Open Loop Control, 37
Closed Loop System, 37
CO_2 Utilization Focus Area, 248
CO_2 Utilization Goals, 249
CO_2 Utilization Technologies, 249
Coal and Biomass to Liquids, 247
Coal to Liquids, 246
Combining Neurons into Neural Networks, 278

Common Topologies, 85
Compiling the Array, 263
Composition Control: Ratio Control, 95
Composition Sensors, 159
Comprehensive Charts, 177
Conditions for Cascade Control, 54
Conductive Level Sensing, 131
Conductivity Cells, 167
Cone and Plate, 190
Constant Pressure Drop, 222
Contact Sensors, 98
Control Valve Gain, 222
Controlled by Steam Flowrate, 88
Controlled by Steam Pressure, 87
Controller Tuning and Constraints, 81
Controlling the Cool Side Stream, 82
Controlling the Hot Side Stream, 83
Cooling Down The Process Stream, 84
Coriolis Mass Flow Meter, 148
Corrosion and Rusting, 6
Cross-Limiting Override Control, 68

D

Decouple, 254
Dehydration Reaction, 13
Dehydrogenation, 7
Density and Specific Gravity, 175
Detailed Gasification Chemistry, 228
Dew Point Measuring Devices, 194
Diagram of Ratio Dependent System, 61
Diaphragms, 114
Differential Pressure Cells, 121
Difficulties with Ratio Controllers, 62
Direct Mass Measurement, 176
Dispersive Photometers, 161
Displacement – Hydrometer, 176
Disturbances to CSTRs, 86
Disturbances to PFRs, 86
Diverter Valves, 130

Doppler Meters, 151
Dual Composition Control, 80
Dynamic Compensation, 41

E

Efficiency Benefits, 243
Elastic Distortion, 113
Electric Sensors, 115
Electrical Level Sensors, 131
Electrical Methods, 113
Electrometric Analysis, 167
Electrophilic Alkylating Agents, 24
Endothermic Reactors, 87
Environment, 112
Environmental Benefits, 243
Environmental Concerns, 21
Epoxides to Glycol, 11
Esters and Amides, 9
Example of Cascade Control, 48

F

Falling Ball, 190
Feedback and Feed-Forward, 85
Feedback Control, 31
Feed-forward Control, 38
Feed-Forward Design Procedure, 46
Filled System Thermometer, 99
First-Order System Example, 272
Float Type Level Sensors, 126
Flow Characteristics, 206
Flow Control: Flow Meters, 97
Flow Fraction Controller, 64
Flow Nozzle, 143
Flow Profile Distortion, 157
Fluorination, 15
Free Radical Halogenation, 14
Frequency, 149
Fundamentals, 226
Fusion-Type (Wax-Filled)Systems, 110

G

Gasification of Wood Chips, 236
Gasification Technology R&D, 242
Gas-solid Chromatography (GSC), 172
Gear Flow Meter, 155
General Cascade Control Schematic, 51
General Model, 271
Globe Valves, 199

H

Halogenation, 13
Heat Exchanger Control, 81
Height of Liquid in Column, 112
Hot Chamber Systems, 110
How Gasification Works, 238
Humidity Sensors, 194
Hydration of other substrates, 12
Hydration Reaction, 11
Hydraulic, 215
Hydrolysis, 8
Hygrometers, 194
Hysteresis, 213

I

Implementing MPC using Excel, 275
Inductive, 116
Industrial MPC Applications, 274
Inferential Temperature Control, 77
Infrared (IR) Spectrophotometers, 164
Infrared Sensors, 195
Inhibitors of Nitrification, 19
Inorganic and Materials Chemistry, 13
Inorganic Chemistry, 17
Interpreting the RGA, 266
Introduction to pH, 180
Introduction to Viscosity, 186
Iodination, 16
Ion Selective Electrodes, 168

Ionization Chambers, 195
Ionization Gauges, 123

L

Law of Homogenous Material, 105
Law of Intermediate Materials, 105
Laws for Thermocouples, 105
Learning Process, 280
Level Control Basics, 70
Level Control: Level Switches, 96
Level Measurement Noise, 71
Level Sensors, 124
Limitations of MPC, 273
Liquid Column, 175
Liquid Flow, 207
Liquid Level Control Model, 72
Liquid Pressure Control Model, 72
Liquid-Chromatography (LC), 172
Liquid-Filled Systems, 110

M

Magnetic Flow Meter, 154
Mass Spectrometry, 173
Metal Aqua Ions, 11
Metal Displacement, 5
Methane Inhibition, 21
Microbiology and Ecology, 17
Microphones, 195
Microwave / Radar Level Sensors, 134
Miscellaneous Sensors, 194
Model Predictive Control, 268
Model Predictive Control Example, 271
More on Exothermic Reactors, 91
Motivation, 270
MS Components, 173

N

Negative Feedback, 33
Neural Networks, 276

Neurons, 277
NI Analysis with RGA, 267
Nitrification, 17
Noncontact Sensors, 99
Non-dispersive Photometers, 161
Nuclear Level Sensors, 135
Nucleophilic Alkylating Agents, 23

O

Off-line Instruments, 188
Oil Refining, 25
On-line Instruments, 192
Opacity Monitors, 167
Open Loop and Closed Loop, 54
Open Loop System, 37
Orifice Meter, 139
Oscillating Cylinder, 192
Oxidizers, 3
Oxidizing and Reducing Agents, 3
Oxychlorination, 15

P

Paddle Wheel Sensors, 148
Photometric Analysis, 161
Photometry Using Visible Light, 165
Photopolymerization, 29
Physical Property Measurements, 175
Piezoelectric, 117
Pilot Assisted Safety Relief Valve, 202
Pneumatic, 215
Polarographic Sensors, 169
Polymerization, 27
Polysaccharides, 10
Positive Feedback, 33
Potentiometric, 118
Pressure Control Basics, 69
Pressure Control: Pressure Switch, 94
Pressure Measuring Methods, 112
Pressure Range, 112

Pressure Relief Valves, 200
Pressure Sensors, 111
Pressure-Flowrate Relationship, 207
Primary and Secondary Loops, 50
Process Description, 235
Process Model Available, 264
Propeller Flow Meter, 148
Psychrometers, 194

R

Radiation, 165
Radiation-based Level Sensors, 133
Radiation-Density Gauges, 176
Rangeability, 223
Ratio Relay Controller, 63
Ratio Relay with Remote Input, 65
Reactions and Transformations, 227
Redox Reactions in Industry, 6
Reference Half-Cell, 182
Refractometry, 165
Regulator Operation, 108
Regulator Structure, 108
Relative Gain Array, 259
Research and Development Needs, 245
Resistance Temperature Detectors, 100

S

Select Elements in Ratio Control, 66
Sensing Half-Cell, 182
Sensor Selection Criteria, 111
Sight Tube Indicators, 125
Single Composition Control, 78
Single Select Override Control, 66
Singular Value Decomposition, 256
Smoke Sensors, 195
Solid Oxide Fuel Cells, 244
Sound Sensors, 195
Spectrophotometers, 162
Staged gasification, 233

Standards and Calibration Curves, 160
Starting up a Cascade System, 56
Steady State Issues, 62
Steam Traps, 203
Strain Gauge, 119
Summary on Control Architectures, 68
Syngas Composition Variability, 231
Syngas Composition, 230

T

Table of Flow Meters, 156
Table of Viscometer Uses, 193
Temperature Compensation, 184
Temperature Control in Distillation, 76
Temperature Control Loops, 73
Temperature Detecting Elements, 109
Temperature Regulators, 107
Temperature Sensors, 98
Thermal Conductivity for Gases, 174
Thermal Conductivity Gauges, 122
Thermal Flow Meters, 156
Thermocouple Operation, 104
Thermocouple Structure, 102
Thermocouples, 102
Thermodynamics and Kinetics, 229
Thermometers, 99
Turbidimeters, 166
Turbine Meter, 147
Turndown Ratio, 157
Types of Chromatography, 171
Types of Photometers, 161

Types of Sensors, 113
Types of Temperature Regulators, 109
Typical pH Sensor Construction, 181
Tyrosine Sulfation, 22

U

Ultrasonic (Sonic) Level Sensors, 134
Ultrasonic Flow Meters, 150
Unit Processing in Chemical Process, 2
Upstream Pressure Increase, 213

V

Vacuum Sensors, 122
Valve Modeling, 218
Valve Types Selection, 195
Valve-based Level Sensors, 129
Vapour-Filled Systems, 110
Variable Pressure Drop, 223
Venturi Meter, 142
Vibrating Element, 120
Vibrating Rod or Cylinder, 193
Visual Level Sensors, 125
Vortex Shedding Flow Meter, 149

W

Wabash River Syngas Composition, 231
What is RGA?, 260
Why is pH Relevant for Chemical Engineering?, 181
Why is Viscosity Relevant for Chemical Engineering?, 187

This page left intentionally blank.